Toxicity Evaluation of Nanoparticles

Toxicity Evaluation of Nanoparticles

Guest Editors

Bing Yan
Rongrong Liu

MDPI

Basel • Beijing • Wuhan • Barcelona • Belgrade • Novi Sad • Cluj • Manchester

Guest Editors

Bing Yan
Institute of Environmental Research at Greater Bay
Guangzhou University
Guangzhou
China

Rongrong Liu
Institute of Environmental Research at Greater Bay
Guangzhou University
Guangzhou
China

Editorial Office
MDPI AG
Grosspeteranlage 5
4052 Basel, Switzerland

This is a reprint of the Special Issue, published open access by the journal *Nanomaterials* (ISSN 2079-4991), freely accessible at: https://www.mdpi.com/journal/nanomaterials/special_issues/nano_toxicity.

For citation purposes, cite each article independently as indicated on the article page online and as indicated below:

Lastname, A.A.; Lastname, B.B. Article Title. *Journal Name* **Year**, *Volume Number*, Page Range.

ISBN 978-3-7258-7695-2 (Hbk)
ISBN 978-3-7258-7696-9 (PDF)
https://doi.org/10.3390/books978-3-7258-7696-9

Contents

About the Editors . . . vii

Rongrong Liu and Bing Yan
Predictive and Mechanism-Based Toxicity Evaluation of Engineered Nanoparticles
Reprinted from: *Nanomaterials* **2026**, *16*, 185, https://doi.org/10.3390/nano16030185 . . . 1

Jialong Wu, Yanyan Yu and Gaoxing Su
Safety Assessment of 2D MXenes: In Vitro and In Vivo
Reprinted from: *Nanomaterials* **2022**, *12*, 828, https://doi.org/10.3390/nano12050828 . . . 4

Valérie Forest
Experimental and Computational Nanotoxicology—Complementary Approaches for Nanomaterial Hazard Assessment
Reprinted from: *Nanomaterials* **2022**, *12*, 1346, https://doi.org/10.3390/nano12081346 . . . 19

Imran Aslam and Maarten B. J. Roeffaers
Carbonaceous Nanoparticle Air Pollution: Toxicity and Detection in Biological Samples
Reprinted from: *Nanomaterials* **2022**, *12*, 3948, https://doi.org/10.3390/nano12223948 . . . 45

In Young Kim, Minjeong Kwak, Jaeseok Kim, Tae Geol Lee and Min Beom Heo
Comparative Study on Nanotoxicity in Human Primary and Cancer Cells
Reprinted from: *Nanomaterials* **2022**, *12*, 993, https://doi.org/10.3390/nano12060993 . . . 72

Chi Be Hlaing, Arpamas Chariyakornkul, Chalermchai Pilapong, Charatda Punvittayagul, Somdet Srichairatanakool and Rawiwan Wongpoomchai
Assessment of Systemic Toxicity, Genotoxicity, and Early Phase Hepatocarcinogenicity of Iron (III)-Tannic Acid Nanoparticles in Rats
Reprinted from: *Nanomaterials* **2022**, *12*, 1040, https://doi.org/10.3390/nano12071040 . . . 85

Tara Lyons-Darden, Jason L. Blum, Mark W. Schooley, Melissa Ellis, Jennifer Durando, Daniel Merrill and Adriana R. Oller
An Assessment of the Oral and Inhalation Acute Toxicity of Nickel Oxide Nanoparticles in Rats
Reprinted from: *Nanomaterials* **2023**, *13*, 261, https://doi.org/10.3390/nano13020261 . . . 103

Joëlle Auclair, Patrice Turcotte, Christian Gagnon, Caroline Peyrot, Kevin James Wilkinson and François Gagné
Form-Dependent Toxicity of Silver Nanomaterials in Rainbow Trout Gills
Reprinted from: *Nanomaterials* **2023**, *13*, 1356, https://doi.org/10.3390/nano13081356 . . . 111

Elena Maria Scalisi, Roberta Pecoraro, Antonio Salvaggio, Fabiano Capparucci, Cosimo Gianluca Fortuna, Massimo Zimbone, et al.
Titanium Dioxide Nanoparticles: Effects on Development and Male Reproductive System
Reprinted from: *Nanomaterials* **2023**, *13*, 1783, https://doi.org/10.3390/nano13111783 . . . 125

Yang Liu, Xuepeng Zhao, Yuhui Ma, Wanqin Dai, Zhuda Song, Yun Wang, et al.
Interaction of Cerium Oxide Nanoparticles and Ionic Cerium with Duckweed (*Lemna minor* L.): Uptake, Distribution, and Phytotoxicity
Reprinted from: *Nanomaterials* **2023**, *13*, 2523, https://doi.org/10.3390/nano13182523 . . . 147

Yuan Chen, Jiaxing Wang, Daidi Chen, Boxi Wang, Jinchuan Wu, Rongrong Liu and Qingxin Li
The Pd (II) Reduction Mechanisms in *Bacillus megaterium* Y-4 Revealed by Proteomic Analysis
Reprinted from: *Nanomaterials* **2024**, *14*, 512, https://doi.org/10.3390/nano14060512 . . . 160

About the Editors

Bing Yan

Bing Yan is a Fellow of the Royal Society of Chemistry and a Cheung Kong Scholar Professor at Guangzhou University, focusing his research on nanotoxicology and environmental health. He earned his Ph.D. in Chemistry from Columbia University in 1990, followed by postdoctoral research at the Department of Physiology, University of Cambridge, U.K., and the Department of Microbiology and Molecular Genetics, University of Texas Medical School, Houston, from 1990 to 1993. From 1993 to 2006, he worked in drug discovery at Novartis Biomedical Research Institute and Bristol-Myers Squibb before serving as a Professor of Pharmaceutics and Director of the High-Throughput Analytical Center at St. Jude Children's Research Hospital in Memphis from 2007 to 2012. Since 2005, he has been a Cheung Kong Scholar Professor, first at Shandong University and later at Guangzhou University, China. He has authored or edited 11 books and published over 380 peer-reviewed papers. He also serves as Editor-in-Chief of *Ecotoxicology and Environmental Safety* and as Associate Editor of *NanoImpact*.

Rongrong Liu

Rongrong Liu is an Associate Professor at Guangzhou University. She earned her Ph.D. in Chemical Biology from Nanyang Technological University, Singapore. She worked as a Research Scientist at the Agency for Science, Technology and Research (A*STAR) in Singapore and received the Dedicated Service Award. In 2019, she joined the Institute of Environmental Research at Greater Bay Area and Key Laboratory for Water Quality and Conservation of the Pearl River Delta (Ministry of Education) at Guangzhou University through the Overseas Young Talents Program. Her research interests focus on environmental microorganism and nanotoxicity. Since 2020, she has served as a council member of the Guangdong Society of Environmental Mutagens.

nanomaterials

Editorial

Predictive and Mechanism-Based Toxicity Evaluation of Engineered Nanoparticles

Rongrong Liu * and Bing Yan

Key Laboratory for Water Quality and Conservation of the Pearl River Delta, Institute of Environmental Research at the Greater Bay Area, Ministry of Education, Guangzhou University, Guangzhou 510006, China; drbingyan@yahoo.com
* Correspondence: liurr@gzhu.edu.cn

The rapid expansion of nanotechnology has driven the widespread use of engineered nanomaterials in energy, electronics, medicine, and consumer products [1,2]. Inevitably, however, these materials are released into the environment during their production, use, and disposal, raising increasing concern about their potential ecological and human health risks [3,4]. Owing to their nanoscale dimensions, large specific surface area, and highly reactive surface chemistries, nanoparticles (NPs) interact with biological systems in ways fundamentally different from their bulk counterparts, often giving rise to novel bioavailability, transport, and toxicity profiles [5]. Consequently, the development of scientifically rigorous, reliable, and predictive frameworks for NP toxicity evaluation has become a prerequisite for the sustainable and responsible advancement of nanotechnology.

Nanotoxicology has evolved from early assessments of acute cytotoxicity toward a more comprehensive understanding of long-term, low-dose, and system-level biological effects. The contributions to this Special Issue reflect this transition and illustrate important progress on multiple fronts. With respect to toxicological outcomes, it is now evident that NP toxicity is strongly dependent on both material properties and exposure context. For example, carbonaceous nanoparticles—a major component of fine atmospheric particulate matter—are closely associated with adverse effects on the respiratory, cardiovascular, and nervous systems [6]. Likewise, silver nanomaterials exhibit pronounced shape-dependent toxicity, with spherical, cubic, and prismatic forms eliciting markedly different biological responses [7].

In parallel, this Special Issue highlights the diversification and increasing sophistication of toxicity assessment models. Studies span conventional in vitro systems (e.g., A549 and HepG2 cells) to whole-organism models, including rats, zebrafish, rainbow trout, and aquatic plants such as *Lemna minor* L. [7–11]. Importantly, model choice has emerged as a critical determinant of toxicological outcomes. Differences in nanoparticle uptake, intracellular processing, and stress responses between primary cells and immortalized cancer cell lines underscore the necessity of selecting biologically relevant systems [12]. Moreover, the integration of experimental data with computational nanotoxicology approaches offers a powerful strategy for screening and prioritizing emerging nanomaterials [13].

Despite these advances, several fundamental challenges remain. First, the intrinsic heterogeneity of NPs, encompassing their size, shape, surface chemistry, and colloidal stability, combined with highly variable exposure conditions severely limits cross-study comparability and predictive extrapolation. For example, the toxicity of MXenes is closely linked to their environmental stability and degradation pathways [14]. Second, the accurate identification, quantification, and speciation of NPs and their transformation products within complex biological matrices remain technically demanding, as illustrated by the

 https://doi.org/10.3390/nano16030185

in situ reduction of Ce(IV) to Ce(III) in plant tissues following CeO_2 NP exposure [10]. Third, although oxidative stress, inflammatory signaling, and endocrine disruption are frequently implicated, a mechanistically integrated and system-level understanding of the effects of NP-induced toxicity—such as TiO_2 NP-mediated impairment of zebrafish gonadal function—has yet to be established [11].

In future, nanoparticle toxicity evaluation must transition toward more precise, mechanism-based, and predictive paradigms. This will require the systematic integration of experimental toxicology with advanced computational tools, including QSAR, machine learning, and deep learning models, to enable intelligent screening, hazard ranking, and exposure-effect prediction. Equally important, risk assessment frameworks must adopt a life-cycle perspective, accounting not only for pristine nanomaterials but also for their environmental transformation products and dynamic behavior in complex biological and ecological systems [15]. Only through such integrative and forward-looking approaches can society fully realize the benefits of nanotechnology while ensuring the protection of human health and environmental safety.

Author Contributions: Conceptualization, R.L. and B.Y.; Writing—original draft preparation, R.L.; writing—review and editing, R.L. and B.Y.; Funding acquisition, R.L. and B.Y. All authors have read and agreed to the published version of the manuscript.

Funding: This work was supported by the Guangdong Basic and Applied Basic Research Foundation (2024A1515011251, 2022A1515010366) and the Introduced Innovative R&D team project under the "The Pearl River Talent Recruitment Program" of Guangdong Province (2019ZT08L387).

Acknowledgments: We would like to express our sincere gratitude to all the authors of this Special Issue for their contributions and to the editorial staff of *Nanomaterials* for their professional support and guidance.

Conflicts of Interest: The authors declare no conflicts of interest.

References

1. Gupta, D.; Boora, A.; Thakur, A.; Gupta, T.K. Green and sustainable synthesis of nanomaterials: Recent advancements and limitations. *Environ. Res.* **2023**, *231*, 116316. [CrossRef] [PubMed]
2. Smita, S.; Gupta, S.K.; Bartonova, A.; Dusinska, M.; Gutleb, A.C.; Rahman, Q. Nanoparticles in the environment: Assessment using the causal diagram approach. *Environ. Health* **2012**, *11*, S13. [CrossRef] [PubMed]
3. Sengul, A.B.; Asmatulu, E. Toxicity of metal and metal oxide nanoparticles: A review. *Environ. Chem. Lett.* **2020**, *18*, 1659–1683. [CrossRef]
4. Wang, Y.-L.; Lee, Y.-H.; Chou, C.-L.; Chang, Y.-S.; Liu, W.-C.; Chiu, H.-W. Oxidative stress and potential effects of metal nanoparticles: A review of biocompatibility and toxicity concerns. *Environ. Pollut.* **2024**, *346*, 123617. [CrossRef] [PubMed]
5. Yan, X.; Yue, T.; Winkler, D.A.; Yin, Y.; Zhu, H.; Jiang, G.; Yan, B. Converting Nanotoxicity Data to Information Using Artificial Intelligence and Simulation. *Chem. Rev.* **2023**, *123*, 8575–8637. [CrossRef] [PubMed]
6. Aslam, I.; Roeffaers, M.B.J. Carbonaceous Nanoparticle Air Pollution: Toxicity and Detection in Biological Samples. *Nanomaterials* **2022**, *12*, 3948. [CrossRef] [PubMed]
7. Auclair, J.; Turcotte, P.; Gagnon, C.; Peyrot, C.; Wilkinson, K.J.; Gagné, F. Form-Dependent Toxicity of Silver Nanomaterials in Rainbow Trout Gills. *Nanomaterials* **2023**, *13*, 1356. [CrossRef] [PubMed]
8. Hlaing, C.B.; Chariyakornkul, A.; Pilapong, C.; Punvittayagul, C.; Srichairatanakool, S.; Wongpoomchai, R. Assessment of Systemic Toxicity, Genotoxicity, and Early Phase Hepatocarcinogenicity of Iron (III)-Tannic Acid Nanoparticles in Rats. *Nanomaterials* **2022**, *12*, 1040. [CrossRef] [PubMed]
9. Lyons-Darden, T.; Blum, J.L.; Schooley, M.W.; Ellis, M.; Durando, J.; Merrill, D.; Oller, A.R. An Assessment of the Oral and Inhalation Acute Toxicity of Nickel Oxide Nanoparticles in Rats. *Nanomaterials* **2023**, *13*, 261. [CrossRef] [PubMed]
10. Liu, Y.; Zhao, X.; Ma, Y.; Dai, W.; Song, Z.; Wang, Y.; Shen, J.; He, X.; Yang, F.; Zhang, Z. Interaction of Cerium Oxide Nanoparticles and Ionic Cerium with Duckweed (*Lemna minor* L.): Uptake, Distribution, and Phytotoxicity. *Nanomaterials* **2023**, *13*, 2523. [CrossRef] [PubMed]

11. Scalisi, E.M.; Pecoraro, R.; Salvaggio, A.; Capparucci, F.; Fortuna, C.G.; Zimbone, M.; Impellizzeri, G.; Brundo, M.V. Titanium Dioxide Nanoparticles: Effects on Development and Male Reproductive System. *Nanomaterials* **2023**, *13*, 1783. [CrossRef] [PubMed]
12. Kim, I.Y.; Kwak, M.; Kim, J.; Lee, T.G.; Heo, M.B. Comparative Study on Nanotoxicity in Human Primary and Cancer Cells. *Nanomaterials* **2022**, *12*, 993. [CrossRef] [PubMed]
13. Forest, V. Experimental and Computational Nanotoxicology—Complementary Approaches for Nanomaterial Hazard Assessment. *Nanomaterials* **2022**, *12*, 1346. [CrossRef] [PubMed]
14. Wu, J.; Yu, Y.; Su, G. Safety Assessment of 2D MXenes: In Vitro and In Vivo. *Nanomaterials* **2022**, *12*, 828. [CrossRef] [PubMed]
15. Chen, Y.; Wang, J.; Chen, D.; Wang, B.; Wu, J.; Liu, R.; Li, Q. The Pd (II) Reduction Mechanisms in Bacillus megaterium Y-4 Revealed by Proteomic Analysis. *Nanomaterials* **2024**, *14*, 512. [CrossRef] [PubMed]

https://doi.org/10.3390/nano16030185

nanomaterials

Review

Safety Assessment of 2D MXenes: In Vitro and In Vivo

Jialong Wu [1], Yanyan Yu [2,*] and Gaoxing Su [2,*]

[1] Key Laboratory for Water Quality and Conservation of the Pearl River Delta, Ministry of Education, Institute of Environmental Research at Greater Bay, Guangzhou University, Guangzhou 510006, China; 2111904062@e.gzhu.edu.cn
[2] School of Pharmacy, Nantong University, Nantong 226001, China
* Correspondence: yuyanyan@ntu.edu.cn (Y.Y.); sugaoxing@ntu.edu.cn (G.S.)

Abstract: MXenes, representing a new class of two-dimensional nanomaterial, have attracted intense interest in a variety of fields as supercapacitors, catalysts, and sensors, and in biomedicine. The assessment of the safety of MXenes and related materials in biological systems is thus an issue that requires significant attention. In this review, the toxic effects of MXenes and their derivatives are summarized through the discussion of current research into their behaviors in mammalian cells, animals and plants. Numerous studies have shown that MXenes have generally low cytotoxicity and good biocompatibility. However, a few studies have indicated that MXenes are toxic to stem cells and embryos. These in vitro and in vivo toxic effects are strongly associated with the dose of material, the cell type, the mode of exposure, and the specific type of MXene. In addition, surface modifications alter the toxic effects of MXenes. The stability of MXenes must be considered during toxicity evaluation, as degradation can lead to potentially toxic byproducts. Although research concerning the toxicity of MXenes is limited, this review provides an overview of the current understanding of interactions of MXenes with biological systems and suggests future research directions.

Keywords: MXenes; 2D materials; toxicology; safety assessments; in vitro; in vivo

1. Introduction

Due to the remarkable properties of graphene, two-dimensional (2D) materials have attracted extensive scientific and engineering interest. In 2011, a novel class of inorganic 2D materials named MXenes was discovered by scientists at Drexel University [1]. MXenes are represented by the general chemical formula $M_{n+1}X_nT_x$, in which M represents an early transition metal (e.g., Ti, V, Sc, Nb, Zr, Mo, Hf, or Ta), X is C and/or N, n = 1 to 6, and T_x stands for the surface termination groups (typically, –O, –OH, –F, –Cl, –Br, or –S). To date, more than 30 stoichiometric MXenes have been experimentally produced, and over 70 MXenes have been theoretically predicted [2]. Among them, $Ti_3C_2T_x$ is the most widely studied.

MXenes are fabricated by etching layers of A-elements including Al, Si, Ga, and Ge, containing the parent transition metal carbides and/or nitrides in the so-called MAX phase. The A layers are selectively etched without destroying the M−X bonds because of the stronger chemical activity of M−A bonds relative to M−X bonds. As the resulting 2D M-X materials have graphene-like morphologies, they have been termed "MXenes".

MXenes exhibit favorable properties for use in electronics, mechanics, magnetism, and optics because the layers are only a few atoms thick [3]. The use of MXenes in multiple fields, especially in biomedical applications, is facilitated by abundant polar surface functional groups that make MXenes hydrophilic. In addition, the preparation of MXenes is relatively easy and can be readily scaled up. These properties make MXenes valuable to a variety of fields, including supercapacitors [4], catalysis [5], sensors [6], electrochemical energy storage [2], optoelectronics [7], environmental remediation [8,9], electromagnetic interference shielding [10] and biomedicine [11,12].

With the rapid growth in the production and application of MXenes, the release of these materials into the environment is inevitable, and potential toxicities of MXenes should be carefully evaluated. Interactions with physiological systems is especially important due to the increasing interest in biomedical applications of MXenes, including in anticancer therapy [13], antibacterial therapy [14], biosensing [15,16], bioimaging [17], radioprotection [18], combined therapy and theranostics [11,19]. Before these materials are approved for clinical use, the basic interactions with physiological systems should be understood, and evaluations of biocompatibility are essential. The assessment of the safety of MXenes is important both regarding the use of MXenes in nanomedicine and regarding implications for the synthesis and widespread use of MXenes for the environment.

Once nanomaterials come in contact with cells, they could be internalized into cells or adhered to cell membranes. Intracellular homeostasis may then be disrupted due to the physical presence of the nanomaterial and its interactions with organelles, cell membranes, proteins, lipids, and other cellular structures. This disruption may lead to apoptosis [20], autophagy [21], neurosis [20], immunostimulation [22], and other deleterious effects. Two-dimensional materials with ultra-thin nanostructures are likely to enter into multiple cell types [23], and cytotoxicity has been observed experimentally [24]. In vivo studies have demonstrated that 2D materials can accumulate in organs including liver, spleen, kidneys and lungs. Numerous in vivo and in vitro investigations into mechanisms and consequences exposure to 2D ultrathin nanostructures have been reported [24,25]. However, since MXenes are relatively new 2D materials, a limited number of studies regarding the toxicity of this material have been carried out. Some of these studies focus on the biomedical applications of MXenes, and evaluations of toxic effects are merely tangential aspects of the work. Generally, MXenes have been reported as having no obvious toxicity in these cell-based or short-term experiments. Thus far, deep investigations about physiological influences and basic interactions between MXenes and physiological environments have been rare, but they are in high demand.

In this review, we first introduce the synthesis of MXenes, paying particular attention to important hazards related to the synthetic procedures. Next, recent studies regarding interactions between MXenes and various classes of mammalian cells and tissues are summarized, while noting that the toxicology and mechanisms of toxicity vary depending on cell type and route and extent of exposure. The toxicological profiles also depend on the physicochemical properties of the MXenes, including stability and surface chemistry. Finally, we conclude this review with future perspectives for the assessment of the safety production and application of MXenes.

2. Synthesis of MXenes and Safety Concerns

Among the multiple bottom-up and top-down approaches to MXene synthesis, etching of A-element layers from the MAX phase is the most popular strategy (Figure 1) [1,26,27]. In this process, the MAX phase powder is stirred in aqueous hydrofluoric acid (HF), and after an appropriate amount of time, the solid is separated from the supernatant by centrifugation or filtration. After washing, the dense MAX particles are converted to a loosely packed accordion-like structure, which is referred to as multilayer MXenes. Subsequent gentle sonication results in the exfoliation of MXene nanosheets, and 10 or fewer layers of terminated MXenes, with thicknesses of approximately 10 nm, are ultimately obtained.

As the production of MXene materials is scaled up to commercial levels, it is important to analyze the safety of the synthetic procedure. Dust ignition, runaway reactions, and toxic chemical exposure are the most important potentially hazardous factors in MXene synthesis (Figure 2) [28]. Firstly, in MAX phase synthesis, hazards associated with dust production are present in both reactant handling and postsynthesis processing procedures. Using alternative milling methods may increase safety.

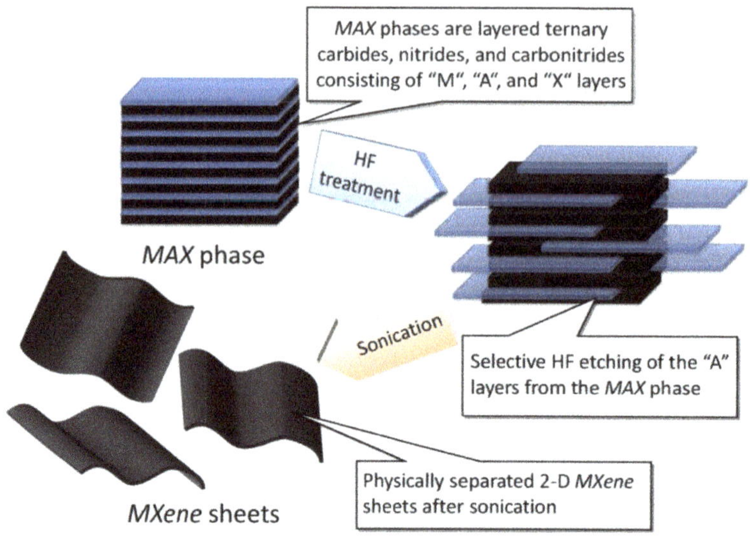

Figure 1. Synthesis of MXenes from the corresponding MAX phases. Reproduced with permission from [27]. Copyright American Chemical Society, 2012.

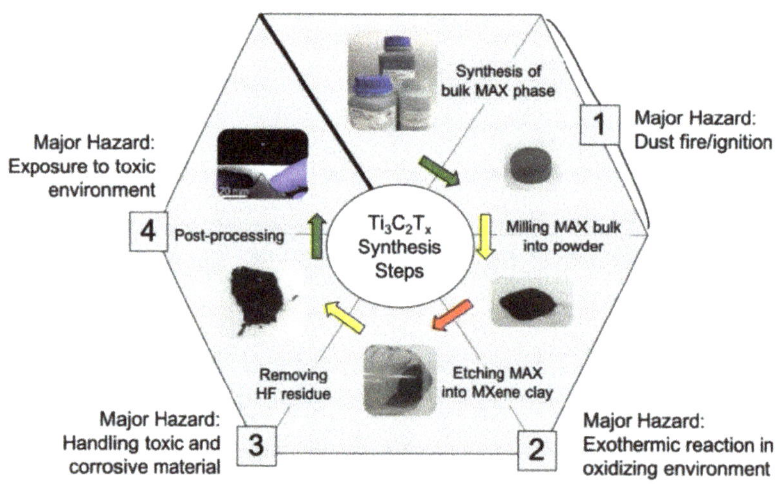

Figure 2. Safety concerns in the synthesis of $Ti_3C_2T_x$ from raw materials. Reproduced with permission from [28]. Copyright American Chemical Society, 2019.

Next, a significant hazard exists in the process of etching the Al layer in the MAX phase. If the Ti_3AlC_2 MAX phase is added too quickly, corrosive materials can leave containment, and toxic fumes can be produced. In particular, the most widely used etching methods employ HF, which is extremely toxic and containment of HF represents a significant challenge to the industry. When using HF, HF-specific personal protective equipment should be worn to minimize HF exposure [29]. Additionally, if the $Ti_3C_2T_x$ clay is not properly washed, HF-related hazards persist throughout the procedure, even into postprocessing, and this durability adds to the toxicity found through in vitro and in vivo studies.

Due to the hazardous nature of HF, researchers have explored an alternative milder etching processes, including electrochemistry [30], alkali treatment [31], hydrothermal methods [32] and the use of Lewis acidic molten salts [33]. In addition to using sonication for exfoliation, organic solvents including dimethyl sulfoxide, *n*-butylamine, tetrabutylammonium hydroxide and choline hydroxide have been investigated for the generation of MXenes [3,34]. Furthermore, multilayer MXenes can be immersed in such solvents to swell, which with further sonication would result in the formation of colloidal particles. After centrifugation, homogeneous MXene nanosheets were prepared.

3. In Vitro Toxic Effects of MXenes

Toxicity concerns are not only associated with the production of MXenes, in addition, the use of MXenes, especially in medical related application, may be associated with safety issues. Cell-based assays are usually initially employed to evaluate the toxic effects of engineered nanomaterials in vitro (Table 1). As $Ti_3C_2T_x$ nanosheets are the most widely applied MXenes, the cytotoxicity of this material was the first to be evaluated. Zhang et al. revealed that there was no obvious acute cytotoxicity to human umbilical vein endothelial cells (HUVECs) after treatment with two different concentrations of $Ti_3C_2T_x$ nanosheets (100 and 500 $\mu g\ mL^{-1}$) [35]. While scanning electron microscopy (SEM) images of MXene-treated cells showed that $Ti_3C_2T_x$ nanosheets contacted cell membranes and formed spherical structures, the ratios of living, apoptotic, and necrotic cells after treatment were similar to those of the control group. However, at a high dose, a significant metabolic shift was observed in HUVECs, which was reflected as the enhancement of glycolysis, fatty acid biosynthesis, and lipid accumulation. On the other hand, in experiments using HUVECs as an in vitro model, Gu et al. found that Ti_3C_2 quantum dots (QDs) exhibited higher cytotoxic effects than did Nb_2C QDs [36]. After treatment with 25 $\mu g\ mL^{-1}$ of Ti_3C_2 QDs, the cellular viability was greater than 70%, but higher concentrations (50 and 100 $\mu g\ mL^{-1}$) of Ti_3C_2 QDs led to significantly lower cellular viability as compared with Nb_2C QDs. For Nb_2C QDs, there was no significant cytotoxicity after being treated with these three concentrations. Due to the different cytotoxicities of MXene QDs relative to other forms of MXene, it is likely that the MXene structure has an important impact on its toxicity.

Treating human mesenchymal stem cells (hMSCs) with a high dose of $Ti_3C_2T_x$ nanosheets (>50 $\mu g\ mL^{-1}$) led to obvious cytotoxicity, as shown in [37]. At a dose of 100 $\mu g\ mL^{-1}$, hMSCs displayed a significant decrease of cell proliferation early in this study. As for cells treated with 50 $\mu g\ mL^{-1}$ of MXenes, the first decrease of cell proliferation was noted after 5 days of incubation.

In a similar finding from another lab [38], $Ti_3C_2T_x$ nanosheets were shown to induce dose-dependent cytotoxicity to primary neural stem cells (NSCs) and NSC-derived differentiated cells. At a dose of 12.5 $\mu g\ mL^{-1}$, $Ti_3C_2T_x$ nanosheets exhibited no observable adverse effects on NSCs or NSCs-derived differentiated cells. However, at 25 $\mu g\ mL^{-1}$, $Ti_3C_2T_x$ nanosheets caused significant cytotoxicity and were internalized into the NSCs upon compromising of the cell membrane. Subsequent biological assays revealed that the 25 $\mu g\ mL^{-1}$ dose significantly increased the rate of apoptosis of the NSCs. Furthermore, 198 differently expressed genes (DEGs), which were mainly associated with the extracellular regions, were identified. The characteristics of these DEGs suggested that the decrease of viability of NSCs might due to the disruption of the cell membrane upon internalization of nanosheets at high doses.

In addition to being dependent on the dose, the in vitro toxic effects of MXenes are also dependent on the target cell types. Jastrzebska et al. investigated the cytotoxicity in two normal (MRC-5 and HaCaT) and two cancerous (A549 and A375) cell lines [39]. For $Ti_3C_2T_x$ nanosheets, the toxic effects were higher against the cancerous cells in comparison to normal cells. Scheibe et al. also found that the short-term cytotoxicities of different $Ti_3C_2T_x$ MXene structures and their precursors (TiC, Ti_2AlC, and Ti_3AlC_2) were higher when they were applied to cancer-derived HeLa cells than when they were applied to normal human fibroblasts [40]. The higher ROS generation and stronger adhesion to cell membranes brought about higher cytotoxicity to cancer cells. In addition, for HeLa cells, MXene structures showed higher cell viability in comparison to MAX phases.

Previous studies have demonstrated that oxidative stress was the most probable cause of the cytotoxicity of 2D materials. However, images from confocal laser scanning microscopy demonstrated higher generation of reactive oxygen species (ROS) upon treatment with MXenes in comparison to their precursor MAX phases. As MXenes tend to be more toxic, this finding indicated that cytotoxicity pathways are not strongly dependent on the ROS production, but rather on direct physicochemical interactions of larger particles with the cell membranes. These mechanisms were further confirmed through the evaluation of the cytotoxicity of the Ti_2NT_x MXene [41]. In this experiment, Ti_2NT_x nanosheets also showed higher toxicity towards cancerous cell lines in comparison to normal ones, but intracellular ROS production only increased in A375 cells. Meanwhile, in each normal cell line, and in MCF-7 cells, their ROS-specific fluorescence remained constant over time and relative to controls. Therefore, the mechanism of cytotoxicity of Ti_2NT_x nanosheets was not solely dependent on the generation of ROS, but instead, internalization of the 2D sheets and cell membrane disruption were also the mechanisms leading to the toxic effects of MXenes.

Notably, the cytotoxicity of MXenes has been applied in a therapeutic way. Due to the differential toxic effects of MXenes towards normal cells, and because MXenes have been shown to have photothermal effects, MXenes have been developed as cancer phototherapeutic agents [10]. Two cancer cell lines, breast 4T1 cancer cells and glioma U87 cancer cells, were treated with polyvinyl pyrrolidone (PVP)-coated Nb_2CT_x nanosheets at various concentrations from 0 to 200 μg mL^{-1}. After a 24 h incubation, Nb_2CT_x nanosheets showed negligible effects on the survival of both 4T1 and U87 cells, even the highest dose. However, after incubation with 40 μg mL^{-1} Nb_2CT_x nanosheets for 4 h followed by infrared laser irradiation (λ = 808 or 1064 nm), the cancer cells exhibited significant growth inhibition. The same lab also performed similar photothermal therapy using soybean phospholipid-modified $Ti_3C_2T_x$ nanosheets. These nanomaterials showed negligible effect on the viability of 4T1 cells, even at a dose of 400 μg mL^{-1} [42]. When $Ti_3C_2T_x$ nanosheets and near infrared laser irradiation were applied the majority of 4T1 cells were killed, suggesting that the increased intracellular temperature promoted cancer cell ablation.

In addition, Ti_2N QDs have also been explored as photothermal therapeutic agents [43]. At a dose of 80 μg mL^{-1} Ti_2N QDs exhibited no cytotoxicity to all tested cells, including 293T human embryonic kidney cells, 4T1 murine mammary carcinoma cells, and U87 human malignant glioma cells (Figure 3). However, supplementing this treatment with near infrared laser irradiation for 5 min led to almost all of the cancer cells being killed. These studies suggested tremendous potential for the use of MXenes in cancer treatment.

Furthermore, 3D cell culture systems have also been used to evaluate the cytotoxicity of Ti_2CT_x MXene in vitro [44]. Using microscopy, it was observed that MXenes possessed a strong binding capacity to the cellular membranes of HeLa cells due to the electrostatic interactions between the negatively charged MXenes and positively charged proteins on the cell surface. Accordingly, cell viability decreased from 98% to 43% as the dose of Ti_2CT_x MXenes increased from 50 to 300 μg mL^{-1}, indicating that the cytotoxicity was dose dependent. In the 3D cell culture system, the cytotoxicity of MXenes was evaluated by dispersing the Ti_2CT_x MXenes at doses of 50 μg mL^{-1} and 500 μg mL^{-1}, among HeLa cells encapsulated in alginate microbeads. Using fluorescence microscopy, dead cells were observed in the area where the agglomerates of Ti_2CT_x MXenes directly bound to the surface of the 3D microtissues. The direct adhesion of Ti_2CT_x MXenes to the cell membranes of microtissues limited the growth of microtissues and cell division, resulting in lower cell viability. Importantly, oxidative stress assays showed that there was no significant difference in ROS level between control cells and Ti_2CT_x MXene-treated cells, meaning that the cytotoxicity pathways of Ti_2CT_x MXene were not associated with oxidative stress, but instead were influenced by the direct physicochemical interactions of Ti_2CT_x MXene with the cell membrane.

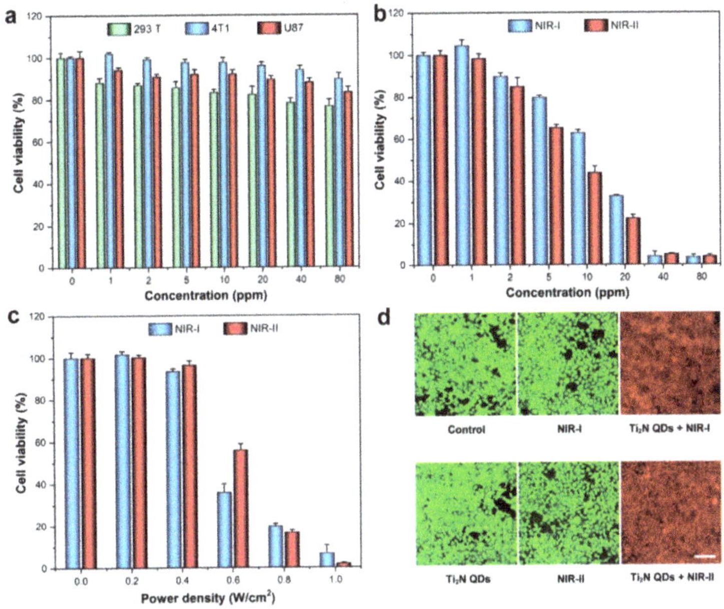

Figure 3. (**a**) Cell viability of 293T, 4T1, and U87 cells after 24 h incubation with different concentrations of Ti_2N QDs. (**b**) Cell viability of 4 T1 cells co-cultured with different concentrations of Ti_2N QDs for 4 h after 5 min irradiation with 808 or 1064 nm lasers. (**c**) Cell viability of 4T1 cells after incubation with Ti_2N QDs (40 μg mL^{-1}) irradiated with 808 or 1064 nm laser at different power densities for 5 min. (**d**) Confocal fluorescence images of calcein-AM (green, live cells) and PI (red, dead cells) co-stained cells after the treatment (scale bars: 100 μM). Reproduced with permission from [43]. Copyright Elsevier, 2020.

As in vitro cytotoxicity evaluations are time-consuming and expensive, computational machine learning methods have recently been applied to predict the interactions between MXenes and cells. Marchwiany et al. constructed the predictive model of 2D MXenes and identified that the most relevant highly surface-specific features that might be responsible for the cytotoxic behavior of MXenes [45]. The machine learning models predicted that two types of 2D MXenes could exhibit cytotoxic properties with high probabilities, while other MXenes studied were predicted to be non-toxic. In the non-toxic MXenes, there was no presence of M_xO_y on the surface, suggesting that the presence of M_xO_y was the key toxicity-generating feature. This study also identified the presence of surface Li^+ ions as an important cytotoxicity-generating feature. Overall, these predictions indicated that the crucial factor for cytotoxicity of MXenes is not the structure of the materials, but the surface properties.

Interactions between MXenes and immune cells were also explored. Rafieerad et al. employed MXene (Ti_3C_2) QDs for immunomodulation and tissue repair [46]. In a simulated lymphocyte population, Ti_3C_2 QDs was shown to selectively reduce the activation of $CD4^+IFN\text{-}\gamma^+$ T-lymphocytes, while promoting the expansion of immunosuppressive regulatory T-cells. After being incorporated into a chitosan-based hydrogel, the constructed platform enhanced the stem cell delivery and tissue repair. The immunomodulatory properties of MXenes remind us to pay attention to the immunotoxicity of MXenes. Overall, the cytotoxicity of MXene has been found to depend on dose, material type, and cell type. At low doses, MXenes exhibits low cytotoxicity and good biocompatibility. Meanwhile, MXenes showed higher toxicity to cancer cells as compared to normal cells, and are more toxic to stem cells than to differentiated cells. Ultrasmall MXene QDs tend to exhibit higher cytotoxicity than nanosheet materials. Unlike other 2D nanomaterials, the cytotoxicity of MXenes is not strongly associated with the intracellular ROS, but instead with the stable binding of MXenes to cell membranes. Upon near-infrared (NIR) irradiation, certain MXenes showed strong photothermal therapy effects in promoting cancer cell ablation. Very recently, artificial intelligence methods were also used to verify and even to predict the cytotoxicity, and surface properties of MXene were identified as the crucial factors leading to cytotoxicity. Computer-based modeling thus represents a new way to gain additional knowledge regarding the in vitro toxicity of MXenes.

Table 1. Summary of in vitro toxicity of MXenes.

Type of MXenes	Type of Cells	Dose	Toxicity Effects	Reference
$Ti_3C_2T_x$	human umbilical vein endothelial cells (HUVECs)	100 and 500 μg mL^{-1}, 48 h	No obvious acute cytotoxicity. The ratios of living, apoptotic, and necrotic cells exhibited patterns similar to those of the control group.	[35]
Ti_3C_2 QDs, Nb_2C QDs	HUVECs	0–100 μg mL^{-1}, 24 h	At 25 μg mL^{-1} of Ti_3C_2 QDs, the cellular viability was larger than 70%. While at 50 and 100 μg mL^{-1}, Ti_3C_2 QDs led to significantly lower cellular viability compared with Nb_2C QDs. For Nb_2C QDs, there was no significant cytotoxicity after treated with these three concentrations.	[36]
$Ti_3C_2T_x$	human mesenchymal stem cells (hMSCs)	0–100 μg mL^{-1}, 7 days	>50 μg mL^{-1}, obvious cytotoxicity was shown	[37]
$Ti_3C_2T_x$	neural stem cells (NSCs) and NSCs-derived differentiated cells	12.5–100 μg mL^{-1}, 24 h	At 25 μg mL^{-1}, $Ti_3C_2T_x$ nanosheets caused significant cytotoxicity to NSCs. 198 differently expressed genes (DEGs) which were mainly associated with the extracellular regions were identified.	[38]
$Ti_3C_2T_x$	A549, MRC-5, A375, and HaCaT cells	0–500 μg mL^{-1}, 24 h	Concentration dependent cytotoxicity. Toxic effects were higher against cancerous cells in comparison to normal ones.	[39]
$Ti_3C_2T_x$ MXene structures and their precursors (TiC, Ti_2AlC, and Ti_3AlC_2)	HeLa cells and normal fibroblasts (MSU1.1)	10–400 μg mL^{-1}, 24–48 h	Concentration dependent and cell-type dependent cytotoxicity. MXene structures showed higher cell viability in comparison to MAX phases.	[40]
Ti_2NT_x	human skin malignant melanoma cells, human immortalized keratinocytes, human breast cancer cells, and normal human mammary epithelial cells	62.5–500 μg mL^{-1}, 24 h	Higher toxicity towards cancerous cell lines in comparison to normal ones.	[41]
Nb_2CT_x	breast 4T1 cancer cells and glioma U87 cancer cells	0–200 μg mL^{-1}, 24 h	Negligible effect on the cell viability at 200 μg mL^{-1}. After exposed to 808 or 1064 nm laser, cancer cells were killed significantly with the increase of laser intensity.	[13]

Table 1. *Cont.*

Type of MXenes	Type of Cells	Dose	Toxicity Effects	Reference
Ti_2N QDs	293T, 4T1, and U87	0–80 μg mL^{-1}, 24 h	Adding irradiating with 808 or 1064 nm lasers for 5 min after incubation with the Ti_2N QDs, cancer cells were almost completely killed.	[43]
Ti_2CT_x	3D HeLa cell culture system	0–500 μg mL^{-1}, 24 h	The direct adhesion of Ti_2CT_x MXenes to the cell membrane of microtissues limited the growth of microtissues and cell division, resulting in lower cell viability.	[44]

4. Toxic Effects of MXenes In Vivo

Zebrafish (*Danio rerio*) are the most often used model animals in in vivo investigations of the toxicity of nanomaterials. The utility of zebrafish in this regard is due to their high genetic homology to humans, small size, rapid reproduction and development and their transparency during early growth stages [47]. Nasrallah et al. employed a zebrafish embryo model to evaluate the potential in vivo acute toxicity of $Ti_3C_2T_x$ nanosheets [48]. No cumulative mortality was observed at a dose of 50 μg mL^{-1}. At 100 μg mL^{-1}, however, the cumulative mortality rate was 21%, and this dose represented the lowest concentration with an observed effect. According to a dose curve generated at 96 h post fertilization, the calculated dose that led to 50% lethality (LC_{50}) was 257.46 μg mL^{-1}. In addition to lethality, other assays revealed that $Ti_3C_2T_x$-treated embryos exhibited a normal locomotion behavior at the lowest dose (50 μg mL^{-1}), nor did $Ti_3C_2T_x$ nanosheets at this concentration influence neuron or muscle activity of the zebrafish embryos. Therefore, according to the Acute Toxicity Rating Scale by the Fish and Wildlife Service (FWS), $Ti_3C_2T_x$ nanosheets can be categorized as the "practically nontoxic" group.

Alhussain et al. employed chicken embryos to investigate the developmental toxicity of MXenes [49]. At a dose of 30 μg per embryo, after 5 days incubation, $Ti_3C_2T_x$ nanosheets provoked a significant toxicity on the chicken embryos, as approximately 46% of embryos died shortly after exposure. Moreover, at the tested dose, a significant inhibition of angiogenesis in the chorioallantoic membrane was also observed. These toxic effects may be attributed to the downregulation of several genes that control cell proliferation, survival, cell death and angiogenesis, including *ATF3*, *FOXA2*, *INHBA*, *SERPINA3* and *VEGF-C*. These findings demonstrated that MXenes potentially exert development toxicity on embryos. However, only one dose was tested in this study, and the method of exposure was debatable.

The recent reports of the use of MXenes for photothermal conversion and cancer photothermal therapy using near infrared radiation has led to a need for the in vivo evaluation of toxicity in a mouse model [13]. In Kunning mice, normal hematological parameters and standard blood biochemical indexes were confirmed and no statistical significance in these assays was observed in groups with 20 mg kg^{-1} Nb_2CT_x at 1, 7, and 28 days post-injection in comparison to the control group. These results indicated that no significant inflammation was caused by exposure to Nb_2CT_x nanosheets.

Hematoxylin and eosin staining assays were performed to evaluate histological changes of major organs, including the heart, liver, spleen, lung, and kidney. No significant histological abnormality was found during the treatment period after exposure to Nb_2CT_x nanosheets with or without near infrared light irradiation. These results revealed that MXenes are biocompatible and thus appropriate for in vivo photothermal ablation of cancer.

The biocompatibility of $Ti_3C_2T_x$ MXenes films was also evaluated when they were used for guided bone regeneration [50]. After implanting the films in subcutaneous sites and sites of calvarial defects in rats, $Ti_3C_2T_x$ MXene films was found to have no obvious inflammatory or toxic side effects, indicating their safety for in vivo uses.

To evaluate the impacts of MXenes on organ function, Sui et al. investigated the distribution and accumulation of $Ti_3C_2T_x$ nanosheets in major organs [51]. ICR mice were exposed to $Ti_3C_2T_x$ nanosheets by intravenous injection at a dose of 20 mg kg^{-1}. At 1, 3, 7, 14 and 28 days, the blood, lungs, heart, kidneys, liver, spleen and intestines were harvested, and Ti contents were measured. The researchers found that $Ti_3C_2T_x$ nanosheets were primarily distributed in the lung and liver (Figure 4). Those in the lungs were internalized by endothelial and alveolar epithelial cells, resulting in the decrease of surfactant protein B content in alveolar epithelial cells and causing an increase in Penh, an indicator of lung dysfunction. A decrease of respiratory rate was also observed. These results indicated that when $Ti_3C_2T_x$ nanosheets crossed the lung, they potentially exerted toxicity that led to respiratory disorders. $Ti_3C_2T_x$ nanosheets that accumulated in liver were endocytosed by sinusoidal endothelial cells and were gradually excreted via bile canaliculi. During the entire exposure period, inflammatory responses and pathological lesions, including within the small and large airways of the lung, were observed as not obviously abnormal. Several issues with the experimental design in this case mean that results should be interpreted with caution. For example, the stability of the material was not investigated. In addition, while the diameter of the $Ti_3C_2T_x$ nanosheets was stated to 100 to 200 nm, no statistical analysis of the tunneling electron microscope images of the nanosheets was presented.

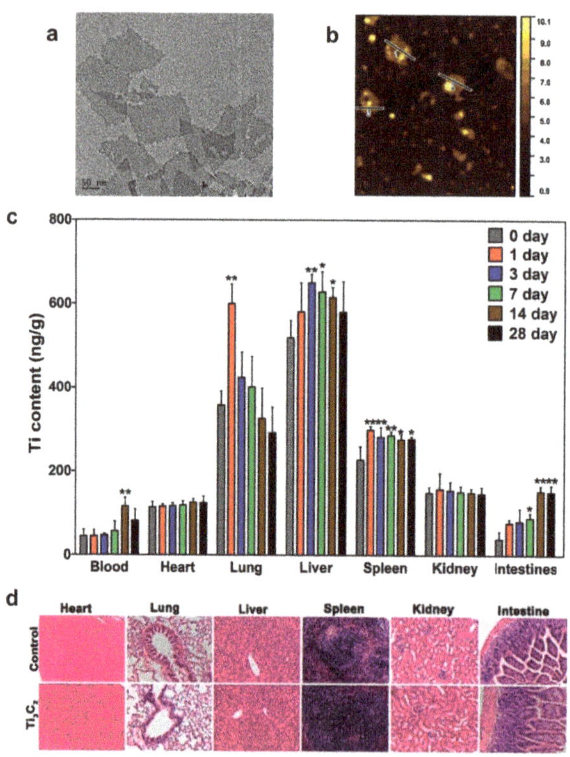

Figure 4. (**a**) TEM image of $Ti_3C_2T_x$ nanosheets. (**b**) AFM image of $Ti_3C_2T_x$ nanosheets. (**c**) Biodistribution of $Ti_3C_2T_x$ nanosheets in blood, heart, lung, liver, spleen, kidney, and intestines at 1, 3, 7, 14, 21 and 28 days after intravenous injection into mice. * $p < 0.05$, ** $p < 0.01$. (**d**) Representative sections from mice tissues after treatment of $Ti_3C_2T_x$ nanosheets. Reproduced with permission from [51]. Copyright Elsevier, 2021.

As environmental release of MXenes or byproducts of synthesis is inevitable, toxicity studies in plants are also important. MXenes are fabricated by etching the A-layer of MAX phase, and MAX phase materials are wildly used in the fields of ceramics, catalysis, environment remediation, and biomedicine. Therefore, seed germination and growth of rice were employed to assess the potential phytotoxicity of MAX phase materials [52]. After foliar spraying at doses of 100 and 1000 $\mu g \cdot mL^{-1}$, Ti_3AlC_2 nanosheets inhibited the growth of seedlings. In contrast to previous studies of toxicity of MXenes, the inhibition in this case was found to be due to the generation of ROS. At a dose of 100 $\mu g \cdot mL^{-1}$, the stomatal aperture was increased to 78.6%. Meanwhile, the number of trichomes was increased by 100%. These responses were mainly caused by the entrance of titanium ions released from the Ti_3AlC_2 nanosheets. To reduce the entrance of external ions, rice changed the structure and function of the leaves so as to activate the immune system and enhance the activity of antioxidant enzymes. However, at high doses, the plant was not able resist abiotic stress, the cell membranes were damaged, and biomass and chlorophyll content were decreased. These eventually led the rice growth inhibition.

There are few studies on the in vivo toxicity of MXenes. The iv vivo toxicity evaluation of MXenes mentioned above was summarized in Table 2. In particular, barely any studies use environmental doses and environmentally relevant modes of exposure to evaluate the consequences of long-term exposure. Although in vivo toxicity can be observed at high doses in some studies, these assays do not provide enough evidence to suggest the safety of MXenes in biomedical applications.

Table 2. Summary of the in vivo toxicity of MXenes.

Type of MXenes	Type of Models	Dose	Toxicity Effects	Reference
$Ti_3C_2T_x$	Zebrafish embryos	25–200 $\mu g\ mL^{-1}$	The calculated LC_{50} was 257.46 $\mu g\ mL^{-1}$. At the concentration of 50 $\mu g\ mL^{-1}$, no acutoxicity or neurotoxicity was observed.	[48]
$Ti_3C_2T_x$	Chicken embryos	30 μg per embryo, 5 days incubation	Potential toxicity on the early stage of embryogenesis; down regulation of several controller genes of cell proliferation, survival, cell death and angiogenesis; inhibition of blood vessel development.	[49]
Nb_2CT_x	Kunming mice	20 $mg\ kg^{-1}$	No significant inflammation was caused. No significant histological abnormality was found. Nb_2CT_x are highly biocompatible.	[13]
$Ti_3C_2T_x$	ICR mice	20 $mg\ kg^{-1}$	$Ti_3C_2T_x$ nanosheets could accumulate in the liver lungs. Those in the lung might influence respiratory function via the downregulation of surfactant protein B in alveolar epithelial cells.	[51]
Ti_3AlC_2	Rice	0.1–1000 $\mu g\ mL^{-1}$	At the doses of 100 and 1000 $\mu g \cdot mL^{-1}$, Ti_3AlC_2 nanosheets inhibited the growth of seedlings due to the generation of ROS. At the dose of 100 $\mu g \cdot mL^{-1}$, the stomatal aperture was increased to 78.6%. Meanwhile, the number of trichomes was increased to 100%.	[52]

5. Properties Associated with the Safety of MXenes

5.1. Surface Functionalization

To improve the biocompatibility of MXenes, collagen has been conjugated to the surfaces of $Ti_3C_2T_x$ and Ti_2CT_x nanosheets [53]. The interactions between the MXenes and collagen mainly involved electrostatic forces, and zeta potential measurements were used to characterize and control the adsorption and desorption of collagen. The viability of A375 (human skin malignant melanoma cells), HaCaT (human immortalized keratinocytes), MCF-7 (human breast cancer cells) and MCF-10A (mammary epithelial cells) were evaluated upon treatment with surface-modified and unmodified nanosheets. It was found that the surface collagen modification significantly decreased the cytotoxicity, as well as reduced oxidative stress. The mechanism of increasing the biocompatibility of MXenes by collagen modification should be further investigated.

Rashid et al. conjugated polypropylene glycol (PPG) and polyethylene glycol (PEG) to the surfaces of $Ti_3C_2T_x$ nanosheets [54] and assessed their cytotoxicity towards normal (HaCaT and MCF-10A) and cancerous (MCF-7 and A375) cells. PEG-modified $Ti_3C_2T_x$ nanosheets were the most toxic to normal and cancer cell lines, followed by PPG-modified nanosheets, and the least toxic were bare MXenes. As photothermal therapy agents, PPG-modified nanosheets were safer to normal cells as compared to PEG-modified nanosheets. As polymer coating typically reduces the toxicity of nanomaterials, conclusions from this study should be further evaluated with additional in vitro or in vivo assays.

To integrate the photothermal therapy capabilities of MXenes with targeting and imaging functions, Hussein et al. fabricated two plasmonic $Ti_3C_2T_x$-based nanocomposites, i.e., Au/MXene and Au/Fe_3O_4/MXene nanocomposites [55]. In a human breast cancer cell lines MCF7, these nanocomposites showed similar photothermal therapy efficiencies. However, hybrid nanocomposites showed less in vivo toxicity than did pure MXene. The in vivo acute toxicity assays were conducted using zebrafish embryos. Au/MXene and Au/Fe_3O_4/MXene led to less embryonic mortality with LC_{50} values greater than 1000 μg mL^{-1} than pure MXene, with an LC_{50} value of 257.46 μg mL^{-1}. No explanations for this phenomenon were proposed and the mechanisms should be further explored.

In addition, Ti_2CT_x nanosheets can be surface modified by post-delamination using mild heating at 60 °C [56]. After thermal treatment, the cytotoxicity of Ti_2CT_x nanosheets showed selectively toxic to cancer cells compared to normal cells. At a concentration of 62.5 μg mL^{-1}, post-treatments, Ti_2CT_x nanosheets were lethal to A375 cells, but not to normal HaCaT cells. This difference was because of a Ti_2O_3 layer that was formed on the Ti_2CT_x nanosheets after mild oxidation at 60 °C. The presence of the Ti_2O_3 layer caused higher generation of ROS as compared to non-oxide Ti_2CT_x nanosheets.

5.2. Stability and Degradation

The dispersion state and sizes of the aggregates are important factors that influence cellular uptake and subsequent cytotoxicity [57]. It is thus important to evaluate aggregation states before performing toxicology evaluations. Nasrallah et al. tested the stability of $Ti_3C_2T_x$ nanosheets in seawater and E3 medium by monitoring the scattering signal of nanosheet aggregates [48]. It was observed that, at 50 μg mL^{-1}, a suspension of $Ti_3C_2T_x$ nanosheets with diameters of 200 nm was more stable than at higher concentrations. There was no significant change in the scattering signal over the 18 h incubation both in seawater and E3 medium. Xie et al. systematically investigated the colloidal properties and stability of $Ti_3C_2T_x$ nanosheets in aqueous solutions by adjusting the pH, ionic types, and ionic strength [58]. It was found the stability of $Ti_3C_2T_x$ nanosheets varies with ionic types and ionic strength after a 24 h incubation in different electrolytes (Figure 5). The effects of the electrolytes in destabilizing $Ti_3C_2T_x$ nanosheets follows the order of $CaCl_2$ > $MgCl_2$ > NaCl > Na_2HPO_4 > $NaHCO_3$. When the electrolyte reached a critical concentration, $Ti_3C_2T_x$ nanosheet suspensions became unstable and formed visible agglomerates. The aggregation states were accompanied by an increasing of zeta potentials and hydrodynamic diameters. Since the MXenes were suspended in culture medium for longer than 24 h, it would have been important to evaluate the stability of the MXene suspensions in the culture medium.

MXenes are susceptible to environmental oxidation and can be decomposed to transition metal oxides and carbide-derived carbon. The decomposed byproducts can also induce toxic effects in vitro and in vivo. Jastrzebska et al. studied the oxidation of the most unstable V_2CT_x nanosheets in cell culture media and the resulting toxicity [59]. V_2CT_x nanosheets in cell culture medium were found to undergo a rapid oxidation, leading to decomposition and entry into the mixture of vanadium oxides and carbide-derived-carbon after 48 h. The formed vanadium oxides were highly cytotoxic and the decreasing of cell viability was oxidation dependent. The mechanisms of toxicity were related to the disruption of cell cycle and cell membrane disintegration through contact with oxidized flakes.

In addition, biodegradation of MXenes was also investigated by incubation of Nb_2CT_x nanosheets with human myeloperoxidase (hMPO) and H_2O_2 in PBS [13]. The suspension turned translucent in 24 h, indicating the degradation of Nb_2CT_x nanosheets. However, the decomposed products and their toxicity were not studied in this report.

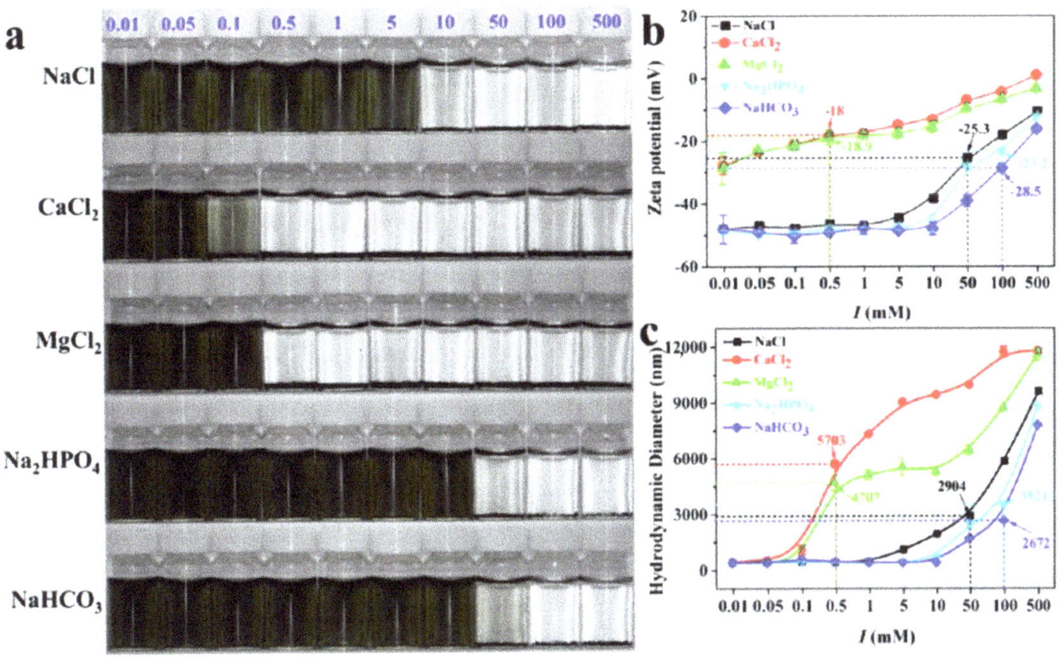

Figure 5. (**a**) Photo illustration of the effects of cations and anions with different concentrations (mM) on the aggregation behaviors of $Ti_3C_2T_x$ nanosheets. (**b**) Zeta potential of $Ti_3C_2T_x$ nanosheets in the presence of different electrolytes. (**c**) Hydrodynamic diameter of $Ti_3C_2T_x$ nanosheets in the presence of different electrolytes. Reproduced with permission from [58]. Copyright American Chemical Society, 2020.

6. Conclusions and Future Perspectives

Understanding the fundamental interactions between MXenes and biological systems and the toxic effects of MXenes in vitro and in vivo are of utmost importance for further development of MXene-based nanomedicines. These insights are also required for the development of safety guidelines for researchers working with this new class of 2D nanomaterial. Like other 2D nanomaterials, MXenes can enter cells and induce cytotoxicity. The available in vivo data suggests that MXenes exhibit potential pulmonary, reproductive, and developmental toxicities in laboratory animal and plant models. However, the information about the toxicity of MXenes remains very limited, especially when taking into account the many different types of MXenes and their potential modifications. The composition, size, surface chemistry, dose, stability, cell types, and means of exposure are all factors determining the in vitro and in vivo toxicity of MXenes. Oxidative stress and cell membrane disruption may be involved in the mechanisms of the toxicity of MXenes.

Toxicology evaluation methods have been well-established. However, we should pay particular attention to material characterization before using these methods, as toxicity varies with the physiochemical properties of MXenes. Key parameters, including size, morphology, layer number, zeta potential, surface chemistry, and dispersion states, influence the interactions between MXenes and biological systems and can lead to different toxicities. Many studies cited in this review did not provide full information regarding the physiochemical properties of the MXenes that were tested. In addition, while protein corona formation is known as a key factor affecting the interactions between MXenes and biological systems [60–62], no studies have considered this factor in toxicology studies. The influence of protein corona composition on MXenes and their behaviors in biological systems needs to be clarified so as to further the understanding of toxicological mechanisms.

This review demonstrated that studies have tended to focus on acute toxicological effects, but no long-term toxicology investigations at environmental doses have been performed. In addition to intravenous injection, oral and respiratory routes of exposure should also be considered. To completely understand the toxicity of MXenes, studies should be conducted using various types of well-characterized MXenes. In addition, toxicities to dermal, ocular, genomic and immunology systems should be performed. Considering the large number of MXene types and derivatives and the time consuming and expensive nature of toxicology evaluations, computer-aided toxicology methods, especially artificial intelligence models, should be employed as powerful tools for toxicology prediction and mechanism discovery [63,64]. Furthermore, the in vivo degradation, metabolization and excretion behaviors of MXenens also need to be established in future research. In short, more systematic investigations are still in high demand in order to address safety concerns before the wide practical application of MXenes.

Author Contributions: J.W., Y.Y. and G.S. designed this work of review. J.W. and G.S. performed the literature search of the databases. J.W., Y.Y. and G.S. wrote the manuscript. Y.Y. and G.S. revised the manuscript. All authors have read and agreed to the published version of the manuscript.

Funding: This work was supported by the National Natural Science Foundation of China (22076085), the Natural Science Foundation of Jiangsu Province for the Excellent Young Scholars (BK20211604), Qing-Lan Projects of Jiangsu Province, the Science and Technology Plan Projects of Nantong (JC2020031, XG202008-4, MS12021051).

Conflicts of Interest: The authors declare no conflict of interest.

References

1. Naguib, M.; Kurtoglu, M.; Presser, V.; Lu, J.; Niu, J.; Heon, M.; Hultman, L.; Gogotsi, Y.; Barsoum, M.W. Two-Dimensional Nanocrystals Produced by Exfoliation of Ti_3AlC_2. *Adv. Mater.* **2011**, *23*, 4248–4253. [CrossRef] [PubMed]
2. Anasori, B.; Lukatskaya, M.R.; Gogotsi, Y. 2D metal carbides and nitrides (MXenes) for energy storage. *Nat. Rev. Mater.* **2017**, *2*, 1–17. [CrossRef]
3. Lei, J.-C.; Zhang, X.; Zhou, Z. Recent advances in MXene: Preparation, properties, and applications. *Front. Phys.* **2015**, *10*, 276–286. [CrossRef]
4. Hu, M.; Zhang, H.; Hu, T.; Fan, B.; Wang, X.; Li, Z. Emerging 2D MXenes for supercapacitors: Status, challenges and prospects. *Chem. Soc. Rev.* **2020**, *49*, 6666–6693. [CrossRef]
5. Morales-García, A.N.; Calle-Vallejo, F.; Illas, F. MXenes: New horizons in catalysis. *ACS Catal.* **2020**, *10*, 13487–13503. [CrossRef]
6. Sajid, M. MXenes: Are they emerging materials for analytical chemistry applications?–A review. *Anal. Chim. Acta* **2021**, *1143*, 267–280. [CrossRef]
7. Jiang, X.; Kuklin, A.V.; Baev, A.; Ge, Y.; Ågren, H.; Zhang, H.; Prasad, P.N. Two-dimensional MXenes: From morphological to optical, electric, and magnetic properties and applications. *Phys. Rep.* **2020**, *848*, 1–58. [CrossRef]
8. Khandelwal, N.; Darbha, G.K. A decade of exploring MXenes as aquatic cleaners: Covering a broad range of contaminants, current challenges and future trends. *Chemosphere* **2021**, *279*, 130587. [CrossRef]
9. Rasool, K.; Pandey, R.P.; Rasheed, P.A.; Buczek, S.; Gogotsi, Y.; Mahmoud, K.A. Water treatment and environmental remediation applications of two-dimensional metal carbides (MXenes). *Mater. Today* **2019**, *30*, 80–102. [CrossRef]
10. Liu, J.; Zhang, H.B.; Sun, R.; Liu, Y.; Liu, Z.; Zhou, A.; Yu, Z.Z. Hydrophobic, flexible, and lightweight MXene foams for high-performance electromagnetic-interference shielding. *Adv. Mater.* **2017**, *29*, 1702367. [CrossRef]
11. Sivasankarapillai, V.S.; Somakumar, A.K.; Joseph, J.; Nikazar, S.; Rahdar, A.; Kyzas, G.Z. Cancer theranostic applications of MXene nanomaterials: Recent updates. *Nano-Struct. Nano-Objects* **2020**, *22*, 100457.
12. Lu, B.; Zhu, Z.; Ma, B.; Wang, W.; Zhu, R.; Zhang, J. 2D MXene Nanomaterials for Versatile Biomedical Applications: Current Trends and Future Prospects. *Small* **2021**, *17*, 2100946. [CrossRef]
13. Lin, H.; Gao, S.; Dai, C.; Chen, Y.; Shi, J. A Two-Dimensional Biodegradable Niobium Carbide (MXene) for Photothermal Tumor Eradication in NIR-I and NIR-II Biowindows. *J. Am. Chem. Soc.* **2017**, *139*, 16235–16247. [CrossRef]
14. Xu, X.; Wang, S.; Wu, H.; Liu, Y.; Xu, F.; Zhao, J. A multimodal antimicrobial platform based on MXene for treatment of wound infection. *Colloids Surf. B Biointerfaces* **2021**, *207*, 111979. [CrossRef] [PubMed]

15. Sinha, A.; Zhao, H.; Huang, Y.; Lu, X.; Chen, J.; Jain, R. MXene: An emerging material for sensing and biosensing. *TrAC Trends Anal. Chem.* **2018**, *105*, 424–435. [CrossRef]
16. Wu, L.; You, Q.; Shan, Y.; Gan, S.; Zhao, Y.; Dai, X.; Xiang, Y. Few-layer $Ti_3C_2T_x$ MXene: A promising surface plasmon resonance biosensing material to enhance the sensitivity. *Sens. Actuators B: Chem.* **2018**, *277*, 210–215. [CrossRef]
17. Huang, M.; Gu, Z.; Zhang, J.; Zhang, D.; Zhang, H.; Yang, Z.; Qu, J. MXene and Black Phosphorus Based 2D Nanomaterials in Bioimaging and Biosensing: Progress and Perspective. *J. Mater. Chem. B* **2021**, *9*, 5195–5220. [PubMed]
18. Ren, X.; Huo, M.; Wang, M.; Lin, H.; Zhang, X.; Yin, J.; Chen, Y.; Chen, H. Highly catalytic niobium carbide (MXene) promotes hematopoietic recovery after radiation by free radical scavenging. *ACS Nano* **2019**, *13*, 6438–6454. [CrossRef]
19. Wang, Y.; Feng, W.; Chen, Y. Chemistry of two-dimensional MXene nanosheets in theranostic nanomedicine. *Chin. Chem. Lett.* **2020**, *31*, 937–946. [CrossRef]
20. Andón, F.T.; Fadeel, B. Programmed cell death: Molecular mechanisms and implications for safety assessment of nanomaterials. *Acc. Chem. Res.* **2013**, *46*, 733–742. [CrossRef]
21. Zhou, X.; Jia, J.; Luo, Z.; Su, G.; Yue, T.; Yan, B. Remote Induction of Cell Autophagy by 2D MoS2 Nanosheets via Perturbing Cell Surface Receptors and mTOR Pathway from Outside of Cells. *Acs Appl. Mater. Interfaces* **2019**, *11*, 6829–6839. [CrossRef]
22. Wei, M.; Chen, N.; Li, J.; Yin, M.; Liang, L.; He, Y.; Song, H.; Fan, C.; Huang, Q. Polyvalent immunostimulatory nanoagents with self-assembled CpG oligonucleotide-conjugated gold nanoparticles. *Angew. Chem.* **2012**, *124*, 1228–1232. [CrossRef]
23. Mu, Q.; Su, G.; Li, L.; Gilbertson, B.O.; Yu, L.H.; Zhang, Q.; Sun, Y.-P.; Yan, B. Size-Dependent Cell Uptake of Protein-Coated Graphene Oxide Nanosheets. *Acs Appl. Mater. Interfaces* **2012**, *4*, 2259–2266. [CrossRef]
24. Seabra, A.B.; Paula, A.J.; de Lima, R.; Alves, O.L.; Durán, N. Nanotoxicity of Graphene and Graphene Oxide. *Chem. Res. Toxicol.* **2014**, *27*, 159–168. [CrossRef]
25. Huang, H.; Feng, W.; Chen, Y. Two-dimensional biomaterials: Material science, biological effect and biomedical engineering applications. *Chem. Soc. Rev.* **2021**, *50*, 11381–11485. [CrossRef]
26. Naguib, M.; Mochalin, V.N.; Barsoum, M.W.; Gogotsi, Y. 25th Anniversary Article: MXenes: A New Family of Two-Dimensional Materials. *Adv. Mater.* **2014**, *26*, 992–1005. [CrossRef]
27. Naguib, M.; Mashtalir, O.; Carle, J.; Presser, V.; Lu, J.; Hultman, L.; Gogotsi, Y.; Barsoum, M.W. Two-Dimensional Transition Metal Carbides. *ACS Nano* **2012**, *6*, 1322–1331. [CrossRef]
28. Lakhe, P.; Prehn, E.M.; Habib, T.; Lutkenhaus, J.L.; Radovic, M.; Mannan, M.S.; Green, M.J. Process safety analysis for $Ti_3C_2T_x$ MXene synthesis and processing. *Ind. Eng. Chem. Res.* **2019**, *58*, 1570–1579. [CrossRef]
29. Shuck, C.E.; Ventura-Martinez, K.; Goad, A.; Uzun, S.; Shekhirev, M.; Gogotsi, Y. Safe Synthesis of MAX and MXene: Guidelines to Reduce Risk During Synthesis. *ACS Chem. Health Saf.* **2021**, *28*, 326–338. [CrossRef]
30. Li, X.; Li, M.; Yang, Q.; Liang, G.; Huang, Z.; Ma, L.; Wang, D.; Mo, F.; Dong, B.; Huang, Q. In situ electrochemical synthesis of MXenes without Acid/Alkali usage in/for an aqueous zinc ion battery. *Adv. Energy Mater.* **2020**, *10*, 2001791. [CrossRef]
31. Li, T.; Yao, L.; Liu, Q.; Gu, J.; Luo, R.; Li, J.; Yan, X.; Wang, W.; Liu, P.; Chen, B. Fluorine-free synthesis of high-purity $Ti_3C_2T_x$ (T = OH, O) via alkali treatment. *Angew. Chem. Int. Ed.* **2018**, *57*, 6115–6119. [CrossRef]
32. Peng, C.; Wei, P.; Chen, X.; Zhang, Y.; Zhu, F.; Cao, Y.; Wang, H.; Yu, H.; Peng, F. A hydrothermal etching route to synthesis of 2D MXene (Ti_3C_2, Nb_2C): Enhanced exfoliation and improved adsorption performance. *Ceram. Int.* **2018**, *44*, 18886–18893. [CrossRef]
33. Li, Y.; Shao, H.; Lin, Z.; Lu, J.; Liu, L.; Duployer, B.; Persson, P.O.; Eklund, P.; Hultman, L.; Li, M. A general Lewis acidic etching route for preparing MXenes with enhanced electrochemical performance in non-aqueous electrolyte. *Nat. Mater.* **2020**, *19*, 894–899. [CrossRef]
34. Wu, Y.; Nie, P.; Wang, J.; Dou, H.; Zhang, X. Few-layer MXenes delaminated via high-energy mechanical milling for enhanced sodium-ion batteries performance. *ACS Appl. Mater. Interfaces* **2017**, *9*, 39610–39617. [CrossRef]
35. Zhang, D.; Zheng, W.; Li, X.; Li, A.; Ye, N.; Zhang, L.; Liu, Y.; Liu, X.; Zhang, R.; Wang, M.; et al. Investigating the effect of Ti_3C_2 (MXene) nanosheet on human umbilical vein endothelial cells via a combined untargeted and targeted metabolomics approach. *Carbon* **2021**, *178*, 810–821. [CrossRef]
36. Gu, M.; Dai, Z.; Yan, X.; Ma, J.; Niu, Y.; Lan, W.; Wang, X.; Xu, Q. Comparison of toxicity of Ti_3C_2 and Nb_2C Mxene quantum dots (QDs) to human umbilical vein endothelial cells. *J. Appl. Toxicol.* **2021**, *41*, 745–754. [CrossRef]
37. Jang, J.H.; Lee, E.J. Influence of MXene Particles with a Stacked-Lamellar Structure on Osteogenic Differentiation of Human Mesenchymal Stem Cells. *Materials* **2021**, *14*, 4453. [CrossRef]
38. Wu, W.; Ge, H.; Zhang, L.; Lei, X.; Yang, Y.; Fu, Y.; Feng, H. Evaluating the Cytotoxicity of Ti_3C_2 MXene to Neural Stem Cells. *Chem. Res. Toxicol.* **2020**, *33*, 2953–2962. [CrossRef] [PubMed]
39. Jastrzebska, A.M.; Szuplewska, A.; Wojciechowski, T.; Chudy, M.; Ziemkowska, W.; Chlubny, L.; Rozmyslowska, A.; Olszyna, A. In vitro studies on cytotoxicity of delaminated Ti_3C_2 MXene. *J. Hazard Mater.* **2017**, *339*, 1–8. [CrossRef] [PubMed]
40. Scheibe, B.; Wychowaniec, J.K.; Scheibe, M.; Peplinska, B.; Jarek, M.; Nowaczyk, G.; Przysiecka, L. Cytotoxicity Assessment of Ti-Al-C Based MAX Phases and $Ti_3C_2T_x$ MXenes on Human Fibroblasts and Cervical Cancer Cells. *ACS Biomater. Sci. Eng.* **2019**, *5*, 6557–6569. [CrossRef] [PubMed]
41. Szuplewska, A.; Rozmyslowska-Wojciechowska, A.; Pozniak, S.; Wojciechowski, T.; Birowska, M.; Popielski, M.; Chudy, M.; Ziemkowska, W.; Chlubny, L.; Moszczynska, D.; et al. Multilayered stable 2D nano-sheets of Ti_2NT_x MXene: Synthesis, characterization, and anticancer activity. *J. Nanobiotechnology* **2019**, *17*, 114. [CrossRef]

42. Lin, H.; Wang, X.; Yu, L.; Chen, Y.; Shi, J. Two-Dimensional Ultrathin MXene Ceramic Nanosheets for Photothermal Conversion. *Nano Lett.* **2017**, *17*, 384–391. [CrossRef]
43. Shao, J.; Zhang, J.; Jiang, C.; Lin, J.; Huang, P. Biodegradable titanium nitride MXene quantum dots for cancer phototheranostics in NIR-I/II biowindows. *Chem. Eng. J.* **2020**, *400*, 126009. [CrossRef]
44. Lim, G.P.; Soon, C.F.; Jastrzębska, A.M.; Ma, N.L.; Wojciechowska, A.R.; Szuplewska, A.; Wan Omar, W.I.; Morsin, M.; Nayan, N.; Tee, K.S. Synthesis, characterization and biophysical evaluation of the 2D Ti_2CT_x MXene using 3D spheroid-type cultures. *Ceram. Int.* **2021**, *47*, 22567–22577. [CrossRef]
45. Marchwiany, M.E.; Birowska, M.; Popielski, M.; Majewski, J.A.; Jastrzebska, A.M. Surface Related Features Responsible for Cytotoxic Behavior of MXenes Layered Materials Predicted with Machine Learning Approach. *Materials* **2020**, *13*, 3083. [CrossRef]
46. Rafieerad, A.; Yan, W.; Sequiera, G.L.; Sareen, N.; Abu-El-Rub, E.; Moudgil, M.; Dhingra, S. Application of Ti_3C_2 MXene quantum dots for immunomodulation and regenerative medicine. *Adv. Healthc. Mater.* **2019**, *8*, 1900569. [CrossRef]
47. Liu, W.; Huang, G.; Su, X.; Li, S.; Wang, Q.; Zhao, Y.; Liu, Y.; Luo, J.; Li, Y.; Li, C. Zebrafish: A promising model for evaluating the toxicity of carbon dot-based nanomaterials. *ACS Appl. Mater. Interfaces* **2020**, *12*, 49012–49020. [CrossRef]
48. Nasrallah, G.K.; Al-Asmakh, M.; Rasool, K.; Mahmoud, K.A. Ecotoxicological assessment of $Ti_3C_2T_x$ (MXene) using a zebrafish embryo model. *Environ. Sci. Nano* **2018**, *5*, 1002–1011. [CrossRef]
49. Alhussain, H.; Augustine, R.; Hussein, E.A.; Gupta, I.; Hasan, A.; Al Moustafa, A.-E.; Elzatahry, A. MXene nanosheets may induce toxic effect on the early stage of embryogenesis. *J. Biomed. Nanotechnol.* **2020**, *16*, 364–372. [CrossRef]
50. Zhang, J.; Fu, Y.; Mo, A. Multilayered titanium carbide MXene film for guided bone regeneration. *Int. J. Nanomed.* **2019**, *14*, 10091. [CrossRef]
51. Sui, B.; Liu, X.; Sun, J. Biodistribution, inter-/intra-cellular localization and respiratory dysfunction induced by Ti_3C_2 nanosheets: Involvement of surfactant protein down-regulation in alveolar epithelial cells. *J. Hazard. Mater.* **2021**, *402*, 123562. [CrossRef] [PubMed]
52. Jiang, H.; Li, Y.; Jin, Q.; Yang, D.; Wu, C.; Cui, J. Physiological and biochemical effects of Ti_3AlC_2 nanosheets on rice (*Oryza sativa* L.). *Sci. Total Environ.* **2021**, *770*, 145340. [CrossRef]
53. Rozmyslowska-Wojciechowska, A.; Szuplewska, A.; Wojciechowski, T.; Pozniak, S.; Mitrzak, J.; Chudy, M.; Ziemkowska, W.; Chlubny, L.; Olszyna, A.; Jastrzebska, A.M. A simple, low-cost and green method for controlling the cytotoxicity of MXenes. *Mater. Sci. Eng. C Mater. Biol. Appl.* **2020**, *111*, 110790. [CrossRef] [PubMed]
54. Rashid, B.; Anwar, A.; Shahabuddin, S.; Mohan, G.; Saidur, R.; Aslfattahi, N.; Sridewi, N. A Comparative Study of Cytotoxicity of PPG and PEG Surface-Modified 2-D Ti_3C_2 MXene Flakes on Human Cancer Cells and Their Photothermal Response. *Materials* **2021**, *14*, 4370. [CrossRef] [PubMed]
55. Hussein, E.A.; Zagho, M.M.; Rizeq, B.R.; Younes, N.N.; Pintus, G.; Mahmoud, K.A.; Nasrallah, G.K.; Elzatahry, A.A. Plasmonic MXene-based nanocomposites exhibiting photothermal therapeutic effects with lower acute toxicity than pure MXene. *Int. J. Nanomed.* **2019**, *14*, 4529–4539. [CrossRef]
56. Jastrzębska, A.; Szuplewska, A.; Rozmysłowska-Wojciechowska, A.; Chudy, M.; Olszyna, A.; Birowska, M.; Popielski, M.; Majewski, J.; Scheibe, B.; Natu, V. On tuning the cytotoxicity of Ti_3C_2 (MXene) flakes to cancerous and benign cells by post-delamination surface modifications. *2d Materials* **2020**, *7*, 025018. [CrossRef]
57. Zook, J.M.; MacCuspie, R.I.; Locascio, L.E.; Halter, M.D.; Elliott, J.T. Stable nanoparticle aggregates/agglomerates of different sizes and the effect of their size on hemolytic cytotoxicity. *Nanotoxicology* **2011**, *5*, 517–530. [CrossRef]
58. Xie, Y.; Gao, Y.; Ren, X.; Song, G.; Alsaedi, A.; Hayat, T.; Chen, C. Colloidal behaviors of two-dimensional titanium carbide in natural surface waters: The role of solution chemistry. *Environ. Sci. Technol.* **2020**, *54*, 3353–3362. [CrossRef]
59. Jastrzębska, A.M.; Scheibe, B.; Szuplewska, A.; Rozmysłowska-Wojciechowska, A.; Chudy, M.; Aparicio, C.; Scheibe, M.; Janica, I.; Ciesielski, A.; Otyepka, M. On the rapid in situ oxidation of two-dimensional V2CTz MXene in culture cell media and their cytotoxicity. *Mater. Sci. Eng. C* **2021**, *119*, 111431. [CrossRef]
60. Bai, X.; Wang, J.; Mu, Q.; Su, G. In vivo Protein Corona Formation: Characterizations, Effects on Engineered Nanoparticles' Biobehaviors, and Applications. *Front. Bioeng. Biotechnol.* **2021**, *9*, 263. [CrossRef]
61. Mahmoudi, M.; Bertrand, N.; Zope, H.; Farokhzad, O.C. Emerging understanding of the protein corona at the nano-bio interfaces. *Nano Today* **2016**, *11*, 817–832. [CrossRef]
62. Qin, M.; Zhang, J.; Li, M.; Yang, D.; Liu, D.; Song, S.; Fu, J.; Zhang, H.; Dai, W.; Wang, X. Proteomic analysis of intracellular protein corona of nanoparticles elucidates nano-trafficking network and nano-bio interactions. *Theranostics* **2020**, *10*, 1213. [CrossRef]
63. Basile, A.O.; Yahi, A.; Tatonetti, N.P. Artificial intelligence for drug toxicity and safety. *Trends Pharmacol. Sci.* **2019**, *40*, 624–635. [CrossRef]
64. Wang, J.; Su, G.; Yan, X.; Zhang, W.; Jia, J.; Yan, B. Predicting cytotoxicity of binary pollutants towards a human cell panel in environmental water by experimentation and deep learning methods. *Chemosphere* **2022**, *287*, 132324. [CrossRef]

nanomaterials

Review

Experimental and Computational Nanotoxicology—Complementary Approaches for Nanomaterial Hazard Assessment

Valérie Forest

Mines Saint-Etienne, Univ Lyon, Univ Jean Monnet, Etablissement Français du Sang, INSERM, U1059 Sainbiose, Centre CIS, F-42023 Saint-Etienne, France; vforest@emse.fr

Abstract: The growing development and applications of nanomaterials lead to an increasing release of these materials in the environment. The adverse effects they may elicit on ecosystems or human health are not always fully characterized. Such potential toxicity must be carefully assessed with the underlying mechanisms elucidated. To that purpose, different approaches can be used. First, experimental toxicology consisting of conducting in vitro or in vivo experiments (including clinical studies) can be used to evaluate the nanomaterial hazard. It can rely on variable models (more or less complex), allowing the investigation of different biological endpoints. The respective advantages and limitations of in vitro and in vivo models are discussed as well as some issues associated with experimental nanotoxicology. Perspectives of future developments in the field are also proposed. Second, computational nanotoxicology, i.e., in silico approaches, can be used to predict nanomaterial toxicity. In this context, we describe the general principles, advantages, and limitations especially of quantitative structure–activity relationship (QSAR) models and grouping/read-across approaches. The aim of this review is to provide an overview of these different approaches based on examples and highlight their complementarity.

Keywords: nanomaterials; toxicity; experimental models; computational models; in vitro; in vivo; in silico

1. Introduction

In addition to ultra-fine particles produced naturally (as by-products of wildfires, volcanic eruptions, and other natural processes) or unintentionally released as a consequence of human activities (present in polluting emissions, such as welding fumes, cigarette smoke, aircraft waste gas, or diesel exhaust), engineered nanomaterials are produced purposefully to take advantage of their unique physical, chemical, and/or biological characteristics associated with their nanostructure. The remarkable development and applications of nanomaterials lead to an increasing release of these materials in the environment, and, consequently, ecosystems and humans are increasingly exposed to them, while the adverse effects they may elicit are not always fully characterized. By January 2022, the Nanodatabase [1] counted 5224 products containing nanoparticles available on the European market.

Besides the huge amount and variety of nanomaterials, we should keep in mind that nanomaterials are highly heterogeneous in terms of physicochemical features, which represents a challenge for their hazard assessment and the generalization of conclusions [2,3]. As a result, a thorough characterization of nanomaterials is a prerequisite to the evaluation of their potential toxicity [4,5]. In particular, the ISO/TR 13014:2012 guideline [6] recommends the systematic consideration of parameters such as particle size and particle size distribution, aggregation/agglomeration state, shape, surface area, composition, surface chemistry, surface charge, and solubility/dispersibility.

The toxicity of nanomaterials should be carefully assessed as a prerequisite for their safe and successful use in any application. To that purpose, many models and approaches are available, each having their advantages and limitations. The nanotoxicology field was first introduced in 2005 by Oberdörster et al. [7] and has evolved ever since. It is now also referred to as "nanosafety" [8] and includes in vitro and in vivo studies, and it aims to establish a dose/effect relationship and thresholds below which no biological effects are observed. It also studies the influence of specific properties of nanomaterials on biological responses; this interdisciplinary approach involves physicochemists as well as biologists. Furthermore, nanotoxicology includes the investigation of the interactions of nanomaterials with living organisms and the study of the mechanisms of action potentially responsible for adverse effects. Although nanotoxicology is a specific field due to the specific nature of nanomaterials, some methodologies and approaches can be derived from standard toxicology. Indeed, the toxicity of metals and other elements either naturally occurring or produced by anthropogenic activities has already been extensively studied in the context of environmental pollution [9].

We can roughly distinguish two types of approaches to evaluate nanomaterial hazards: (i) empirical toxicology consisting of conducting in vitro or in vivo experiments and (ii) in silico approaches relying on computational studies. The aim of this review is twofold: (i) to provide an overview of these different fields based on recent examples of applications and (ii) highlight their complementarity. Indeed, while the literature is rich in reviews focused either on experimental or computational approaches (sometimes involving slightly different scientific communities), very few studies deal with both. Thus, the present paper aims to bridge this gap by exploring a balanced combination of wet and dry lab approaches that are available for the toxicity assessment of nanomaterials, arguing for cooperative projects.

2. Experimental Nanotoxicology/Empirical Approaches

Experimental toxicology includes in vitro and in vivo studies. Basically, the principle is to expose individuals, tissues, or cells to the substance of study, observe the effects induced, and compare them with the response observed in a control group (treated in the same conditions but unexposed to the substance). Clinical studies can also be considered as part of experimental nanotoxicology.

The main parameters used to evaluate nanotoxicity include the study of the targets of the nanomaterials (biodistribution and cellular uptake), the assessment of the induction of cytotoxicity (cell damage, cell death), the pro-inflammatory response, oxidative stress, and genotoxic effects as these events have been reported to be the mechanisms of action of nanomaterial toxicity [10–12].

As mentioned before, a prerequisite to the assessment of the toxicity of nanomaterials is their thorough physicochemical characterization as some features can deeply influence the biological outcomes [7,13,14].

2.1. In Vitro Studies

2.1.1. In Vitro Models

A wide variety of in vitro models exist: from cell lines to primary cultures (developed from tissue biopsies) and tissues. Primary cultures are quite challenging to establish due to the lack of tissue availability, the specific handling they require, and donor-specific variations [13,15]. Usually, cells lines are preferred because of their homogeneity and stability, resulting in reproducible results. However, they are either cancer cells or cells artificially immortalized, and although their high proliferative rate makes them easily cultivable, available in large quantities, and inexpensive, they exhibit altered pathways compared to normal cells. Consequently, they do not answer exactly as healthy cells would do, making them a poorly reliable representation of what really occurs in vivo. In addition, if they are used for long periods of time, cell de-differentiation and thus a change in phenotype can occur [13,15]. This is why some scientists question their relevance. However, Verdon et al. observed a similar pattern of response between the human neutrophil-like

HL-60 cell line and human primary neutrophils after exposure to various nanomaterials [16]. They concluded that although some neutrophil functions were compromised in the cell line, it still represented a valuable model that can be used in a first attempt, especially for screening purposes, and primary cells can be used for more focused assessments.

Additionally, different cell types can be chosen depending on the scientific question to address: cells from the immune system, in particular macrophages, epithelial cells, fibroblasts, cells from the lung, cells from the gastrointestinal tract, neurons, etc. For instance, while fibroblasts are commonly used to assess pro-fibrotic signals, macrophages are used because they are the first line of defense of the organism. Additionally, lung and gastrointestinal cells are often used as they represent the primary target organs after the inhalation and ingestion of nanomaterials, two major exposure pathways [13].

Cell lines also offer the possibility to work with different species. While human cell lines are supposed to more closely mimic human responses in comparison to rodent cell lines, the latter are particularly useful for comparison of the results with animal data [13].

It should be taken into account that cell type-dependent effects have been reported, and this is why it is recommended to use several cell types to assess nanomaterial toxicity before drawing firm conclusions [15].

Cell culture conditions should also be considered carefully as they can impact the reproducibility and accuracy of the results. Drasler et al. reviewed and made recommendations on the main experimental parameters involved in in vitro assays to achieve a reliable assessment of nanomaterial–cell response [13]. For instance, the concentration of nanomaterials used in such experiments should be realistic and range from 1 to 100 μg/mL, with the lower and higher limits at 0.125 and 200 μg/mL, respectively. It is also recommended to include negative and positive controls to allow comparison between studies, both intra- and interlaboratory [13].

The use of cell lines in monoculture systems is very common and is indeed recommended for the first stage of nanomaterial hazard evaluation for the sake of simplicity [13]. However, it is acknowledged that they do not reproduce a tissue or organ which possesses a defined three-dimensional (3D) structure and do not include the complex cross-talks between cells [13–15,17–19]. To overcome these shortcomings, models able to more closely mimic a physiological reality have been developed. These models of increasing complexity are referred to as advanced in vitro models and include co-cultures and 3D cultures [18,20].

Multiple-cell cultures, also known as multicellular systems, 3D models, or "oid" cultures (e.g., organoids, spheroids), represent more physiologically relevant systems, especially for mechanistic studies, and have been described to have the potential to be more predictive in toxicology testing, thus filling a gap between 2D systems and animal experiments [8,13]. On the other hand, advanced culture systems are more expensive, time-consuming, technically more demanding, more difficult to standardize, and usually less suitable for high-throughput analyses [18,21].

Co-culture systems can be more or less complex; they comprise two or more cell types and are not necessarily 3D. For example, in the study of the effects induced by inhaled nanomaterials, co-cultures of alveolar cells and macrophages are interesting as the latter cell type can provide information on an indirect toxicity mechanism where nanomaterial internalization by macrophages leads to an inflammatory response and subsequent tissue damage [14,17]. In an attempt to simulate the alveolar epithelial barrier, Barosova et al. [22] developed a human alveolar cell co-culture model using alveolar epithelial type II cells and two types of immune cells (human monocyte-derived macrophages and dendritic cells). Another type of 3D lung co-culture model was developed to assess the hazard potential of multi-walled carbon nanotubes, consisting of a mix of alveolar epithelial cells, fibroblasts, and macrophages [23].

Similarly, different models of the 3D gastrointestinal tract were developed to assess the hazard of nanomaterials [24–26]. While Saez-Tenorio et al. used a Caco-2/HT29 co-culture, Bredeck et al. used a triple culture consisting of Caco-2, HT29-MTX-E12, and THP-1 cells [24,26].

Usually, a higher sensitivity of co-culture models has been reported compared to that of monocultures. As an example, after exposure to nanomaterials, biological adverse effects were observed in A549+THP-1 co-cultures but not in A549 monocultures [27–29].

Multicellular spheroids and organoids even more closely resemble in vivo organs [14,15,30,31]. These tissues have enhanced morphologies and show increased functional activity [17]. The difference between spheroid and organoid lies in the fact that in spheroids, cells are free to form aggregates into a tight 3D ball, while in organoids, cells are embedded in a matrix and confer a more ordered configuration, serving as a scaffolding support [19].

Such advanced 3D models have been developed to mimic the liver, pancreas, or lung [31–34]. For instance, Kabadi et al. developed 3D lung microtissues using human lung epithelial cells, fibroblasts, and macrophages [32]. Upon exposure to multi-walled carbon nanotubes and asbestos fibers, they observed responses similar to those previously observed in vivo in a rodent model, validating this advanced in vitro model.

Further, the incorporation of tissues or organs on a microfluidic platform, so-called organ-on-a-chip, enables the evaluation of nanomaterial toxicity in highly dynamic conditions in vitro [35,36]. A lung-on-a-chip based on human vascular endothelial cells and alveolar epithelial cells has recently been developed for the assessment of the pulmonary toxicity of TiO_2 and ZnO nanoparticles [37]. The same model was used to evaluate the pulmonary effect of air pollutant fine particulate matter [38].

Ready-to-use commercial solutions are also available such as Epithelix, EpiAirway, MatTek, and MucilAir. They are 3D human reconstructed epithelia of the respiratory tract. These cultures are made of primary airway epithelial cells isolated from different parts of the respiratory tract of healthy donors or patients with chronic lung diseases, asthma, or chronic obstructive pulmonary disease. However, these models are expensive, and the donor variation has to be taken into account [39–42].

Besides more physiologically relevant cell models, progress has been made to more realistically reproduce cell or organ exposure to nanomaterials. For instance, when considering airborne nanomaterials, air–liquid interface (ALI) systems better recapitulate the physiological exposure of the lungs and seem better suited than conventional submerged exposure assays to predict lung toxicity [40,43]. Indeed, when exposed to ceria and titania nanoparticles, human A549 lung epithelial cells and differentiated THP-1 macrophages cells were more sensitive at the ALI compared to under classical submerged conditions [43]. Similarly, Loret et al. reported that the biological responses of A549/THP-1 cells exposed to TiO_2 and CeO_2 nanoparticles were usually observed at lower doses at the ALI than in submerged conditions [27]. Others have also found a higher sensitivity of cells exposed at the ALI compared to submerged exposure experiments [41,44,45]. Therefore, exposure of cells at the air–liquid interface has been shown to be a valid and sensitive method to assess the toxicity of several poorly soluble nanomaterials in monocultures and co-cultures and could bridge the gap between traditional 2D in vitro assays and animal models of airway exposure [4,27,39,44,46–51].

2.1.2. Main Biological Endpoints Considered

Nanomaterials can induce toxicity both in vitro and in vivo through various mechanisms such as oxidative stress, cell death mechanisms (apoptosis, autophagy, and necrosis), genotoxicity, and immunological responses [10–12,19,52]. Different assays allow evaluating these biological endpoints to report on the nanotoxicity; they can be subdivided into cytotoxicity assessment (including cell viability, proliferation, and cell death assays), pro-inflammatory response, oxidative stress, and genotoxicity [11]. To draw firm conclusions, a battery of tests should be employed, with each assay being considered depending on the cell type and nanomaterials used [13].

Cytotoxicity can be measured based on the observation of cells' morphological alteration, measurement of cells' viability, or the ability of cells to proliferate and form colonies. Cell viability assays are generally based on the use of colorimetric or fluorimetric assays [13], e.g., MTT, XTT, WST-1, Alamar blue, neutral red uptake, or evaluation of membrane integrity (LDH, trypan blue assays), whereas cell division or proliferation can be evaluated through clonogenic assays or cell counting, respectively. Moreover, apoptosis assays (e.g., TUNEL assay, caspase activation) or apoptosis/necrosis (Annexin V/PI) assays can be used to characterize the type of cell death induced by nanomaterials [10,11,52–56].

A widely used approach to assess the pro-inflammatory response consists of the analysis of the soluble factors such as cytokines or chemokines secreted by cells and assessed in the cell culture supernatant by an enzyme-linked immunosorbent assay (ELISA) [13].

Oxidative stress is acknowledged to be a major mechanism of nanomaterial toxicity [11]. Indeed, exposure to nanomaterials leads to the production of reactive oxygen species (ROS) and reactive nitrogen species (RNS) [11]. Cells generate ROS to maintain normal metabolism/homeostasis, but their overproduction can interfere with a variety of signal transduction pathways and even induce cell apoptosis, pro-inflammation, and DNA damage [5,11,13,56]. ROS can be detected using electron paramagnetic resonance (EPR) or fluorescent probes [11].

Genotoxicity describes the potential damage induced by nanomaterials to the DNA [13]. Primary genotoxicity can be distinguished from secondary genotoxicity: primary genotoxicity is defined as genetic damage induced by nanomaterials themselves, whereas in secondary genotoxicity, nanomaterials do not interact directly with the target cell but produce an inflammatory response in neighboring cells, resulting in the oxidative damage of DNA by ROS. Please note that primary genotoxicity can also be mediated by ROS formation formed in cells by the interaction of nanomaterials, e.g., with mitochondria [13,57]. To assess genotoxicity, different markers can be used evaluating DNA damage, gene mutations, and chromosomal damage [21]. They include the in vitro mammalian cell gene mutation assay and the in vitro mammalian cell micronucleus assay. The latter determines the chromosome breakage leading to the formation of an additional nucleus (micronucleus) during cellular division [13,33]. The micronucleus assay also assesses chromosomal loss, i.e., aneugenic effects. In addition, the comet assay is widely used to detect the genotoxic potential of nanomaterials, allowing the detection of single- and double-stranded DNA breaks in individual cells [11,58,59]. Other genotoxicity assays include γ-H2AX foci formation as a marker for DNA double-strand breaks but with limited mechanistic insight for investigations of more precise mechanisms of genotoxicity [13].

2.1.3. Advantages of In Vitro Approaches

In vitro models offer many advantages. They permit different levels of study: organ, tissue, cell (one or several populations); they allow large screening of effects with a very small amount of test material; and they are very well adapted for the study of mechanisms, mainly for short-term studies [53,60]. As they are performed under controlled testing conditions, they allow a reduction in variability between experiments [53]. They were initially developed to apply the "3R" rule introduced in 1959 and aiming to Reduce, Refine, and Replace animal experiments [61], which is now widely encouraged by international legislation [21]. In 2010, the European Commission requested the partial and even full replacement of animal studies [18,62]. In addition to preventing ethical issues related to animal testing, such methods are easier, faster, and cheaper [11].

In particular, in nanotoxicology, in vitro models allow a better understanding of nanomaterial uptake by cells and translocation through biological barriers, cytotoxicity and cellular effects, and genotoxicity induced by nanomaterials in regard to their physicochemical features. Thus, such in vitro studies have allowed evidencing the involvement of a specific physicochemical parameter in the toxicity of some nanomaterials. For instance, correlations were observed between nanomaterial toxicity and their shape (cerium nanoparticles) [63], their surface charge (silica nanoparticles) [64,65], their particle size and surface chemistry (silicon carbide nanoparticles) [66,67], their agglomeration size (boehmite nanoparticles) [68], and their size, shape, and agglomeration (TiO_2 nanoparticles) [28].

This type of data is very useful, especially in the context of safer-by-design approaches. Indeed, when the physicochemical parameters responsible for the toxicity of a nanomaterial are identified, we can alter them at the early stages of the nanomaterial development to produce safer materials [14].

2.1.4. Limitations of In Vitro Approaches

Although very convenient, in vitro models have the major disadvantage of being too simplistic. They do not reflect the complexity of a whole organism and do not reproduce the toxicokinetics. Therefore, they may not be predictive of what really occurs in vivo. In addition, chronic effects cannot be tested with these models due to difficulties of performing long-term simulation [14,20,55,56,60]. This is why, despite its limitation due to ethical reasons, the use of animal models still remains necessary.

2.2. In Vivo Models

The main advantage of in vivo models is that they are able to better mimic physiopathological processes and allow the systemic evaluation of the effects triggered by nanomaterials. They also enable studying the defense mechanisms that could counterbalance a biological response (e.g., antioxidant defense, tissue repair mechanism, clearance by cells from the immune system). Indeed, whole organisms remain the most scientifically relevant models as they are able to capture effects of nanomaterials after they have entered the body and have been distributed and processed [69]. As a matter of fact, a major advantage of in vivo models is that they allow the assessment of the kinetics of nanomaterials through absorption, distribution, metabolism, and excretion (ADME) [14]. In addition, while the long-term effects of inhaled nanomaterials, for instance, cannot be studied in vitro, they can be observed in vivo as they are retained in the body for longer periods [60]. Finally, in vivo models could allow establishing modes of action of nanomaterials as well as studies on the second generation, for instance, in the case of reproductive toxicity assays [70,71] or embryotoxicity evaluation [72]. Another asset of in vivo models is that they allow considering the biopersistence of nanomaterials in the organism. This is of paramount importance in the sense that it is correlated with toxicity. Indeed, the most persistent particles can accumulate in tissues where they can elicit adverse effects [73]. For instance, compared to silver ions, silver nanoparticles were shown to be more persistent in the circulating system and organs in terms of overall body distribution [72].

In the nanotoxicity field, the most frequently used in vivo models are mammal models, in particular small rodents such as mice, rats, and rabbits because of their close resemblance to humans. In addition, they are cheaper and easier to be maintained than larger animals such as pigs, which are genetically very close to humans [11,14,15]. Dogs and primates show the highest similarity to the human respiratory system; however, they are rarely used for toxicity studies due to ethical issues and experimental costs. On the other hand, guinea pigs have been used for sensitization to inhaled antigens since their airways show similar sensitivity to mediators to human airways, while rodent lungs are less sensitive [60].

To assess nanomaterial toxicity in vivo, different parameters can be evaluated, including the biodistribution of the nanomaterials in the organism (localization in tissues or organs), clearance (examination of excretion and metabolism of nanomaterials at various time points after exposure), hematology, serum chemistry (examination of changes), and histopathology (evaluation of alterations and potential damage to tissues) [11,14,15,19]. The LC_{50} (concentration of nanomaterial that causes the death of 50% of the population), LOEC (lowest concentration that causes a noticeable effect on the organism), and NOEC (maximum concentration at which no effect is observed on the organism) can also be determined [14,19]. The genotoxicity of nanomaterials has also been widely investigated in rodent models with the added value of evaluating the response of a whole organism upon nanomaterial exposure. In addition, with an appropriate study design, several endpoints can be analyzed in a single animal, reducing the number of animals and the costs of the in vivo experiments. For instance, the in vivo genotoxic potential of nanofibrillated cellulose, TiO_2, SiO_2, or Al_2O_3 nanomaterials administered to mice was evaluated using several genotoxicity endpoints [74–77].

Regarding the exposure of animals to nanomaterials, different routes are possible such as inhalation, intravenous or intraperitoneal injections, ingestion, intratracheal instillation, dermal administration, or gavage, to be chosen depending on the question to address [14,60]. Many studies have focused on inhalation as it is a main exposure route. Whole-body exposure can be performed by placing animals in inhalation chambers. While it best corresponds to pulmonary exposure, it is difficult to perform because usually nanomaterials are contained in dry powder and thus need to be dispersed in airflow, which can generate high concentrations of nanomaterials and thus aggregation [60]. Moreover, it is challenging to precisely quantify the inhaled dose as nanomaterials are not only inhaled but also ingested (through licking of the coat where they are also deposited). In addition, the animals can avoid exposure by huddling together or burying their noses in corners of cages or in the fur of another animal [60]. The "nose/head-only" system has been developed to prevent some of these drawbacks where animals are immobilized in a tube, and only the nose or head is exposed to the source of nanomaterials. However, such an immobilization is stressful for the animals and can therefore produce biological impact, thus inducing biased results when monitoring biological parameters. Inhalation studies require specialized equipment and are more difficult and expensive to carry out than oral administration [69].

Another widely used exposure pathway of animals to nanomaterials is through intratracheal instillation [60,69]. Nanomaterials are suspended in a physiological solution which is deposited directly in the respiratory tract through a cannula inserted in the trachea. It is easy to perform and allows controlling the dose effectively inserted into the lungs, but this technique can cause local tissue damage and an uneven distribution of the test substance in the lungs [60]. Moreover, it is not a physiological exposure, and when nanomaterials are suspended in a liquid, a more or less homogeneous solution is produced, resulting in a lack of reproducibility of the experiments.

The exposure route should be chosen depending on the availability of materials (whole-body exposure requires high amounts of material), the technical expertise of the personnel (intratracheal instillation is technically challenging), and the exposure duration [60]. Anyway, the exposure route should be considered carefully as it has been shown to have a significant impact on the results. As an example, inhaled single-walled carbon nanotubes in mice elicited a greater effect than instilled particles [78], while the opposite was found for titanium dioxide nanoparticles [79]. However, Warheit et al. observed similar results with both exposure routes [80].

The high doses tested and route of administration used in inhalation studies are not always relevant to human exposure scenarios and can result in so-called "overload" of the test system. This is why short-term in vivo inhalation studies (STIS) have been developed to reduce the need for 90-day inhalation studies [69].

Intravenous injection of nanomaterials is also commonly used, especially for the study of toxicity mechanisms, to mimic translocation through the blood barrier. Similarly, other pathways such as oral administration or skin contact also have their pros and cons and should be considered depending on the scientific question to address.

Besides the ethical reasons and legislations [62] that tend to limit animal testing to that which is strictly necessary, in vivo models have intrinsic limitations. As mentioned before, from a practical point of view, such models are expensive, and facilities and personnel are necessary for animal housing, feeding, care, etc., with increasing needs with the increasing size of the animal model. Experiments with animals can be long and complex to handle. Additionally, due to the huge amount and variety of nanomaterials, the systematic assessment of their toxicity profile in vivo is simply unfeasible and unsustainable [81]. In addition, while they remain the gold standard for toxicology testing, questions remain on the relevance of in vivo models when assessing hazard implications towards human health [8]. Indeed, because of anatomical, physiological, and biochemical differences between species, one may wonder if in vivo approaches are really predictive for humans [18,60,82]. As a matter of fact, some models may also not be sensitive enough. For instance, it has been reported that little pleural reactivity was induced in rats even with particles known to be very carcinogenic (e.g., asbestos). This is due to the rat lifespan being too short to assess diseases that can occur 30 to 40 years after exposure in humans [60]. Similarly, the drug thalidomide was shown to be inert within rodents, while it induced significantly detrimental effects on human fetuses [8].

2.3. Clinical Studies

Biological monitoring or biomonitoring is defined as "the repeated, controlled measurement of chemical or biological markers in fluids, tissues, or other accessible samples from subjects exposed or exposed in the past or to be exposed to chemical, physical or biological risk factors in the workplace and/or the general environment" [83].

Biomonitoring has been widely used in pulmonology, especially in the case of pneumoconiosis, and has uncovered critical information on the relationship between exposure to a harmful substance and biological and even pathological effects. For instance, the assessment of asbestosis bodies in patient lung tissues or in broncho-alveolar lavage (BAL) fluids has allowed defining values specific to diseases [84–86]. It has also been shown that BAL from patients suffering from idiopathic pulmonary fibrosis had a different chemical composition to that of patients with other interstitial lung diseases or healthy subjects [87]. In the context of health risk assessment, the biomonitoring of nanoparticles in human biological samples could be a particularly interesting approach to gain new insights into the role of inhaled biopersistent nanoparticles in the etiology/development of some respiratory diseases. Although technically challenging [88,89], this strategy appears promising. Indeed, the biomonitoring of nanoparticles in human lung tissues or fluids could fill a gap between exposure to nanomaterials (evaluated by the external dose assessed by ambient monitoring) and the biological effects and even diseases induced by these nanomaterials [88,90–93] (Figure 1).

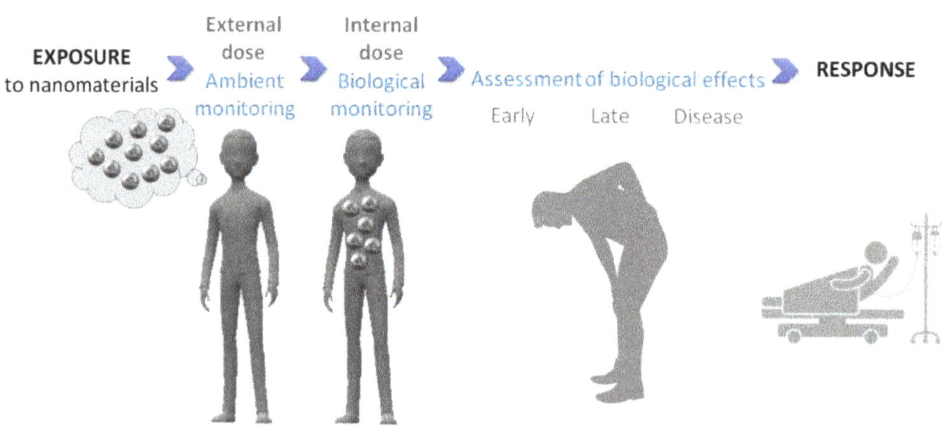

Figure 1. Biological monitoring of nanomaterials in human samples could fill a gap and help better understand the relationship between exposure to nanomaterials and adverse effects through the analysis of both biomarkers of exposure and biomarkers of effects.

We adopted this approach to detect and quantify nanoparticles in various types of clinical samples such as seminal and follicular fluids [94], the colon [95], amniotic fluids [96], or BAL [97–99]. We especially focused our attention on BAL, where we separated micron-sized particles (>1 µm) from submicron (100 nm–1 µm) and nano-sized particles (<100 nm) contained in BAL from a cohort of 100 patients who suffered from interstitial lung diseases (ILDs). We then determined the metal load in each of these size fractions and evidenced a higher concentration of submicron silica particles in patients suffering from sarcoidosis than in patients suffering from other ILDs, suggesting a potential role of these inhaled particles in the etiology and/or development of sarcoidosis [98]. Similarly, we observed a higher concentration of titanium nanoparticles in patients suffering from idiopathic fibrosis than in patients suffering from other ILDs, allowing us to suspect a relationship between titanium nanoparticles and idiopathic pulmonary fibrosis even though, in this case, we had a limited number of patients to reach a satisfactory statistical power to draw firm conclusions.

We recently went one step further to gain a comprehensive vision of the events from exposure to airborne nanoparticles to the biological response induced (Figure 1) and investigated associations between respiratory diseases and occupational exposures [100]. To that purpose, we estimated the exposure to inhaled unintentionally released nanoparticles of patients for each job held in their working life. Most of the patients showed a high probability of exposure to airborne unintentionally released nanoparticles (>50%), suggesting a potential role of inhaled nanoparticles in lung physiopathology. Depending on the respiratory disease, the number of patients likely exposed to unintentionally released nanoparticles was variable (e.g., from 88% for idiopathic pulmonary fibrosis to 54% for sarcoidosis). These findings were consistent with the mineralogical analyses. Further investigations are necessary to draw firm conclusions, but these first results strengthen the array of presumptions on the contribution of some inhaled particles (from nano to submicron size) to some idiopathic lung diseases.

While BAL reflects a diluted sample of the respiratory tract lining fluid, it is invasive, takes time, must be performed in a medical setting, and cannot easily be repeated in short intervals. This is why exhaled breath, which is more easily accessible, has emerged as a promising matrix for biological monitoring [101]. The exhaled air is cooled to condensate, and the obtained exhaled breath condensate (EBC) is collected and analyzed for the presence of nanoparticles.

Finally, of course, large epidemiological studies can be informative, but they are personnel- and time-consuming and require important financial resources.

2.4. Issues Associated with Experimental Nanotoxicology

2.4.1. Lack of Standardized Assays

As described before, a wide variety of assays are available to assess nanomaterial toxicity [10]. However, there are no standardized protocols, and because of the different sensitivities of different cell lines, the influence of cell culture media, and dispersion techniques of nanomaterials, the outcome of in vitro assays may vary, making it difficult to compare results from different studies [14,56,102]. As an example, when reviewing studies from the literature to gain insight into the toxicity of graphene-based materials with regard to their physicochemical features, no firm conclusions could be drawn because of the lack of standardization of toxicity assessment as well as the incomplete nanomaterial physicochemical characterization [103]. There is also a lack of consensus on the dose metric; often, the concentration is expressed as mass per volume, which is not always the most relevant unit [15]. Indeed, the nanoparticle surface area or the nanoparticle number can be used as alternative dose metrics to nanoparticle mass. The more adequate metric should be chosen depending on the question to address; for instance, Oberdörster et al. demonstrated that to evaluate dose–response relationships of nanoparticle-induced pulmonary inflammation, the particle surface area was a more appropriate dose metric than particle mass or particle number [104]. Similarly, the biological endpoints can be expressed using different scales; for instance, cytotoxicity can be expressed using a binary system (toxic/nontoxic), or a (semi)quantitative scale (low/medium/high toxicity), with absolute values, LC_{50}, NOAEL, LOAEL, etc. This heterogeneity makes it challenging to compare studies.

Finally, failure to use standardized methods and appropriate control experiments questions the reliability of nanotoxicological assay results. In particular, the lack of positive controls at the nanoscale is an issue [105,106].

2.4.2. Potential Artifacts

It has been well documented that nanomaterials can interact with different cytotoxicological assays, resulting in inaccurate results and misinterpretations [13,15,53,56]. In particular, due to their large surface area, nanomaterials can adsorb assay reagents, enzymes, substrates, or reporter dyes, inducing a bias in the assay outcome [15,107].

Moreover, as many cytotoxicity assays are based on optical readouts (spectrophotometric or fluorometric), the intrinsic optical properties of the nanomaterials (high absorption or scattering) can interfere with the detection method [15,53,107].

This is why it is highly recommended to evaluate possible interferences prior to any cytotoxicity assay to ensure reliable results. In case of interferences, nanomaterial-specific adaptations of the conventional cytotoxicity assays (e.g., MTT and DCF assays) might eliminate nanomaterial interference with the method [13,107]. Alternatively, after quantification and modeling of the interferences, corrective factors can be applied to obtain accurate results [108].

Kroll et al. characterized the interferences of 24 nanomaterials with 4 standard cytotoxicity assays for oxidative stress, cell viability, cell death, and inflammatory cytokine production (DCF, MTT, LDH, and IL-8 ELISA) and reported concentration-, particle-, and assay-specific interferences [107]. Similarly, it was demonstrated that interferences occurred between carbon nanotubes and the commonly used LDH assay, due to the carbon nanotubes' intrinsic absorbance and their ability to adsorb LDH at their surface [109].

Besides such interferences, biological artifacts can also occur and bias the results. An example is the contamination by endotoxins or lipopolysaccharides (LPSs) that cause acute inflammatory responses in humans [55].

Another important factor inducing variation in nanotoxicity data is the medium in which the nanomaterials are dispersed. This can alter the agglomeration/aggregation state of the nanomaterial, which will subsequently have an impact on the cell uptake and toxicity of the nanomaterial [15]. This is why it is recommended to characterize the physicochemical properties of nanomaterials after dispersion in a cell culture medium.

Consequently, possible nanomaterial–assay interactions should be carefully assessed. Furthermore, each biological endpoint should be evaluated with multiple assays which ideally supply complementary information and have different assay and detection principles, to validate the obtained results [15].

2.4.3. Difficulty to Compare In Vitro and In Vivo Data

While the merit of in vitro models has been acknowledged, concerns remain about their correlation with in vivo data [60,102]. All the more, there is a lack of studies specifically dedicated to the comparison between results obtained from in vitro and in vivo approaches [69]. The in vitro/in vivo gap is an issue that makes the formulation of unambiguous conclusions on nanomaterial toxicity challenging [15].

Some factors can explain the conflicting results that have been observed between in vitro and in vivo data. First, as discussed before, in vitro models, especially standard 2D cell cultures, are too simplistic and do not recapitulate the complexity of whole organisms [15]. The development of advanced in vitro models will hopefully enable reducing this gap. Second, the nanomaterial doses used in vitro and in vivo may significantly differ due to the contribution of clearance [13] or due to the concentrations selected for the different assays. Indeed, while in vitro studies may use unrealistically high nanomaterial concentrations to highlight dose-dependent effects, such dosages would not be applicable in in vivo studies [102]. Third, we have to keep in mind that physicochemical features of the nanomaterials can deeply affect their toxicity, and even small changes in these features can lead to major alterations in nanomaterials' interactions with biological systems [14,15].

However, efforts have been made to fill this gap; for instance, a rat NR8383 alveolar macrophage assay has been developed to assess the in vitro effects of SiO_2 nanomaterials. The results were consistent with the effects observed in vivo, after intratracheal instillation of the nanoparticles in rat lungs [110]. Similarly, Kampfer et al. investigated the cytotoxicity profile of silver and titanium dioxide nanoparticles in four in vitro models of increasing complexity. The results were consistent with the in vivo effects of the same nanoparticles through tissue analysis from 28-day oral exposure studies in mice, especially for advanced in vitro models, underlining the relevance of these models [111].

2.4.4. Formation of a Protein Corona in Biological Media

To fully explore the nanomaterial interactions with biological systems, we should take into account not only the "synthetic identity" of the nanomaterials (corresponding to their pristine, intrinsic nature), but also their "biological identity", acquired in biological media [14,112]. Indeed, upon contact with blood, biological fluids, or cell culture media, biomolecules contained in these fluids adsorb on the nanomaterial surface, forming a crown, a layer mainly composed of proteins, the so-called "protein corona". This new interface with biological systems will deeply impact the nanomaterial biological fate and response [113–117]. The composition of this protein corona is unique for each nanomaterial and influenced by the nanomaterial's intrinsic physicochemical features as well as by biological environmental factors. The protein corona is also characterized by its dynamic nature [14,114,117]. Indeed, its composition can vary over time following what is called the "Vroman effect" where the most abundant proteins adsorb first and are then replaced by proteins of higher affinity [118]. The protein corona composition also varies, especially in vivo, depending on its location in the organism as the nanomaterial passes from one biological compartment to another, each having its own distinct set of proteins that interact in a unique way with the nanomaterial [117].

The characterization of the protein corona may be technically challenging, and a lack of reproducibility can also be mentioned. In addition, despite numerous works, no general rule can be established. Indeed, there is no universal corona; it is unique to each system as its formation depends both on the nanomaterial properties and the biological environment [65,117,119,120].

However, it is of paramount importance to take this component into account as it can deeply influence interactions with cells and the subsequent biological outcomes. Lesniak et al. showed that for identical particles and cells, under identical conditions, the interactions with cells and the biological outcomes could vary greatly in the presence or absence of a preformed corona in serum [121]. The protein corona may also impact the biopersistence of nanomaterials in vivo as it can be recognized by macrophages, resulting in an accelerated clearance [122].

Better knowledge of the impact of the protein corona is critical for nanotoxicology.

2.4.5. Combined Effects of Nanoparticles

We are constantly exposed to numerous and various sources of nanoparticles (natural as well as engineered). It is thus difficult to reproduce such complexity in experiments; therefore, samples used in nanotoxicology assessment are not always fully representative of real life exposure and may not lead to accurate human health risk assessment and management as they overlook the interactions between the contaminants and their resulting combined toxicity [123]. Indeed, the biological effects triggered by nanoparticles are usually assessed focusing on individual nanoparticles [124,125], while their interaction with co-contaminants can deeply impact, either positively or negatively, their biodistribution, fate in the organism, and toxicological profile (additive, synergistic, or antagonistic responses) [123,126]. Even if the toxicities of individual compounds are well characterized, unexpected adverse effects can be induced by the mixing of such compounds [127,128]. Even worse, compounds that do not elicit adverse effects individually can induce significant toxicity when in a mixture, resulting in an underestimation of the risk [129]. While additive or synergistic effects can occur (effects higher than the simple addition of the effects of the individual components) [130], antagonistic effects can also be reported [124,127,131,132].

Different mechanisms can be responsible for combined toxicity [123]. A carrier effect can be observed where the co-pollutant adsorbs at the surface of the nanomaterial that serves as a cargo and eases the entry and accumulation of the co-pollutant inside the cell. Moreover, the co-pollutant may alter the physicochemical features of nanoparticles, especially in relation to surface properties (chemistry, charge, hydrophobicity, agglomeration state, etc.), and consequently result in a modified toxicological profile. In addition, the first pollutant can make cells more sensitive to the second one. As an example, it was shown that ZnO nanoparticles altered cell membranes upon accumulation and therefore enhanced the toxicity of co-exposed Cu nanoparticles [133].

Despite the complexity of the issue (there are countless potential mixtures in our environment) and the technical difficulty to recapitulate real-life conditions, it seems crucial to take into account mixture effects [123]. Furthermore, if nanomaterial exposure is usually well considered during the production and manufacturing stages, it is less systematically evaluated during the use and end-of-life stages [81,112].

2.4.6. Low Doses/Chronic Exposure

In the nanotoxicology field, most studies focus on acute exposure with high nanomaterial doses, while very few studies report on long-term exposures over weeks or months to sub-chronic exposures [13]. While long-term and repeated low-concentration exposure studies are scarce, they are extremely important as they better mimic real-life exposure [134]. Furthermore, the mechanisms involved in a biological response to a high single dose may be significantly different from those related to a repeated low-dose exposure [13]. Whereas in vitro studies can be carried out to evaluate acute effects, they are not adapted to long-term studies and chronic effects [13,135]. Such effects can be rather observed using in vivo models, even if they are time-consuming and expensive and few inhalation laboratories are equipped to carry out sub-chronic (90 days) or chronic (1.5–2 years) inhalation tests. This explains why the in vivo chronic effects of inhaled nanomaterials are poorly documented [69].

2.5. Perspectives and Future Developments for Experimental Nanotoxicology

Besides the development of advanced in vitro models that more closely mimic in vivo conditions, progress is awaited from the development of some approaches, techniques, and their combination. As examples, we can name the omics techniques, high-throughput screening (HTS) analyses, and the adverse outcome pathways (AOPs).

The "omics" approaches include genomics, transcriptomics, proteomics, metabolomics, lipidomics, and toxicogenomics and can be applied both in vitro and in vivo (although more easily in vitro) [56]. They provide a more comprehensive understanding of the biological response to nanomaterials than conventional toxicological assays, providing an insight into the underlying mechanisms such as molecular interactions, cellular responses, tissue/organ changes, and organism responses [56,136–139]. In addition, they may predict toxicity at low levels of nanomaterial exposure, which do not produce toxicity but can stress the cells, thereby emphasizing subtle biological changes that would not be detected by a conventional toxicological assessment [56,139]. On the other hand, they require expensive infrastructure and highly skilled personnel to prepare the samples and to analyze the data [2].

High-throughput screening (HTS) methods are defined as the use of automated tools to facilitate the rapid execution of a large number and variety of biological assays that may include several test substances in each assay [2,21]. They thus allow the rapid screening of large numbers of nanomaterials at different concentrations and under different exposure conditions, saving time and money. In vitro HTS may be used in order to rank nanomaterials' hazard potential and prioritize them for in vivo testing [2,112]. Several HTS methods have been validated and applied for nanomaterial testing, including high-throughput flow cytometry, multiplex analysis of secreted products, and genotoxicity methods (high-throughput comet assay, high-throughput micronucleus assay, and γH2AX assay) [2]. HTS can also be combined with high-content analysis (HCA) or high-content imaging (HCI) to deliver rapid and reliable toxicity assessment of large numbers of nanomaterials in parallel and can combine several endpoint measurements in one experiment [140]. For instance, the human hepatoma HepaRG cell line was treated with a large set of nanomaterials, coatings, and supernatants at different concentrations, and 14 different biological endpoints (including viable cell count, cell membrane permeability, apoptotic cell death, mitochondrial membrane potential, and steatosis) were analyzed using HTS [140]. The HTS method was also used to assess the genotoxicity induced by different graphene-based materials [57] or to evaluate the impact of nanomaterials on different DNA repair pathways [141].

An adverse outcome pathway (AOP) is a systematic process that uses the available mechanistic information concerning a toxicological response and describes causal or mechanistic relationships between a molecular initiating event (MIE), a series of intermediate key events (KEs), and the adverse outcome (AO) [56,142]. In other words, an AOP represents a linear chain of events starting from an MIE and ending with an AO. MIEs and AOs are connected by KEs corresponding to specific biochemical mechanisms which are causal and describe toxicity responses at different levels of biological organization including the cellular, organ, organism, and population levels [8,56,134,142]. The AOP framework provides pragmatic insights to promote the development of alternative testing strategies and can significantly support risk assessment by developing predictive methods that utilize mechanistic and evidence-based data [134,142]. The AOP approach has been applied to nanomaterials. While MIEs are not fully understood yet, inflammation, oxidative stress, and cytotoxicity have been identified as KEs and have led to defining lung fibrosis, lung emphysema, and lung cancer as AOs [56,134,142]. Different AOPs can be of relevance to nanomaterials such as AOP 173, AOP 303, AOP 237, AOP 302, and AOP 1.25, whose AOs are lung fibrosis, lung cancer, atherosclerotic plaque progression, acute inhalation toxicity, and lung emphysema, respectively [142]. However, lung fibrosis is the most widely assessed and reported AO following exposure to nanomaterials. AOPs can also be combined as networks. Indeed, individual linear AOPs allow the simplification of the complex biology, but the combination of AOPs could better describe the intricateness of the disease processes and hence is applicable to real-world scenarios [142].

Because of the respective drawbacks of each approach and model as discussed before, it is very difficult to draw firm conclusions on the hazard potential of nanomaterials. In addition, due to the multitude and variety of nanomaterials, it is impossible to assess in vitro or in vivo their toxicity profile on a case-by-case basis [143]. This is why computational approaches (in silico models) have been developed to complement experimental testing and represent useful alternatives proposing models able to predict the potential hazard of nanomaterials, without testing, thereby reducing the time and cost of nanosafety assessments and allowing the prioritization of nanomaterials that deserve further investigations [18,143].

3. Computational Nanotoxicology/In Silico Approaches

In silico toxicology refers to the use of computational methods to analyze, simulate, visualize, or predict the toxicity of chemicals and nanomaterials [54,144]. To be accurately predictive of nanomaterial toxicity, models should carefully consider the complexity of nanomaterials (requiring a complete physicochemical characterization) and their diverse environments (exposure route) [143,145]. Concretely, the aim of computational approaches is to model the relationships between nanomaterials' structure, properties, and biological effects [143]. Therefore, reliable in silico models can be used for the supplementation of data in the first step of nanomaterial hazard assessment, as recommended by the European Chemicals Agency (ECHA), or assist in the second step of hazard assessment (categorization and labeling of nanomaterials) by directly classifying nanomaterials into groups of different hazards [146].

The main advantage of computational approaches is that they can reduce experimental (including animal) testing, thus saving time and money. They are fast and allow the screening of larger numbers of nanomaterials [147]. On the other hand, the development of in silico models has to face some challenges; in particular, to be robust, predictive, and broadly applicable, models should be developed based on large amounts of high-quality and complete experimental data. They have to take into account the wide heterogeneity of published literature data in terms of nanomaterial characterization, exposure, and hazard data reported [103,143]. Missing values or data gaps and differing data quality (numbers of replicates, signal-to-noise ratio, relevance of endpoints, different experimental conditions, etc.) can be a hurdle for modeling.

Despite these limitations, different types of models of variable complexity have been successfully developed to predict nanomaterial toxicity. They mainly use statistical and machine learning (ML) algorithms to model relationships between a nanomaterial's structure, molecular properties, or other parameters and its biological effects [143]. The main machine learning algorithms most commonly used include regression, decision trees, support vector machines, artificial neural networks, partial least squares, and principal component analysis [144,148,149]. Molecular dynamics (MD) has also been used to simulate and predict nanomaterial interactions with biological environments [150].

3.1. Molecular Docking

Molecular docking studies consist of computational simulations that allow predicting how nanomaterials interact with biomolecules based on 3D structural knowledge [20]. Molecular docking has also been used to determine the potential toxicity of CuO, TiO_2, Fe_3O_4, Au, Ag, ZnO, Mn_2O_3, and Fe_3O_4 nanoparticles with biological macromolecules [151].

3.2. Quantitative Structure–Activity Relationship (QSAR)

The most successful computational models able to predict the biological properties of nanomaterials in diverse and complex environments are based on the quantitative structure–activity relationship (QSAR) method [143]. The basic aim of the QSAR model is to define an appropriate function that has a reasonable relationship between the chemical structure and biological activity. This has the potential to further summarize the physiochemical and biological information in order to predict toxicity effects or develop ideal nanomaterials [20]. In other words, and as schematically summarized in Figure 2, QSAR models are based on defining mathematical dependencies between the biological activity of nanomaterials (or toxicity, i.e., the endpoint) and their molecular descriptors (such as physicochemical features) [20,148,152]. They are usually used for a series of often related compounds: datasets.

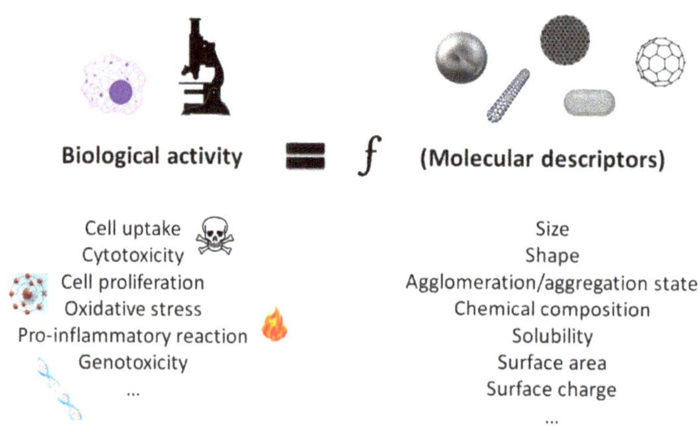

Figure 2. Basic principle of QSAR models for nanomaterials.

This approach assumes that nanomaterials with similar molecular structures will have similar biological effects [153]. Consequently, such models are able to predict a nanomaterial's behavior and effects in biological environments based on its physicochemical properties and molecular descriptors without experimental testing, which allows saving time and money [56,148,154]. Such predictions could also permit researchers to streamline and prioritize toxicological assays and focus on the most promising (i.e., safer) nanomaterials [114,154–157].

The concept of QSAR was proposed a century ago for chemical compounds [158,159] and has been extensively used in the areas of drug discovery and chemical toxicity modeling, but its application to nanomaterials is not without challenges due to their specificities. Indeed, it requires the development of nano-specific descriptors and curated experimental datasets [147,150,153,154,160]. As an example, a molecular descriptor can be experimentally determined or calculated [148]. In the latter case, we should ensure that the calculated descriptors are correlated with experimental values; otherwise, they could have a strong impact on the validity of the model. Therefore, descriptors should be developed and tailored using the material's specific information provided by the physicochemical characterization (experimentally determined), rather than using idealized structures [154,160]. Regarding the datasets necessary to develop QSAR models, they should contain a sufficient number of diverse nanomaterials. These two conditions (diversity and a sufficiently high number of samples) are currently difficult to meet, except when HTS methods are used [154]. Indeed, there are only small databases on experimentally measured basic endpoints [161]. This is partly due to the fact that, although many nanotoxicological studies are available, they are not standardized, resulting in heterogeneous data, and the few nanomaterial databases available are often not suitable for building models [161]. In addition, nanomaterials represent very structurally diverse groups of chemicals, making it difficult to build a significant dataset of structurally related nanomaterials [152]. To sum up, a large quantity of experimental data of a high quality should be available to develop a reliable nanoQSAR model [150,152,156,162–166]. Stronger collaborations between experimentalists and modelers should result in more useful datasets and allow QSAR predictive models to improve their potential [154,161,167], as shown in Figure 3.

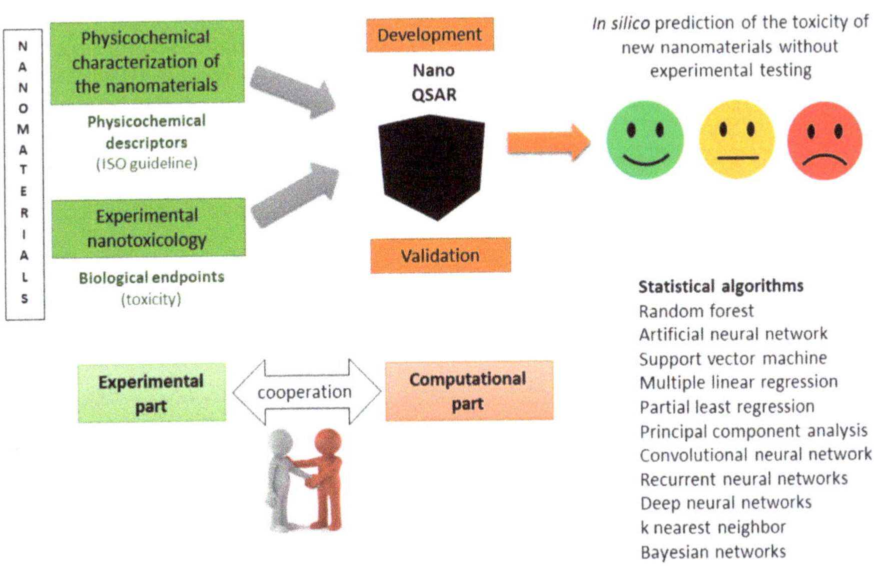

Figure 3. The development of QSAR models for the prediction of nanomaterial toxicity can use various statistical algorithms and is based on nanodescriptors (physicochemical features of nanomaterials) and biological endpoints (toxicity), both experimentally determined. This approach argues for strong collaboration between experimenters and modelers.

As recommended by the OECD [168], five principles should be met to validate a QSAR model: (i) it should be associated with a defined endpoint, (ii) show an unambiguous algorithm, (iii) a defined domain of applicability, and (iv) appropriate measures of goodness of fit, robustness, and predictivity, and (v) possibly provide a mechanistic interpretation. The first criterion indicates that, as a given endpoint could be determined by different experimental protocols and under different experimental conditions, it is therefore crucial to precisely identify the experimental system that is being modeled by the QSAR [154]. The intent of having an unambiguous algorithm is to ensure transparency in the model algorithm that generates predictions of an endpoint. An applicability domain must be defined in the sense that QSARs are reductionist and thus are inevitably associated with limitations in terms of the types of chemical structures, physicochemical properties, and mechanisms of action for which the models can generate reliable predictions. The need for appropriate measures of goodness of fit, robustness, and predictivity indicates that the internal performance (as represented by the goodness of fit and robustness) and predictivity of a model (as determined by external validation) should be carefully assessed [154].

QSAR models have been developed in particular for metal and metal oxide nanomaterials [69,157,169]. Using a dataset of 17 metallic oxide particles differing by their chemical composition, it was found that their toxicity to *Escherichia coli* bacteria could be predicted on the basis of their chemical nature only thanks to calculated or experimental descriptors such as metal electronegativity and cation charge [155,170]. Surprisingly, dimensional or shape descriptors did not play any role. These results were confirmed in a model of mammalian cells, the HaCaT cell line, regarding the toxicity of 18 metal oxide nanoparticles [152,171]. More recently, QSAR models for the prediction of the inflammatory potential of metal oxide nanoparticles were developed based on a dataset of 30 samples [20]. Similarly, a QSAR model able to predict the cellular uptake of 109 functionalized metal oxide nanoparticles to pancreatic cancer cells (PaCa2) has been recently developed [172]. A predictive QSAR model for the cellular association of gold nanoparticles based on their physicochemical properties and protein corona fingerprints was developed and validated using a dataset of 105 nanoparticles [173].

Regarding the statistical methods used for QSAR models, different types of machine learning algorithms can be used (Figure 3), including principal component analysis (PCA), multiple linear regression (MLR), partial least squares (PLS) methods, random forests (RFs), support vector machines (SVMs), k-nearest neighbors, Bayesian networks, and artificial neural networks (ANNs) [20,149,166,172,174–184].

Despite the successful examples of application of nanoQSAR models, they still have some limitations. For instance, they did not prove very useful with regard to in vitro–in vivo extrapolation. Indeed, in QSAR models, the molecular descriptors usually refer to the pristine nanomaterials, while upon contact with biological media, these features are altered [69]. Similarly, because a nanomaterial can exert variable biological effects under different biochemical conditions (cell line, cell species, etc.), QSAR modeling is difficult [143]. Finally, it should be mentioned that one shortcoming of QSAR models lies in the fact that nanomaterial physicochemical features are studied independently while we know that they are directly correlated with each other (e.g., size and surface area). Such an assumption is, however, necessary to manage the complexity and large amount of data.

An alternative to QSAR, especially when data are significantly limited, is the read-across approach [152,165,185].

3.3. Grouping/Read-Across

To fill data gaps without performing additional experimental studies, the concepts of grouping and read-across have been developed. The aim is to predict the effects of an unknown substance (target material) based on the data available from substances classified in the same group because they are considered similar in some way (source material). Similarity is usually based on structural and/or physicochemical properties [13,145,150,152,165,186]. In other words, this approach is based on the assumption that materials that are considered similar may have a comparable toxicological behavior. Consequently, experimental available toxicological properties from a source material can be used to derive the toxicological properties of a target analogue with no (or limited) toxicological experimental data [187].

One may wonder if such an approach initially proposed for chemical compounds is suitable for nanomaterials due to their heterogeneity as previously discussed [13]. However, grouping/read-across methods are considered valid analytical tools to enable a more efficient hazard identification and assessment of nanomaterials by bringing together substances with similar hazardous profiles [145,146,186,188]. A major advantage of grouping/read-across methods is that they do not require a large amount of data to identify groups of similar compounds [152]. On the other hand, the criteria by which similar chemicals are selected for read-across should be carefully justified for the acceptability of the model predictions [183]. For this purpose, the European Chemicals Agency (ECHA) issued a guidance document on information requirements and chemical safety assessment specific to the application of QSAR and grouping approaches to nanomaterials [189].

As an example, following these guidelines, Lamon et al. used grouping/read-across to predict the in vitro comet assay results of TiO_2 nanoforms. They identified the physicochemical properties that define the groups and the similarities between analogues of the same category, using chemoinformatic techniques such as hierarchical clustering (HC), principal component analysis (PCA), and random forest variable selection [190]. The level of similarity can also be determined with distance measures (Euclidean distance, Pearson correlation distance, Manhattan distance, etc.), and then the activity of the target nanomaterial(s) can be predicted based on the activity of the nearest neighbors (i.e., the most similar chemicals) [185]. Another way to justify (or reject) the grouping of a target nanomaterial and a source material is through integrated approaches to testing and assessment (IATAs) [145]. IATAs set out a tiered testing strategy, which reflects the different information needed and levels of uncertainty acceptable for different grouping purposes. The substantiation of a grouping decision is underpinned by the demonstration of similarity between group members, which helps the user to assess whether a target nanomaterial is sufficiently similar to a source material to allow grouping and to assume the target nanomaterial will induce similar toxicity to the source material [145].

Some examples from the literature illustrate the success of the grouping/read-across techniques. For instance, they allow estimating the cytotoxicity of metal oxide nanoparticles with a similar level of accuracy to that provided by QSAR models [152].

4. Conclusions

As schematically summarized in Figure 4, in vitro/in vivo approaches as well as in silico models for nanotoxicology have their respective advantages and limitations.

No gold standard exists to assess the toxicity of a nanomaterial; instead, combinations of several techniques are necessary to adequately describe the material's toxicological profile [55,56]. Similarly, it seems unlikely that one cell-based assay or computational model will ever replace an existing animal test. Thus, the most appropriate methodologies will need to be applied in an integrated assessment and testing strategy [69,191]. Experimental and computational approaches thus appear complementary, and their combination should lead to progress in the nanotoxicology field. Bridging the gap between these approaches implies stronger interactions between experimentalists and developers of models, adopting a multidisciplinary approach. In this context, new approach methodologies (NAMs) have been proposed; according to the ECHA definition, they consist of "a broad context to include in silico approaches, in chemico, and in vitro assays, as well as the inclusion of information from the exposure of chemicals in the context of risk assessment" [192]. As an example, system toxicology, by combining advanced analytical and computational tools, can provide quantitative information on system-wide molecular changes in the context of toxicant exposure, leading to information on how biological networks are perturbed by toxicants. System toxicology aims to change the way in which adverse effects of chemicals or other toxicants are characterized, from isolated empirical endpoints to integrated pathways of toxicity [112]. Similarly, the combination of omics analyses and machine learning has been proposed as a promising approach for analyzing nanotoxicity [139,154,184].

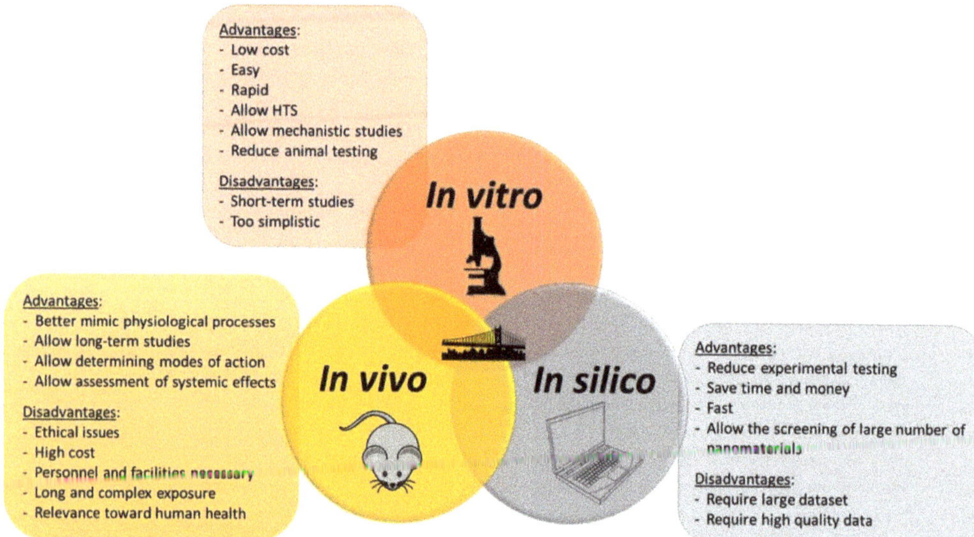

Figure 4. Summary of the advantages and limitations of approaches used in nanotoxicology and perspectives of bridging the gap between these methods.

Funding: This research received no external funding.

Institutional Review Board Statement: Not applicable.

Informed Consent Statement: Not applicable.

Data Availability Statement: Not applicable.

Conflicts of Interest: The author declares no conflict of interest.

References

1. The Nanodatabase. Welcome to The Nanodatabase. Available online: https://nanodb.dk/ (accessed on 28 January 2022).
2. Collins, A.R.; Annangi, B.; Rubio, L.; Marcos, R.; Dorn, M.; Merker, C.; Estrela-Lopis, I.; Cimpan, M.R.; Ibrahim, M.; Cimpan, E.; et al. High throughput toxicity screening and intracellular detection of nanomaterials. *WIREs Nanomed. Nanobiotechnol.* **2017**, *9*, e1413. [CrossRef] [PubMed]
3. Li, Y.; Wang, J.; Zhao, F.; Bai, B.; Nie, G.; Nel, A.E.; Zhao, Y. Nanomaterial libraries and model organisms for rapid high-content analysis of nanosafety. *Natl. Sci. Rev.* **2018**, *5*, 365–388. [CrossRef]
4. Warheit, D.B. Hazard and risk assessment strategies for nanoparticle exposures: How far have we come in the past 10 years? *F1000Research* **2018**, *7*, 376. [CrossRef] [PubMed]
5. Naz, S.; Gul, A.; Zia, M. Toxicity of copper oxide nanoparticles: A review study. *IET Nanobiotechnol.* **2020**, *14*, 1–13. [CrossRef]
6. ISO/TR 13014:2012. Nanotechnologies—Guidance on Physico-Chemical Characterization of Engineered Nanoscale Materials for Toxicologic Assessment. Available online: http://www.iso.org/iso/catalogue_detail?csnumber=52334 (accessed on 28 November 2016).
7. Oberdörster, G.; Oberdörster, E.; Oberdörster, J. Nanotoxicology: An Emerging Discipline Evolving from Studies of Ultrafine Particles. *Environ. Health Perspect.* **2005**, *113*, 823–839. [CrossRef]
8. Clift, M.J.D.; Jenkins, G.J.S.; Doak, S.H. An Alternative Perspective towards Reducing the Risk of Engineered Nanomaterials to Human Health. *Small* **2020**, *16*, e2002002. [CrossRef]
9. Freedman, B. *Environmental Science: A Canadian Perspective*, 6th ed.; Dalhousie University Libraries Digital Editions: Halifax, NS, Canada, 2018.
10. Hillegass, J.M.; Shukla, A.; Lathrop, S.A.; MacPherson, M.B.; Fukagawa, N.K.; Mossman, B.T. Assessing nanotoxicity in cells in vitro. *WIREs Nanomed. Nanobiotechnol.* **2010**, *2*, 219–231. [CrossRef]
11. Kumar, V.; Sharma, N.; Maitra, S.S. In vitro and in vivo toxicity assessment of nanoparticles. *Int. Nano Lett.* **2017**, *7*, 243–256. [CrossRef]

12. Domb, A.J.; Sharifzadeh, G.; Nahum, V.; Hosseinkhani, H. Safety Evaluation of Nanotechnology Products. *Pharmaceutics* **2021**, *13*, 1615. [CrossRef]
13. Drasler, B.; Sayre, P.; Steinhäuser, K.G.; Fink, A.; Rothen-Rutishauser, B. In vitro approaches to assess the hazard of nanomaterials. *NanoImpact* **2017**, *8*, 99–116. [CrossRef]
14. Zielińska, A.; Costa, B.; Ferreira, M.V.; Miguéis, D.; Louros, J.M.S.; Durazzo, A.; Lucarini, M.; Eder, P.; Chaud, M.V.; Morsink, M.; et al. Nanotoxicology and Nanosafety: Safety-By-Design and Testing at a Glance. *Int. J. Environ. Res. Public Health* **2020**, *17*, 4657. [CrossRef] [PubMed]
15. Joris, F.; Manshian, B.B.; Peynshaert, K.; de Smedt, S.C.; Braeckmans, K.; Soenen, S.J. Assessing nanoparticle toxicity in cell-based assays: Influence of cell culture parameters and optimized models for bridging the in vitro–in vivo gap. *Chem. Soc. Rev.* **2013**, *42*, 8339–8359. [CrossRef] [PubMed]
16. Verdon, R.; Gillies, S.L.; Brown, D.M.; Henry, T.; Tran, L.; Tyler, C.R.; Rossi, A.G.; Stone, V.; Johnston, H.J. Neutrophil activation by nanomaterials in vitro: Comparing strengths and limitations of primary human cells with those of an immortalized (HL-60) cell line. *Nanotoxicology* **2021**, *15*, 1–20. [CrossRef] [PubMed]
17. Singh, N.; Wills, J.W.; Doak, S.H. Chapter 10. Three-dimensional Models for In vitro Nanotoxicity Testing. *Nanotoxicology* **2017**, *35*, 248–275. [CrossRef]
18. Fröhlich, E. Comparison of conventional and advanced in vitro models in the toxicity testing of nanoparticles. *Artif. Cells Nanomed. Biotechnol.* **2018**, *46*, 1091–1107. [CrossRef] [PubMed]
19. Prasad, M.; Kumar, R.; Buragohain, L.; Kumari, A.; Ghosh, M. Organoid Technology: A Reliable Developmental Biology Tool for Organ-Specific Nanotoxicity Evaluation. *Front. Cell Dev. Biol.* **2021**, *9*, 696668. [CrossRef] [PubMed]
20. Huang, H.-J.; Lee, Y.-H.; Hsu, Y.-H.; Liao, C.-T.; Lin, Y.-F.; Chiu, H.-W. Current Strategies in Assessment of Nanotoxicity: Alternatives to In Vivo Animal Testing. *Int. J. Mol. Sci.* **2021**, *22*, 4216. [CrossRef] [PubMed]
21. Kohl, Y.; Rundén-Pran, E.; Mariussen, E.; Hesler, M.; El Yamani, N.; Longhin, E.M.; Dusinska, M. Genotoxicity of Nanomaterials: Advanced In Vitro Models and High Throughput Methods for Human Hazard Assessment—A Review. *Nanomaterials* **2020**, *10*, 1911. [CrossRef]
22. Barosova, H.; Drasler, B.; Petri-Fink, A.; Rothen-Rutishauser, B. Multicellular Human Alveolar Model Composed of Epithelial Cells and Primary Immune Cells for Hazard Assessment. *J. Vis. Exp.* **2020**, e61090. [CrossRef]
23. Barosova, H.; Karakocak, B.B.; Septiadi, D.; Petri-Fink, A.; Stone, V.; Rothen-Rutishauser, B. An In Vitro Lung System to Assess the Proinflammatory Hazard of Carbon Nanotube Aerosols. *Int. J. Mol. Sci.* **2020**, *21*, 5335. [CrossRef]
24. Saez-Tenorio, M.; Domenech, J.; García-Rodríguez, A.; Velázquez, A.; Hernández, A.; Marcos, R.; Cortés, C. Assessing the relevance of exposure time in differentiated Caco-2/HT29 cocultures. Effects of silver nanoparticles. *Food Chem. Toxicol.* **2019**, *123*, 258–267. [CrossRef] [PubMed]
25. Ude, V.C.; Brown, D.M.; Stone, V.; Johnston, H.J. Using 3D gastrointestinal tract in vitro models with microfold cells and mucus secreting ability to assess the hazard of copper oxide nanomaterials. *J. Nanobiotechnol.* **2019**, *17*, 70. [CrossRef] [PubMed]
26. Bredeck, G.; Kämpfer, A.A.M.; Sofranko, A.; Wahle, T.; Büttner, V.; Albrecht, C.; Schins, R.P.F. Ingested Engineered Nanomaterials Affect the Expression of Mucin Genes—An In Vitro-In Vivo Comparison. *Nanomaterials* **2021**, *11*, 2621. [CrossRef] [PubMed]
27. Loret, T.; Peyret, E.; Dubreuil, M.; Aguerre-Chariol, O.; Bressot, C.; Le Bihan, O.; Amodeo, T.; Trouiller, B.; Braun, A.; Egles, C.; et al. Air-liquid interface exposure to aerosols of poorly soluble nanomaterials induces different biological activation levels compared to exposure to suspensions. *Part. Fibre Toxicol.* **2016**, *13*, 58. [CrossRef]
28. Kose, O.; Tomatis, M.; Leclerc, L.; Belblidia, N.-B.; Hochepied, J.-F.; Turci, F.; Pourchez, J.; Forest, V. Impact of the Physicochemical Features of TiO2 Nanoparticles on Their In Vitro Toxicity. *Chem. Res. Toxicol.* **2020**, *33*, 2324–2337. [CrossRef]
29. Ventura, C.; Pereira, J.F.S.; Matos, P.; Marques, B.; Jordan, P.; Sousa-Uva, A.; Silva, M.J. Cytotoxicity and genotoxicity of MWCNT-7 and crocidolite: Assessment in alveolar epithelial cells versus their coculture with monocyte-derived macrophages. *Nanotoxicology* **2020**, *14*, 479–503. [CrossRef] [PubMed]
30. Kermanizadeh, A.; Hr, M.L.; Roursgaard, M.; Messner, S.; Gunness, P.; Kelm, J.M.; Møller, P.; Stone, V.; Loft, S. Hepatic toxicology following single and multiple exposure of engineered nanomaterials utilising a novel primary human 3D liver microtissue model. *Part. Fibre Toxicol.* **2014**, *11*, 56. [CrossRef]
31. Broutier, L.; Andersson-Rolf, A.; Hindley, C.J.; Boj, S.F.; Clevers, H.; Koo, B.-K.; Huch, M. Culture and establishment of self-renewing human and mouse adult liver and pancreas 3D organoids and their genetic manipulation. *Nat. Protoc.* **2016**, *11*, 1724–1743. [CrossRef]
32. Kabadi, P.K.; Rodd, A.L.; Simmons, A.E.; Messier, N.J.; Hurt, R.H.; Kane, A.B. A novel human 3D lung microtissue model for nanoparticle-induced cell-matrix alterations. *Part. Fibre Toxicol.* **2019**, *16*, 15. [CrossRef]
33. Llewellyn, S.V.; Conway, G.; Shah, U.-K.; Evans, S.; Jenkins, G.J.; Clift, M.J.; Doak, S.H. Advanced 3D Liver Models for In vitro Genotoxicity Testing Following Long-Term Nanomaterial Exposure. *J. Vis. Exp.* **2020**, e61141. [CrossRef]
34. Llewellyn, S.V.; Conway, G.E.; Zanoni, I.; Jørgensen, A.K.; Shah, U.-K.; Seleci, D.A.; Keller, J.G.; Kim, J.W.; Wohlleben, W.; Jensen, K.A.; et al. Understanding the impact of more realistic low-dose, prolonged engineered nanomaterial exposure on genotoxicity using 3D models of the human liver. *J. Nanobiotechnol.* **2021**, *19*, 193. [CrossRef] [PubMed]
35. Young, E.W.K.; Beebe, D.J. Fundamentals of microfluidic cell culture in controlled microenvironments. *Chem. Soc. Rev.* **2010**, *39*, 1036–1048. [CrossRef] [PubMed]

36. Fierz, M.; Houle, C.; Steigmeier, P.; Burtscher, H. Design, Calibration, and Field Performance of a Miniature Diffusion Size Classifier. *Aerosol Sci. Technol.* **2011**, *45*, 1–10. [CrossRef]
37. Zhang, M.; Xu, C.; Jiang, L.; Qin, J. A 3D human lung-on-a-chip model for nanotoxicity testing. *Toxicol. Res.* **2018**, *7*, 1048–1060. [CrossRef] [PubMed]
38. Xu, C.; Zhang, M.; Chen, W.; Jiang, L.; Chen, C.; Qin, J. Assessment of Air Pollutant PM2.5 Pulmonary Exposure Using a 3D Lung-on-Chip Model. *ACS Biomater. Sci. Eng.* **2020**, *6*, 3081–3090. [CrossRef] [PubMed]
39. Rach, J.; Budde, J.; Möhle, N.; Aufderheide, M. Direct exposure at the air-liquid interface: Evaluation of anin vitroapproach for simulating inhalation of airborne substances. *J. Appl. Toxicol.* **2014**, *34*, 506–515. [CrossRef]
40. Lacroix, G.; Koch, W.; Ritter, D.; Gutleb, A.; Larsen, S.T.; Loret, T.; Zanetti, F.; Constant, S.; Chortarea, S.; Rothen-Rutishauser, B.; et al. Air–Liquid InterfaceIn VitroModels for Respiratory Toxicology Research: Consensus Workshop and Recommendations. *Appl. In Vitro Toxicol.* **2018**, *4*, 91–106. [CrossRef]
41. Kooter, I.; Ilves, M.; Gröllers-Mulderij, M.; Duistermaat, E.; Tromp, P.C.; Kuper, F.; Kinaret, P.; Savolainen, K.; Greco, D.; Karisola, P.; et al. Molecular Signature of Asthma-Enhanced Sensitivity to CuO Nanoparticle Aerosols from 3D Cell Model. *ACS Nano* **2019**, *13*, 6932–6946. [CrossRef]
42. Bessa, M.J.; Brandão, F.; Fokkens, P.; Cassee, F.R.; Salmatonidis, A.; Viana, M.; Vulpoi, A.; Simon, S.; Monfort, E.; Teixeira, J.P.; et al. Toxicity assessment of industrial engineered and airborne process-generated nanoparticles in a 3D human airway epithelial in vitro model. *Nanotoxicology* **2021**, *15*, 542–557. [CrossRef]
43. Diabaté, S.; Armand, L.; Murugadoss, S.; Dilger, M.; Fritsch-Decker, S.; Schlager, C.; Béal, D.; Arnal, M.-E.; Biola-Clier, M.; Ambrose, S.; et al. Air–Liquid Interface Exposure of Lung Epithelial Cells to Low Doses of Nanoparticles to Assess Pulmonary Adverse Effects. *Nanomaterials* **2020**, *11*, 65. [CrossRef]
44. Panas, A.; Comouth, A.; Saathoff, H.; Leisner, T.; Al-Rawi, M.; Simon, M.; Seemann, G.; Dössel, O.; Mülhopt, S.; Paur, H.-R., et al. Silica nanoparticles are less toxic to human lung cells when deposited at the air–liquid interface compared to conventional submerged exposure. *Beilstein J. Nanotechnol.* **2014**, *5*, 1590–1602. [CrossRef] [PubMed]
45. Gohlsch, K.; Mückter, H.; Steinritz, D.; Aufderheide, M.; Hoffmann, S.; Gudermann, T.; Breit, A. Exposure of 19 substances to lung A549 cells at the air liquid interface or under submerged conditions reveals high correlation between cytotoxicity in vitro and CLP classifications for acute lung toxicity. *Toxicol. Lett.* **2019**, *316*, 119–126. [CrossRef] [PubMed]
46. Xie, Y.; Williams, N.G.; Tolic, A.; Chrisler, W.B.; Teeguarden, J.; Maddux, B.; Pounds, J.; Laskin, A.; Orr, G. Aerosolized ZnO Nanoparticles Induce Toxicity in Alveolar Type II Epithelial Cells at the Air-Liquid Interface. *Toxicol. Sci.* **2012**, *125*, 450–461. [CrossRef] [PubMed]
47. Lenz, A.-G.; Karg, E.; Brendel, E.; Hinze-Heyn, H.; Maier, K.L.; Eickelberg, O.; Stoeger, T.; Schmid, O. Inflammatory and Oxidative Stress Responses of an Alveolar Epithelial Cell Line to Airborne Zinc Oxide Nanoparticles at the Air-Liquid Interface: A Comparison with Conventional, Submerged Cell-Culture Conditions. *BioMed Res. Int.* **2013**, *2013*, 652632. [CrossRef]
48. Kim, J.S.; Peters, T.M.; O'Shaughnessy, P.T.; Adamcakova-Dodd, A.; Thorne, P.S. Validation of an in vitro exposure system for toxicity assessment of air-delivered nanomaterials. *Toxicol. Vitr.* **2013**, *27*, 164–173. [CrossRef]
49. Herzog, F.; Loza, K.; Balog, S.; Clift, M.J.D.; Epple, M.; Gehr, P.; Petri-Fink, A.; Rothen-Rutishauser, B. Mimicking exposures to acute and lifetime concentrations of inhaled silver nanoparticles by two different in vitro approaches. *Beilstein J. Nanotechnol.* **2014**, *5*, 1357–1370. [CrossRef]
50. Clippinger, A.J.; Ahluwalia, A.; Allen, D.; Bonner, J.C.; Casey, W.; Castranova, V.; David, R.M.; Halappanavar, S.; Hotchkiss, J.A.; Jarabek, A.M.; et al. Expert consensus on an in vitro approach to assess pulmonary fibrogenic potential of aerosolized nanomaterials. *Arch. Toxicol.* **2016**, *90*, 1769–1783. [CrossRef]
51. Ndika, J.; Ilves, M.; Kooter, I.M.; Gröllers-Mulderij, M.; Duistermaat, E.; Tromp, P.C.; Kuper, F.; Kinaret, P.; Greco, D.; Karisola, P.; et al. Mechanistic Similarities between 3D Human Bronchial Epithelium and Mice Lung, Exposed to Copper Oxide Nanoparticles, Support Non-Animal Methods for Hazard Assessment. *Small* **2020**, *16*, e2000527. [CrossRef]
52. Park, M.V.; Lankveld, D.P.; van Loveren, H.; de Jong, W.H. The status of in vitro toxicity studies in the risk assessment of nanomaterials. *Nanomedicine* **2009**, *4*, 669–685. [CrossRef]
53. Takhar, P.; Mahant, S. In Vitro Methods for Nanotoxicity Assessment: Advantages and Applications. *Arch. Appl. Sci. Res.* **2011**, *3*, 389–403.
54. Gupta, R.; Xie, H. Nanoparticles in Daily Life: Applications, Toxicity and Regulations. *J. Environ. Pathol. Toxicol. Oncol.* **2018**, *37*, 209–230. [CrossRef] [PubMed]
55. Savage, D.T.; Hilt, J.Z.; Dziubla, T.D. In vitro methods for assessing nanoparticle toxicity. In *Nanotoxicity. Methods in Molecular Biology*; Zhang, Q., Ed.; Humana Press: New York, NY, USA, 2019; Volume 1894, pp. 1–29. [CrossRef]
56. Tirumala, M.G.; Anchi, P.; Raja, S.; Rachamalla, M.; Godugu, C. Novel Methods and Approaches for Safety Evaluation of Nanoparticle Formulations: A Focus Towards In Vitro Models and Adverse Outcome Pathways. *Front. Pharmacol.* **2021**, *12*, 612659. [CrossRef] [PubMed]
57. Achawi, S.; Huot, L.; Nesslany, F.; Pourchez, J.; Simar, S.; Forest, V.; Feneon, B. Exploring graphene-based materials' genotoxicity: Inputs of a screening method. *Nanotoxicology* **2021**, *15*, 1279–1294. [CrossRef] [PubMed]
58. El Yamani, N.; Collins, A.; Rundén-Pran, E.; Fjellsbø, L.M.; Shaposhnikov, S.; Zielonddiny, S.; Dusinska, M. In vitrogenotoxicity testing of four reference metal nanomaterials, titanium dioxide, zinc oxide, cerium oxide and silver: Towards reliable hazard assessment. *Mutagenesis* **2017**, *32*, 117–126. [CrossRef] [PubMed]

59. Bajpayee, M.; Kumar, A.; Dhawan, A. The Comet Assay: Assessment of In Vitro and In Vivo DNA Damage. *Methods Mol. Biol.* **2019**, *2031*, 237–257. [CrossRef] [PubMed]
60. Fröhlich, E.; Salar-Behzadi, S. Toxicological Assessment of Inhaled Nanoparticles: Role of in Vivo, ex Vivo, in Vitro, and in Silico Studies. *Int. J. Mol. Sci.* **2014**, *15*, 4795–4822. [CrossRef]
61. Russell, W.M.S.; Burch, R.L. *The Principles of Humane Experimental Technique*; Methuen: London, UK, 1959.
62. European Commission. *European Union Directive 2010/63/EU of the European Parliament and of the Council of 22 September 2010 on the Protection of Animals Used for Scientific Purposes*; European Commission: Brussels, Belgium, 2010.
63. Forest, V.; Leclerc, L.; Hochepied, J.-F.; Trouvé, A.; Sarry, G.; Pourchez, J. Impact of cerium oxide nanoparticles shape on their in vitro cellular toxicity. *Toxicol. In Vitro* **2017**, *38*, 136–141. [CrossRef]
64. Kurtz-Chalot, A.; Klein, J.-P.; Pourchez, J.; Boudard, D.; Bin, V.; Alcantara, G.B.; Martini, M.; Cottier, M.; Forest, V. Adsorption at cell surface and cellular uptake of silica nanoparticles with different surface chemical functionalizations: Impact on cytotoxicity. *J. Nanopart. Res.* **2014**, *16*, 1–15. [CrossRef]
65. Kurtz-Chalot, A.; Villiers, C.; Pourchez, J.; Boudard, D.; Martini, M.; Marche, P.N.; Cottier, M.; Forest, V. Impact of silica nanoparticle surface chemistry on protein corona formation and consequential interactions with biological cells. *Mater. Sci. Eng. C* **2017**, *75*, 16–24. [CrossRef]
66. Pourchez, J.; Forest, V.; Boumahdi, N.; Boudard, D.; Tomatis, M.; Fubini, B.; Herlin-Boime, N.; Leconte, Y.; Guilhot, B.; Cottier, M.; et al. In vitro cellular responses to silicon carbide nanoparticles: Impact of physico-chemical features on pro-inflammatory and pro-oxidative effects. *J. Nanopart. Res.* **2012**, *14*, 1143. [CrossRef]
67. Boudard, D.; Forest, V.; Pourchez, J.; Boumahdi, N.; Tomatis, M.; Fubini, B.; Guilhot, B.; Cottier, M.; Grosseau, P. In vitro cellular responses to silicon carbide particles manufactured through the Acheson process: Impact of physico-chemical features on pro-inflammatory and pro-oxidative effects. *Toxicol. In Vitro* **2014**, *28*, 856–865. [CrossRef] [PubMed]
68. Forest, V.; Pailleux, M.; Pourchez, J.; Boudard, D.; Tomatis, M.; Fubini, B.; Sennour, M.; Hochepied, J.-F.; Grosseau, P.; Cottier, M. Toxicity of boehmite nanoparticles: Impact of the ultrafine fraction and of the agglomerates size on cytotoxicity and pro-inflammatory response. *Inhal. Toxicol.* **2014**, *26*, 545–553. [CrossRef] [PubMed]
69. Burden, N.; Aschberger, K.; Chaudhry, Q.; Clift, M.J.D.; Fowler, P.; Johnston, H.; Landsiedel, R.; Rowland, J.; Stone, V.; Doak, S.H. Aligning nanotoxicology with the 3Rs: What is needed to realise the short, medium and long-term opportunities? *Regul. Toxicol. Pharmacol.* **2017**, *91*, 257–266. [CrossRef] [PubMed]
70. Klein, J.-P.; Boudard, D.; Cadusseau, J.; Palle, S.; Forest, V.; Pourchez, J.; Cottier, M. Testicular biodistribution of 450 nm fluorescent latex particles after intramuscular injection in mice. *Biomed. Microdevices* **2013**, *15*, 427–436. [CrossRef]
71. Leclerc, L.; Klein, J.-P.; Forest, V.; Boudard, D.; Martini, M.; Pourchez, J.; Blanchin, M.-G.; Cottier, M. Testicular biodistribution of silica-gold nanoparticles after intramuscular injection in mice. *Biomed. Microdevices* **2015**, *17*, 66. [CrossRef]
72. Li, Y.; Cummins, E. Hazard characterization of silver nanoparticles for human exposure routes. *J. Environ. Sci. Health Part A* **2020**, *55*, 704–725. [CrossRef]
73. Laux, P.; Riebeling, C.; Booth, A.M.; Brain, J.D.; Brunner, J.; Cerrillo, C.; Creutzenberg, O.; Estrela-Lopis, I.; Gebel, T.; Johanson, G.; et al. Biokinetics of nanomaterials: The role of biopersistence. *NanoImpact* **2017**, *6*, 69–80. [CrossRef]
74. Catalán, J.; Rydman, E.; Aimonen, K.; Hannukainen, K.-S.; Suhonen, S.; Vanhala, E.; Moreno, C.; Meyer, V.; Perez, D.D.S.; Sneck, A.; et al. Genotoxic and inflammatory effects of nanofibrillated cellulose in murine lungs. *Mutagenesis* **2017**, *32*, 23–31. [CrossRef]
75. Balasubramanyam, A.; Sailaja, N.; Mahboob, M.; Rahman, M.F.; Hussain, S.M.; Grover, P. In vivo genotoxicity assessment of aluminium oxide nanomaterials in rat peripheral blood cells using the comet assay and micronucleus test. *Mutagenesis* **2009**, *24*, 245–251. [CrossRef]
76. Louro, H.; Tavares, A.; Vital, N.; Costa, P.M.; Alverca, E.; Zwart, E.; de Jong, W.H.; Fessard, V.; Lavinha, J.; Silva, M.J. Integrated approach to the in vivo genotoxic effects of a titanium dioxide nanomaterial usingLacZplasmid-based transgenic mice. *Environ. Mol. Mutagen.* **2014**, *55*, 500–509. [CrossRef]
77. Maser, E.; Schulz, M.; Sauer, U.G.; Wiemann, M.; Ma-Hock, L.; Wohlleben, W.; Hartwig, A.; Landsiedel, R. In vitro and in vivo genotoxicity investigations of differently sized amorphous SiO_2 nanomaterials. *Mutat. Res. Toxicol. Environ. Mutagen.* **2015**, *794*, 57–74. [CrossRef] [PubMed]
78. Shvedova, A.A.; Kisin, E.; Murray, A.R.; Johnson, V.J.; Gorelik, O.; Arepalli, S.; Hubbs, A.F.; Mercer, R.R.; Keohavong, P.; Sussman, N.; et al. Inhalation vs. aspiration of single-walled carbon nanotubes in C57BL/6 mice: Inflammation, fibrosis, oxidative stress, and mutagenesis. *Am. J. Physiol. Cell. Mol. Physiol.* **2008**, *295*, L552–L565. [CrossRef] [PubMed]
79. Osier, M.; Oberörster, G. Intratracheal Inhalation vs Intratracheal Instillation: Differences in Particle Effects. *Toxicol. Sci.* **1997**, *40*, 220–227. [CrossRef]
80. Warheit, D.B.; Brock, W.J.; Lee, K.P.; Webb, T.R.; Reed, K.L. Comparative Pulmonary Toxicity Inhalation and Instillation Studies with Different TiO_2 Particle Formulations: Impact of Surface Treatments on Particle Toxicity. *Toxicol. Sci.* **2005**, *88*, 514–524. [CrossRef]
81. Burden, N.; Clift, M.J.D.; Jenkins, G.J.S.; Labram, B.; Sewell, F. Opportunities and Challenges for Integrating New In Vitro Methodologies in Hazard Testing and Risk Assessment. *Small* **2021**, *17*, e2006298. [CrossRef]
82. Shanks, N.; Greek, R.; Greek, J. Are animal models predictive for humans? *Philos. Ethics Humanit. Med.* **2009**, *4*, 2. [CrossRef]

83. Manno, M.; Viau, C.; Cocker, J.; Colosio, C.; Lowry, L.; Mutti, A.; Nordberg, M.; Wang, S. Biomonitoring for occupational health risk assessment (BOHRA). *Toxicol. Lett.* **2010**, *192*, 3–16. [CrossRef]
84. De Vuyst, P.; Karjalainen, A.; Dumortier, P.; Pairon, J.-C.; Monsó, E.; Brochard, P.; Teschler, H.; Tossavainen, A.; Gibbs, A. Guidelines for mineral fibre analyses in biological samples: Report of the ERS Working Group. *Eur. Respir. J.* **1998**, *11*, 1416–1426. [CrossRef]
85. Case, B.W.; Abraham, J.L.; Meeker, G.; Pooley, F.D.; Pinkerton, K.E. Applying Definitions of "Asbestos" to Environmental and "Low-Dose" Exposure Levels and Health Effects, Particularly Malignant Mesothelioma. *J. Toxicol. Environ. Health Part B* **2011**, *14*, 3–39. [CrossRef]
86. Mossman, B.T.; Lippmann, M.; Hesterberg, T.W.; Kelsey, K.T.; Barchowsky, A.; Bonner, J.C. Pulmonary Endpoints (Lung Carcinomas and Asbestosis) Following Inhalation Exposure to Asbestos. *J. Toxicol. Environ. Health Part B* **2011**, *14*, 76–121. [CrossRef]
87. Bargagli, E.; Lavorini, F.; Pistolesi, M.; Rosi, E.; Prasse, A.; Rota, E.; Voltolini, L. Trace metals in fluids lining the respiratory system of patients with idiopathic pulmonary fibrosis and diffuse lung diseases. *J. Trace Elem. Med. Biol.* **2017**, *42*, 39–44. [CrossRef] [PubMed]
88. Bergamaschi, E.; Poland, C.; Canu, I.G.; Prina-Mello, A. The role of biological monitoring in nano-safety. *Nano Today* **2015**, *10*, 274–277. [CrossRef]
89. Bitounis, D.; Pourchez, J.; Forest, V.; Boudard, D.; Cottier, M.; Klein, J.-P. Detection and analysis of nanoparticles in patients: A critical review of the status quo of clinical nanotoxicology. *Biomaterials* **2016**, *76*, 302–312. [CrossRef] [PubMed]
90. Groopman, J.D.; Kensler, T.W. The light at the end of the tunnel for chemical-specific biomarkers: Daylight or headlight? *Carcinogenesis* **1999**, *20*, 1–11. [CrossRef] [PubMed]
91. Angerer, J.; Ewers, U.; Wilhelm, M. Human biomonitoring: State of the art. *Int. J. Hyg. Environ. Health* **2007**, *210*, 201–228. [CrossRef]
92. Rinaldo, M.; Andujar, P.; Lacourt, A.; Martinon, L.; Raffin, M.C.; Dumortier, P.; Pairon, J.-C.; Brochard, P. Perspectives in Biological Monitoring of Inhaled Nanosized Particles. *Ann. Occup. Hyg.* **2015**, *59*, 669–680. [CrossRef]
93. Forest, V.; Vergnon, J.-M.; Pourchez, J. Biological Monitoring of Inhaled Nanoparticles in Patients: An Appealing Approach to Study Causal Link between Human Respiratory Pathology and Exposure to Nanoparticles. *Chem. Res. Toxicol.* **2017**, *30*, 1655–1660. [CrossRef]
94. Bitounis, D.; Klein, J.-P.; Mery, L.; El-Merhie, A.; Forest, V.; Boudard, D.; Pourchez, J.; Cottier, M. Ex vivodetection and quantification of gold nanoparticles in human seminal and follicular fluids. *Analyst* **2018**, *143*, 475–486. [CrossRef]
95. Rinaldi, L.; Barabino, G.; Klein, J.-P.; Bitounis, D.; Pourchez, J.; Forest, V.; Boudard, D.; Leclerc, L.; Sarry, G.; Roblin, X.; et al. Metals distribution in colorectal biopsies: New insight on the elemental fingerprint of tumour tissue. *Dig. Liver Dis.* **2015**, *47*, 602–607. [CrossRef]
96. Raia-Barjat, T.; Prieux, C.; Leclerc, L.; Sarry, G.; Grimal, L.; Chauleur, C.; Pourchez, J.; Forest, V. Elemental fingerprint of human amniotic fluids and relationship with potential sources of maternal exposure. *J. Trace Elem. Med. Biol.* **2020**, *60*, 126477. [CrossRef]
97. Forest, V.; Vergnon, J.-M.; Guibert, C.; Bitounis, D.; Leclerc, L.; Sarry, G.; Pourchez, J. Metal load assessment in patient pulmonary lavages: Towards a comprehensive mineralogical analysis including the nano-sized fraction. *Nanotoxicology* **2017**, *11*, 1211–1224. [CrossRef] [PubMed]
98. Forest, V.; Pourchez, J.; Guibert, C.; Bitounis, D.; Leclerc, L.; Sarry, G.; Vergnon, J.-M. Nano to micron-sized particle detection in patients' lungs and its pathological significance. *Environ. Sci. Nano* **2019**, *6*, 1343–1350. [CrossRef]
99. Bitounis, D.; Barnier, V.; Guibert, C.; Pourchez, J.; Forest, V.; Boudard, D.; Hochepied, J.-F.; Chelle, P.; Vergnon, J.-M.; Cottier, M. A method for the quantitative extraction of gold nanoparticles from human bronchoalveolar lavage fluids through a glycerol gradient. *Nanoscale* **2018**, *10*, 2955–2969. [CrossRef] [PubMed]
100. Forest, V.; Pourchez, J.; Pélissier, C.; Durand, S.A.; Vergnon, J.-M.; Fontana, L. Relationship between Occupational Exposure to Airborne Nanoparticles, Nanoparticle Lung Burden and Lung Diseases. *Toxics* **2021**, *9*, 204. [CrossRef]
101. Ljungkvist, G.; Tinnerberg, H.; Löndahl, J.; Klang, T.; Viklund, E.; Kim, J.-L.; Schiöler, L.; Forsgard, N.; Olin, A.-C. Exploring a new method for the assessment of metal exposure by analysis of exhaled breath of welders. *Int. Arch. Occup. Environ. Health* **2022**, *95*, 1–11. [CrossRef]
102. Landsiedel, R.; Sauer, U.G.; Ma-Hock, L.; Schnekenburger, J.; Wiemann, M. Pulmonary toxicity of nanomaterials: A critical comparison of published in vitro assays and in vivo inhalation or instillation studies. *Nanomedicine* **2014**, *9*, 2557–2585. [CrossRef]
103. Achawi, S.; Pourchez, J.; Feneon, B.; Forest, V. Graphene-Based Materials In Vitro Toxicity and Their Structure–Activity Relationships: A Systematic Literature Review. *Chem. Res. Toxicol.* **2021**, *34*, 2003–2018. [CrossRef]
104. Oberdörster, G.; Oberdörster, E.; Oberdörster, J. Concepts of Nanoparticle Dose Metric and Response Metric. *Environ. Health Perspect.* **2007**, *115*, A290. [CrossRef]
105. Stefaniak, A.B.; Hackley, V.; Roebben, G.; Ehara, K.; Hankin, S.; Postek, M.T.; Lynch, I.; Fu, W.-E.; Linsinger, T.P.; Thünemann, A.F. Nanoscale reference materials for environmental, health and safety measurements: Needs, gaps and opportunities. *Nanotoxicology* **2013**, *7*, 1325–1337. [CrossRef]
106. Johnston, L.J.; Gonzalez-Rojano, N.; Wilkinson, K.J.; Xing, B. Key challenges for evaluation of the safety of engineered nanomaterials. *NanoImpact* **2020**, *18*, 100219. [CrossRef]

107. Kroll, A.; Pillukat, M.H.; Hahn, D.; Schnekenburger, J. Interference of engineered nanoparticles with in vitro toxicity assays. *Arch. Toxicol.* **2012**, *86*, 1123–1136. [CrossRef] [PubMed]
108. Pailleux, M.; Boudard, D.; Pourchez, J.; Forest, V.; Grosseau, P.; Cottier, M. New insight into artifactual phenomena during in vitro toxicity assessment of engineered nanoparticles: Study of TNF-α adsorption on alumina oxide nanoparticle. *Toxicol. In Vitro* **2013**, *27*, 1049–1056. [CrossRef] [PubMed]
109. Forest, V.; Figarol, A.; Boudard, D.; Cottier, M.; Grosseau, P.; Pourchez, J. Adsorption of Lactate Dehydrogenase Enzyme on Carbon Nanotubes: How to Get Accurate Results for the Cytotoxicity of These Nanomaterials. *Langmuir* **2015**, *31*, 3635–3643. [CrossRef] [PubMed]
110. Wiemann, M.; Sauer, U.G.; Vennemann, A.; Bäcker, S.; Keller, J.-G.; Ma-Hock, L.; Wohlleben, W.; Landsiedel, R. In Vitro and In Vivo Short-Term Pulmonary Toxicity of Differently Sized Colloidal Amorphous SiO_2. *Nanomaterials* **2018**, *8*, 160. [CrossRef] [PubMed]
111. Kämpfer, A.A.M.; Busch, M.; Büttner, V.; Bredeck, G.; Stahlmecke, B.; Hellack, B.; Masson, I.; Sofranko, A.; Albrecht, C.; Schins, R.P.F. Model Complexity as Determining Factor for In Vitro Nanosafety Studies: Effects of Silver and Titanium Dioxide Nanomaterials in Intestinal Models. *Small* **2021**, *17*, e2004223. [CrossRef] [PubMed]
112. Fadeel, B.; Farcal, L.; Hardy, B.; Vázquez-Campos, S.; Hristozov, D.; Marcomini, A.; Lynch, I.; Valsami-Jones, E.; Alenius, H.; Savolainen, K. Advanced tools for the safety assessment of nanomaterials. *Nat. Nanotechnol.* **2018**, *13*, 537–543. [CrossRef]
113. Cedervall, T.; Lynch, I.; Lindman, S.; Berggard, T.; Thulin, E.; Nilsson, H.; Dawson, K.A.; Linse, S. Understanding the nanoparticle-protein corona using methods to quantify exchange rates and affinities of proteins for nanoparticles. *Proc. Natl. Acad. Sci. USA* **2007**, *104*, 2050–2055. [CrossRef]
114. Winkler, D.A.; Mombelli, E.; Pietroiusti, A.; Tran, L.; Worth, A.; Fadeel, B.; McCall, M.J. Applying quantitative structure–activity relationship approaches to nanotoxicology: Current status and future potential. *Toxicology* **2013**, *313*, 15–23. [CrossRef]
115. Forest, V.; Cottier, M.; Pourchez, J. Electrostatic interactions favor the binding of positive nanoparticles on cells: A reductive theory. *Nano Today* **2015**, *10*, 677–680. [CrossRef]
116. Forest, V.; Pourchez, J. Preferential binding of positive nanoparticles on cell membranes is due to electrostatic interactions: A too simplistic explanation that does not take into account the nanoparticle protein corona. *Mater. Sci. Eng. C* **2017**, *70*, 889–896. [CrossRef]
117. Forest, V. CHAPTER 2: Biological significance of the nanoparticles protein corona. In *Nanoparticle–Protein Corona*; Royal Society of Chemistry: Cambridge, UK, 2019; pp. 31–60.
118. Vroman, L.; Adams, A. Findings with the recording ellipsometer suggesting rapid exchange of specific plasma proteins at liquid/solid interfaces. *Surf. Sci.* **1969**, *16*, 438–446. [CrossRef]
119. Singh, A.V.; Laux, P.; Luch, A.; Sudrik, C.; Wiehr, S.; Wild, A.-M.; Santomauro, G.; Bill, J.; Sitti, M. Review of emerging concepts in nanotoxicology: Opportunities and challenges for safer nanomaterial design. *Toxicol. Mech. Methods* **2019**, *29*, 378–387. [CrossRef] [PubMed]
120. Kose, O.; Stalet, M.; Leclerc, L.; Forest, V. Influence of the physicochemical features of TiO2 nanoparticles on the formation of a protein corona and impact on cytotoxicity. *RSC Adv.* **2020**, *10*, 43950–43959. [CrossRef]
121. Lesniak, A.; Fenaroli, F.; Monopoli, M.P.; Åberg, C.; Dawson, K.A.; Salvati, A. Effects of the Presence or Absence of a Protein Corona on Silica Nanoparticle Uptake and Impact on Cells. *ACS Nano* **2012**, *6*, 5845–5857. [CrossRef]
122. Owens, D.E., III; Peppas, N.A. Opsonization, biodistribution, and pharmacokinetics of polymeric nanoparticles. *Int. J. Pharm.* **2006**, *307*, 93–102. [CrossRef]
123. Forest, V. Combined effects of nanoparticles and other environmental contaminants on human health—An issue often overlooked. *NanoImpact* **2021**, *23*, 100344. [CrossRef]
124. Carlsten, C. Synergistic Environmental Exposures and the Airways Capturing Complexity in Humans. *Chest* **2018**, *154*, 918–924. [CrossRef]
125. Rosário, F.; Bessa, M.J.; Brandão, F.; Costa, C.; Lopes, C.B.; Estrada, A.C.; Tavares, D.S.; Teixeira, J.P.; Reis, A.T. Unravelling the Potential Cytotoxic Effects of Metal Oxide Nanoparticles and Metal(Loid) Mixtures on A549 Human Cell Line. *Nanomaterials* **2020**, *10*, 447. [CrossRef]
126. Abbas, Q.; Yousaf, B.; Amina; Ali, M.U.; Munir, M.A.M.; El-Naggar, A.; Rinklebe, J.; Naushad, M. Transformation pathways and fate of engineered nanoparticles (ENPs) in distinct interactive environmental compartments: A review. *Environ. Int.* **2020**, *138*, 105646. [CrossRef]
127. Silins, I.; Hogberg, J. Combined Toxic Exposures and Human Health: Biomarkers of Exposure and Effect. *Int. J. Environ. Res. Public Health* **2011**, *8*, 629–647. [CrossRef]
128. Zheng, D.; Wang, N.; Wang, X.; Tang, Y.; Zhu, L.; Huang, Z.; Tang, H.; Shi, Y.; Wu, Y.; Zhang, M.; et al. Effects of the interaction of TiO2 nanoparticles with bisphenol A on their physicochemical properties and in vitro toxicity. *J. Hazard. Mater.* **2012**, *199–200*, 426–432. [CrossRef] [PubMed]
129. Jiang, N.; Wen, H.; Zhou, M.; Lei, T.; Shen, J.; Zhang, D.; Wang, R.; Wu, H.; Jiang, S.; Li, W. Low-dose combined exposure of carboxylated black carbon and heavy metal lead induced potentiation of oxidative stress, DNA damage, inflammation, and apoptosis in BEAS-2B cells. *Ecotoxicol. Environ. Saf.* **2020**, *206*, 111388. [CrossRef] [PubMed]
130. Cedergreen, N. Quantifying Synergy: A Systematic Review of Mixture Toxicity Studies within Environmental Toxicology. *PLoS ONE* **2014**, *9*, e96580. [CrossRef]

131. Asweto, C.O.; Wu, J.; Hu, H.; Feng, L.; Yang, X.; Duan, J.; Sun, Z. Combined Effect of Silica Nanoparticles and Benzo[a]pyrene on Cell Cycle Arrest Induction and Apoptosis in Human Umbilical Vein Endothelial Cells. *Int. J. Environ. Res. Public Health* **2017**, *14*, 289. [CrossRef]
132. Liu, Y.; Nie, Y.; Wang, J.; Wang, J.; Wang, X.; Chen, S.; Zhao, G.; Wu, L.; Xu, A. Mechanisms involved in the impact of engineered nanomaterials on the joint toxicity with environmental pollutants. *Ecotoxicol. Environ. Saf.* **2018**, *162*, 92–102. [CrossRef]
133. Li, L.; Fernandez-Cruz, M.L.; Connolly, M.; Conde, E.; Fernández, M.; Schuster, M.; Navas, J.M. The potentiation effect makes the difference: Non-toxic concentrations of ZnO nanoparticles enhance Cu nanoparticle toxicity in vitro. *Sci. Total Environ.* **2015**, *505*, 253–260. [CrossRef]
134. Murugadoss, S. A strategy towards the generation of testable adverse outcome pathways for nanomaterials. *ALTEX* **2021**, *38*, 580–594. [CrossRef]
135. Romeo, D.; Salieri, B.; Hischier, R.; Nowack, B.; Wick, P. An integrated pathway based on in vitro data for the human hazard assessment of nanomaterials. *Environ. Int.* **2020**, *137*, 105505. [CrossRef]
136. Guo, N.L.; Poh, T.Y.; Pirela, S.; Farcas, M.T.; Chotirmall, S.H.; Tham, W.K.; Adav, S.S.; Ye, Q.; Wei, Y.; Shen, S.; et al. Integrated Transcriptomics, Metabolomics, and Lipidomics Profiling in Rat Lung, Blood, and Serum for Assessment of Laser Printer-Emitted Nanoparticle Inhalation Exposure-Induced Disease Risks. *Int. J. Mol. Sci.* **2019**, *20*, 6348. [CrossRef]
137. Bannuscher, A.; Hellack, B.; Bahl, A.; Laloy, J.; Herman, H.; Stan, M.S.; Dinischiotu, A.; Giusti, A.; Krause, B.-C.; Tentschert, J.; et al. Metabolomics profiling to investigate nanomaterial toxicity in vitro and in vivo. *Nanotoxicology* **2020**, *14*, 807–826. [CrossRef]
138. Kinaret, P.A.S.; Ndika, J.; Ilves, M.; Wolff, H.; Vales, G.; Norppa, H.; Savolainen, K.; Skoog, T.; Kere, J.; Moya, S.; et al. Toxicogenomic Profiling of 28 Nanomaterials in Mouse Airways. *Adv. Sci.* **2021**, *8*, 2004588. [CrossRef] [PubMed]
139. Shin, T.H.; Nithiyanandam, S.; Lee, D.Y.; Kwon, D.H.; Hwang, J.S.; Kim, S.G.; Jang, Y.E.; Basith, S.; Park, S.; Mo, J.-S.; et al. Analysis of Nanotoxicity with Integrated Omics and Mechanobiology. *Nanomaterials* **2021**, *11*, 2385. [CrossRef] [PubMed]
140. Joossens, E.; Macko, P.; Palosaari, T.; Gerloff, K.; Ojea-Jimenez, I.; Gilliland, D.; Novak, J.; Torrent, S.F.; Gineste, J.-M.; Römer, I.; et al. A high throughput imaging database of toxicological effects of nanomaterials tested on HepaRG cells. *Sci. Data* **2019**, *6*, 46. [CrossRef] [PubMed]
141. Toprani, S.M.; Bitounis, D.; Huang, Q.; Oliveira, N.; Ng, K.W.; Tay, C.Y.; Nagel, Z.D.; Demokritou, P. High-Throughput Screening Platform for Nanoparticle-Mediated Alterations of DNA Repair Capacity. *ACS Nano* **2021**, *15*, 4728–4746. [CrossRef]
142. Halappanavar, S.; Brule, S.V.D.; Nymark, P.; Gaté, L.; Seidel, C.; Valentino, S.; Zhernovkov, V.; Danielsen, P.H.; de Vizcaya-Ruiz, A.; Wolff, H.; et al. Adverse outcome pathways as a tool for the design of testing strategies to support the safety assessment of emerging advanced materials at the nanoscale. *Part. Fibre Toxicol.* **2020**, *17*, 1–24. [CrossRef]
143. Afantitis, A.; Melagraki, G.; Isigonis, P.; Tsoumanis, A.; Varsou, D.D.; Valsami-Jones, E.; Papadiamantis, A.; Ellis, L.-J.A.; Sarimveis, H.; Doganis, P.; et al. NanoSolveIT Project: Driving nanoinformatics research to develop innovative and integrated tools for in silico nanosafety assessment. *Comput. Struct. Biotechnol. J.* **2020**, *18*, 583–602. [CrossRef]
144. Furxhi, I.; Murphy, F.; Mullins, M.; Arvanitis, A.; Poland, C.A. Nanotoxicology data for in silico tools: A literature review. *Nanotoxicology* **2020**, *14*, 612–637. [CrossRef]
145. Braakhuis, H.M.; Murphy, F.; Ma-Hock, L.; Dekkers, S.; Keller, J.; Oomen, A.G.; Stone, V. An Integrated Approach to Testing and Assessment to Support Grouping and Read-Across of Nanomaterials After Inhalation Exposure. *Appl. In Vitro Toxicol.* **2021**, *7*, 112–128. [CrossRef]
146. Chen, G.; Peijnenburg, W.; Xiao, Y.; Vijver, M.G. Current Knowledge on the Use of Computational Toxicology in Hazard Assessment of Metallic Engineered Nanomaterials. *Int. J. Mol. Sci.* **2017**, *18*, 1504. [CrossRef]
147. Winkler, D.; Burden, F.; Yan, B.; Weissleder, R.; Tassa, C.; Shaw, S.; Epa, V. Modelling and predicting the biological effects of nanomaterials. *SAR QSAR Environ. Res.* **2014**, *25*, 161–172. [CrossRef]
148. Buglak, A.A.; Zherdev, A.V.; Dzantiev, B.B. Nano-(Q)SAR for Cytotoxicity Prediction of Engineered Nanomaterials. *Molecules* **2019**, *24*, 4537. [CrossRef] [PubMed]
149. Singh, A.V.; Ansari, M.H.D.; Rosenkranz, D.; Maharjan, R.S.; Kriegel, F.L.; Gandhi, K.; Kanase, A.; Singh, R.; Laux, P.; Luch, A. Artificial Intelligence and Machine Learning in Computational Nanotoxicology: Unlocking and Empowering Nanomedicine. *Adv. Health Mater.* **2020**, *9*, e1901862. [CrossRef] [PubMed]
150. Winkler, D.A. Recent advances, and unresolved issues, in the application of computational modelling to the prediction of the biological effects of nanomaterials. *Toxicol. Appl. Pharmacol.* **2016**, *299*, 96–100. [CrossRef] [PubMed]
151. Abdelsattar, A.S.; Dawoud, A.; Helal, M.A. Interaction of nanoparticles with biological macromolecules: A review of molecular docking studies. *Nanotoxicology* **2020**, *15*, 66–95. [CrossRef]
152. Gajewicz, A.; Cronin, M.T.; Rasulev, B.; Leszczynski, J.; Puzyn, T. Novel approach for efficient predictions properties of large pool of nanomaterials based on limited set of species: Nano-read-across. *Nanotechnology* **2015**, *26*, 015701. [CrossRef]
153. Barnard, A.S. Computational strategies for predicting the potential risks associated with nanotechnology. *Nanoscale* **2009**, *1*, 89–95. [CrossRef]
154. Burello, E. Review of (Q)SAR models for regulatory assessment of nanomaterials risks. *NanoImpact* **2017**, *8*, 48–58. [CrossRef]
155. Puzyn, T.; Rasulev, B.; Gajewicz, A.; Hu, X.; Dasari, T.P.; Michalkova, A.; Hwang, H.-M.; Toropov, A.; Leszczynska, D.; Leszczynski, J. Using nano-QSAR to predict the cytotoxicity of metal oxide nanoparticles. *Nat. Nanotechnol.* **2011**, *6*, 175–178. [CrossRef]

156. Gajewicz, A.; Rasulev, B.; Dinadayalane, T.C.; Urbaszek, P.; Puzyn, T.; Leszczynska, D.; Leszczynski, J. Advancing risk assessment of engineered nanomaterials: Application of computational approaches. *Adv. Drug Deliv. Rev.* **2012**, *64*, 1663–1693. [CrossRef]
157. Cohen, Y.; Rallo, R.; Liu, R.; Liu, H.H. In Silico Analysis of Nanomaterials Hazard and Risk. *Acc. Chem. Res.* **2013**, *46*, 802–812. [CrossRef]
158. Brown, A.C.; Fraser, T.R. On the Connection between Chemical Constitution and Physiological Action; with special reference to the Physiological Action of the Salts of the Ammonium Bases derived from Strychnia, Brucia, Thebaia, Codeia, Morphia, and Nicotia. *J. Anat. Physiol.* **1868**, *2*, 224–242. [PubMed]
159. Hansch, C.; Maloney, P.P.; Fujita, T.; Muir, R.M. Correlation of Biological Activity of Phenoxyacetic Acids with Hammett Substituent Constants and Partition Coefficients. *Nature* **1962**, *194*, 178–180. [CrossRef]
160. Bañares, M.A.; Haase, A.; Tran, L.; Lobaskin, V.; Oberdörster, G.; Rallo, R.; Leszczynski, J.; Hoet, P.; Korenstein, R.; Hardy, B.; et al. CompNanoTox2015: Novel perspectives from a European conference on computational nanotoxicology on predictive nanotoxicology. *Nanotoxicology* **2017**, *11*, 839–845. [CrossRef] [PubMed]
161. Toropova, A.P.; Toropov, A.A. Nanomaterials: Quasi-SMILES as a flexible basis for regulation and environmental risk assessment. *Sci. Total Environ.* **2022**, *823*, 153747. [CrossRef] [PubMed]
162. Fourches, D.; Pu, D.; Tassa, C.; Weissleder, R.; Shaw, S.Y.; Mumper, R.J.; Tropsha, A. Quantitative Nanostructure–Activity Relationship Modeling. *ACS Nano* **2010**, *4*, 5703–5712. [CrossRef] [PubMed]
163. Nicolotti, O.; Benfenati, E.; Carotti, A.; Gadaleta, D.; Gissi, A.; Mangiatordi, G.F.; Novellino, E. REACH and in silico methods: An attractive opportunity for medicinal chemists. *Drug Discov. Today* **2014**, *19*, 1757–1768. [CrossRef]
164. Dekkers, S.; Oomen, A.G.; Bleeker, E.A.; Vandebriel, R.; Micheletti, C.; Cabellos, J.; Janer, G.; Fuentes, N.; Campos, S.V.; Borges, T.; et al. Towards a nanospecific approach for risk assessment. *Regul. Toxicol. Pharmacol.* **2016**, *80*, 46–59. [CrossRef]
165. Varsou, D.-D.; Tsiliki, G.; Nymark, P.; Kohonen, P.; Grafström, R.; Sarimveis, H. toxFlow: A Web-Based Application for Read-Across Toxicity Prediction Using Omics and Physicochemical Data. *J. Chem. Inf. Model.* **2018**, *58*, 543–549. [CrossRef]
166. Forest, V.; Hochepied, J.-F.; Leclerc, L.; Trouvé, A.; Abdelkebir, K.; Sarry, G.; Augusto, V.; Pourchez, J. Towards an alternative to nano-QSAR for nanoparticle toxicity ranking in case of small datasets. *J. Nanopart. Res.* **2019**, *21*, 95. [CrossRef]
167. Forest, V.; Hochepied, J.-F.; Pourchez, J. Importance of Choosing Relevant Biological End Points to Predict Nanoparticle Toxicity with Computational Approaches for Human Health Risk Assessment. *Chem. Res. Toxicol.* **2019**, *32*, 1320–1326. [CrossRef]
168. OECD. *Guidance Document on the Validation of (Quantitative) Structure-Activity Relationships [(Q)SAR] Models*; OECD: Paris, France, 2007.
169. Pan, Y.; Li, T.; Cheng, J.; Telesca, D.; Zink, J.I.; Jiang, J. Nano-QSAR modeling for predicting the cytotoxicity of metal oxide nanoparticles using novel descriptors. *RSC Adv.* **2016**, *6*, 25766–25775. [CrossRef]
170. Kar, S.; Gajewicz, A.; Puzyn, T.; Roy, K.; Leszczynski, J. Periodic table-based descriptors to encode cytotoxicity profile of metal oxide nanoparticles: A mechanistic QSTR approach. *Ecotoxicol. Environ. Saf.* **2014**, *107*, 162–169. [CrossRef] [PubMed]
171. Gajewicz, A.; Schaeublin, N.; Rasulev, B.; Hussain, S.; Leszczynska, D.; Puzyn, T.; Leszczynski, J. Towards understanding mechanisms governing cytotoxicity of metal oxides nanoparticles: Hints from nano-QSAR studies. *Nanotoxicology* **2015**, *9*, 313–325. [CrossRef] [PubMed]
172. Shi, H.; Pan, Y.; Yang, F.; Cao, J.; Tan, X.; Yuan, B.; Jiang, J. Nano-SAR Modeling for Predicting the Cytotoxicity of Metal Oxide Nanoparticles to $PaCa_2$. *Molecules* **2021**, *26*, 2188. [CrossRef]
173. Afantitis, A.; Melagraki, G.; Tsoumanis, A.; Valsami-Jones, E.; Lynch, I. A nanoinformatics decision support tool for the virtual screening of gold nanoparticle cellular association using protein corona fingerprints. *Nanotoxicology* **2018**, *12*, 1148–1165. [CrossRef]
174. Marvin, H.J.P.; Bouzembrak, Y.; Janssen, E.M.; van der Zande, M.; Murphy, F.; Sheehan, B.; Mullins, M.; Bouwmeester, H. Application of Bayesian networks for hazard ranking of nanomaterials to support human health risk assessment. *Nanotoxicology* **2017**, *11*, 123–133. [CrossRef]
175. Fjodorova, N.; Novic, M.; Gajewicz, A.; Rasulev, B. The way to cover prediction for cytotoxicity for all existing nano-sized metal oxides by using neural network method. *Nanotoxicology* **2017**, *11*, 475–483. [CrossRef]
176. Saini, B.; Srivastava, S. Nanotoxicity prediction using computational modelling—Review and future directions. *IOP Conf. Ser. Mater. Sci. Eng.* **2018**, *348*, 012005. [CrossRef]
177. Findlay, M.R.; Freitas, D.N.; Mobed-Miremadi, M.; Wheeler, K.E. Machine learning provides predictive analysis into silver nanoparticle protein corona formation from physicochemical properties. *Environ. Sci. Nano* **2018**, *5*, 64–71. [CrossRef]
178. Gajewicz, A.; Puzyn, T.; Odziomek, K.; Urbaszek, P.; Haase, A.; Riebeling, C.; Luch, A.; Irfan, M.A.; Landsiedel, R.; Van Der Zande, M.; et al. Decision tree models to classify nanomaterials according to the DF4nanoGrouping scheme. *Nanotoxicology* **2018**, *12*, 1–17. [CrossRef]
179. Ha, M.K.; Trinh, T.X.; Choi, J.S.; Maulina, D.; Byun, H.G.; Yoon, T.H. Toxicity Classification of Oxide Nanomaterials: Effects of Data Gap Filling and PChem Score-based Screening Approaches. *Sci. Rep.* **2018**, *8*, 3141. [CrossRef] [PubMed]
180. Labouta, H.I.; Asgarian, N.; Rinker, K.; Cramb, D.T. Meta-Analysis of Nanoparticle Cytotoxicity via Data-Mining the Literature. *ACS Nano* **2019**, *13*, 1583–1594. [CrossRef] [PubMed]
181. Furxhi, I.; Murphy, F.; Poland, C.A.; Sheehan, B.; Mullins, M.; Mantecca, P. Application of Bayesian networks in determining nanoparticle-induced cellular outcomes using transcriptomics. *Nanotoxicology* **2019**, *13*, 827–848. [CrossRef] [PubMed]

182. Yan, X.; Sedykh, A.; Wang, W.; Yan, B.; Zhu, H. Construction of a web-based nanomaterial database by big data curation and modeling friendly nanostructure annotations. *Nat. Commun.* **2020**, *11*, 2519. [CrossRef]
183. Madden, J.C.; Enoch, S.J.; Paini, A.; Cronin, M.T. A Review of In Silico Tools as Alternatives to Animal Testing: Principles, Resources and Applications. *Altern. Lab. Anim.* **2020**, *48*, 146–172. [CrossRef]
184. Furxhi, I.; Murphy, F. Predicting In Vitro Neurotoxicity Induced by Nanoparticles Using Machine Learning. *Int. J. Mol. Sci.* **2020**, *21*, 5280. [CrossRef]
185. Gajewicz, A. What if the number of nanotoxicity data is too small for developing predictive Nano-QSAR models? An alternative read-across based approach for filling data gaps. *Nanoscale* **2017**, *9*, 8435–8448. [CrossRef]
186. Di Cristo, L.; Oomen, A.G.; Dekkers, S.; Moore, C.; Rocchia, W.; Murphy, F.; Johnston, H.J.; Janer, G.; Haase, A.; Stone, V.; et al. Grouping Hypotheses and an Integrated Approach to Testing and Assessment of Nanomaterials Following Oral Ingestion. *Nanomaterials* **2021**, *11*, 2623. [CrossRef]
187. Lamon, L.; Aschberger, K.; Asturiol, D.; Richarz, A.; Worth, A. Grouping of nanomaterials to read-across hazard endpoints: A review. *Nanotoxicology* **2019**, *13*, 100–118. [CrossRef]
188. Kononenko, V.; Warheit, D.B.; Drobne, D. Grouping of Poorly Soluble Low (Cyto)Toxic Particles: Example with 15 Selected Nanoparticles and A549 Human Lung Cells. *Nanomaterials* **2019**, *9*, 704. [CrossRef]
189. European Chemicals Agency. *Guidance on Information Requirements and Chemical Safety Assessment: Appendix R.6 1 for Nanomaterials Applicable to the Guidance on QSARs and Grouping of Chemicals*; Publications Office of the European Union: Luxembourg, 2017; ISBN 978-92-9495-871-6.
190. Lamon, L.; Asturiol, D.; Richarz, A.; Joossens, E.; Graepel, R.; Aschberger, K.; Worth, A. Grouping of nanomaterials to read-across hazard endpoints: From data collection to assessment of the grouping hypothesis by application of chemoinformatic techniques. *Part. Fibre Toxicol.* **2018**, *15*, 37. [CrossRef] [PubMed]
191. Nel, A.E.; Nasser, E.; Godwin, H.; Avery, D.; Bahadori, T.; Bergeson, L.; Beryt, E.; Bonner, J.C.; Boverhof, D.; Carter, J.; et al. A Multi-Stakeholder Perspective on the Use of Alternative Test Strategies for Nanomaterial Safety Assessment. *ACS Nano* **2013**, *7*, 6422–6433. [CrossRef] [PubMed]
192. Nymark, P.; Bakker, M.; Dekkers, S.; Franken, R.; Fransman, W.; García-Bilbao, A.; Greco, D.; Gulumian, M.; Hadrup, N.; Halappanavar, S.; et al. Toward Rigorous Materials Production: New Approach Methodologies Have Extensive Potential to Improve Current Safety Assessment Practices. *Small* **2020**, *16*, e1904749. [CrossRef] [PubMed]

nanomaterials

MDPI

Review

Carbonaceous Nanoparticle Air Pollution: Toxicity and Detection in Biological Samples

Imran Aslam and Maarten B. J. Roeffaers *

Centre for Membrane Separations, Adsorption, Catalysis, and Spectroscopy for Sustainable Solutions, Department of Microbial and Molecular Systems, KU Leuven Celestijnenlaan 200F, 3001 Leuven, Belgium
* Correspondence: maarten.roeffaers@kuleuven.be

Abstract: Among the different air pollutants, particulate matter (PM) is of great concern due to its abundant presence in the atmosphere, which results in adverse effects on the environment and human health. The different components of PM can be classified based on their physicochemical properties. Carbonaceous particles (CPs) constitute a major fraction of ultrafine PM and have the most harmful effects. Herein, we present a detailed overview of the main components of CPs, e.g., carbon black (CB), black carbon (BC), and brown carbon (BrC), from natural and anthropogenic sources. The emission sources and the adverse effects of CPs on the environment and human health are discussed. Particularly, we provide a detailed overview of the reported toxic effects of CPs in the human body, such as respiratory effects, cardiovascular effects, neurodegenerative effects, carcinogenic effects, etc. In addition, we also discuss the challenges faced by and limitations of the available analytical techniques for the qualitative and quantitative detection of CPs in atmospheric and biological samples. Considering the heterogeneous nature of CPs and biological samples, a detailed overview of different analytical techniques for the detection of CPs in (real-exposure) biological samples is also provided. This review provides useful insights into the classification, toxicity, and detection of CPs in biological samples.

Keywords: carbonaceous nanoparticles; carbon black; black carbon; brown carbon; toxicity; optical detection

1. Introduction

Air pollution has been a major concern for many years due to its harmful effects on the environment and public health [1]. Worldwide, outdoor air pollution is estimated to be the cause of approximately four million premature deaths annually, of which about half a million occur in the European Union (EU). An additional four million premature deaths have been attributed to household air pollution in the World Health Organization's (WHO) report [2–4]. Air pollution is a complex mixture of gaseous components and particulates originating mainly from human activities and natural processes [5]. The composition of the pollutants depends on various factors such as their sources, emission rate, and wind conditions [5]. The major gaseous pollutants are nitrogen oxide (NO_2), ozone (O_3), and carbon monoxide (CO), whereas common particulates in the atmosphere include various particles of natural and anthropogenic origins [6,7]. Particulate matter (PM), or simply particulates, are extremely small particles and liquid droplets suspended in the atmosphere containing a mixture of different salts, organic chemicals, metals, dust particles, and CPs [8,9]. PM pollutants can be classified based on their sizes into coarse PM (PM_{10}; size < 10 µm and >2.5 µm), fine PM ($PM_{2.5}$; size < 2.5 µm), and ultrafine PM ($PM_{0.1}$; size < 100 nm) (Figure 1a). The particulates in PM can either be directly emitted into the atmosphere, i.e., primary PM, or can be formed in the atmosphere from gaseous precursors—the secondary PM [9,10]. Different emission sources from anthropogenic activities and natural processes contribute to the release of the primary components of PM in the environment (Figure 1b) [9]. The major anthropogenic emission sources of PM in the atmosphere are industrial activities, transportation, and the burning of fossil fuel [11,12]. Industrial activities are a major contributor to the presence of metals and engineered nanomaterials (ENMs) in the environment, resulting in elevated levels of metal pollution near sites with industrial activity [13–15]. Transportation is also a major source of CPs, dust, and metals in the atmosphere through exhaust and non-exhaust activities. For example, exhaust gases from the incomplete combustion of fossil fuels emit CPs into the atmosphere. Whereas non-exhaust activities such as the erosion of roads, brakes, and tires are major sources of dust and metals in the atmosphere [16,17]. People living near major highways with heavy traffic are more prone to the adverse effects of CPs [18]. The burning of fossil fuel for heating is another major contributor to CP emissions in the environment [19]. Some examples of the major natural CP sources include volcanic eruptions and forest fires [20]. Secondary particles mostly form in the atmosphere through chemical reactions of gaseous pollutants such as the transformation of nitrogen oxides and sulfur oxides [10].

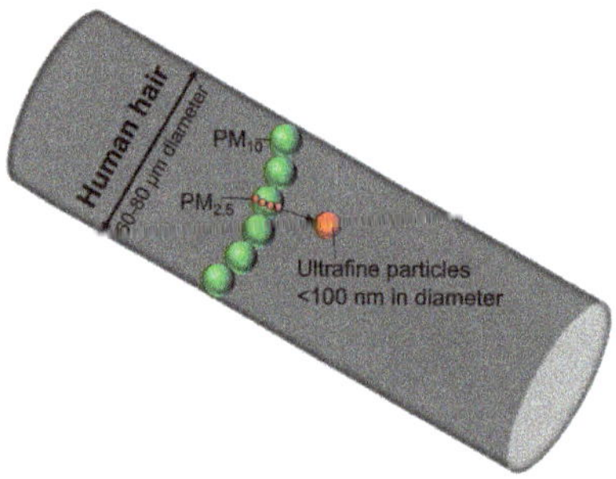

(a)

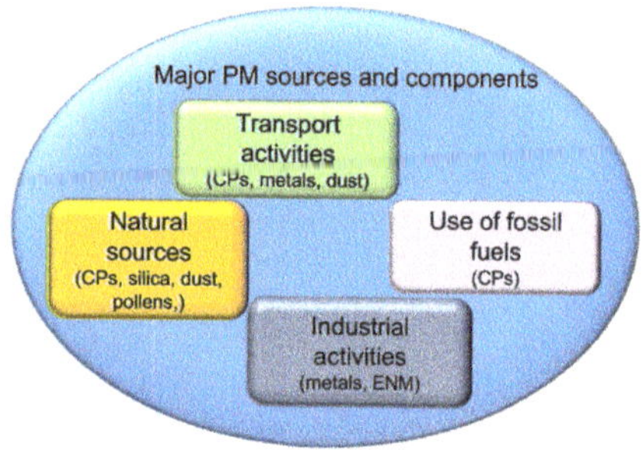

(b)

Figure 1. (**a**) Different components of PM based on size: PM_{10}, $PM_{2.5}$ and $PM_{0.1}$. With a decrease in the size of PM components, their concentration increases exponentially. (**b**) Major sources of PM pollution and different components from major sources.

Despite the considerable improvements in the atmospheric concentrations of PM recently, the environmental and health effects of PM are still manifold [9,21]. The environmental effects of PM depend on the chemical composition and include global warming, damage to crops and forests, and contribution to acidic rain [22]. The impact of PM on public health is very complex due to their variability in size, morphology, and composition [23]. A few well-known effects include, for example, the worsening of many respiratory diseases such as asthma and chronic obstructive pulmonary disease (COPD) caused by short-term exposure to PM_{10} [24], while exposure to $PM_{2.5}$ has been associated with premature death through existing pulmonary diseases and reduced lung function in children [25]. In addition, exposure to $PM_{2.5}$ results in cardiovascular morbidity and mortality through the deterioration of cardiac function [26]. Epidemiological studies indicate that the carcinogenic effects of $PM_{2.5}$ result in increased mortality, for example, due to lung cancer [27,28]. $PM_{0.1}$ is present in the atmosphere in large concentrations and generally enters the body through inhalation and can translocate to different organs through the bloodstream [29]. $PM_{0.1}$ causes pulmonary inflammation and can be retained in the lungs for a long time. In addition, it is also transported through the olfactory nerve into the brain and hence results in autonomic dysfunction [30,31]. $PM_{0.1}$ exposure is also associated with diabetes, cancer, and low birth weight [32,33].

Since PM is a complex mixture of components with different chemistry, variable sizes, and originating from different sources, it is a complicated task to link the adverse health effects of PM to a specific constituent. Specifically, the fine and ultrafine constituents of PM are more toxic compared to others [23,34,35]. CPs constitute a major fraction of $PM_{2.5}$ and $PM_{0.1}$, and their abundant presence in the atmosphere makes them among the most harmful fractions of ambient PM [23,36–38]. Although a few review articles provide insights into the classification of CPs and their detection in environmental samples, this comprehensive review provides useful insights not only into the classification of different CPs but also their toxicity and detection in biological samples.

2. Carbonaceous Particle (CP) Air Pollution

CPs represent a diverse group of materials divided mainly into two major categories: elemental carbon (EC) and organic carbon (OC) [39,40]. EC consists of black carbon (BC), which is directly emitted into the atmosphere through the incomplete combustion of fossil fuels, and carbon black (CB), which is manufactured for use in different applications as a pigment [41]. EC also includes engineered carbon nanomaterials, e.g., carbon nanotubes, fullerenes, and graphene [42]. OC is a complex mixture of different (organic) compounds and can be classified into primary and secondary organic compounds [41]. Among the different components of OC, ambient brown carbon (BrC) is a major fraction resulting from the burning of biomass, the degradation of organic matter, and secondary formation processes in the atmosphere [39,43].

2.1. Classification of Different Components of CPs

The initial chemical composition of CPs depends mainly on their source. For example, some sources produce almost pure EC, while others produce almost 50% (by mass) OC particles [44]. The internal structure of CPs strictly relates to their optical properties, particularly to those relating to UV–Vis absorption; hence, it is used as a diagnostic tool in different fields such as atmospheric chemistry, solid-state physics, materials science, etc. [45–47]. The classification of the different components of CPs based on the most common methods of bulk PM analysis is shown in Figure 2 [39,40]. At the top of this chart, the CB and BC compounds have the strongest optical absorption and lowest volatility. Although CB and BC are inherently complex, their chemical structure and optical properties can be explained by using graphite as an example. Graphite is the most stable (thermodynamically) form of pure EC and is an inert material under atmospheric conditions. Carbon atoms have an sp^2 orbital hybridization in graphitic structures, which results in a hexagonally symmetric planar arrangement of carbon atoms connected by σ-bonds [48–50]. The remaining p-electron is in an orbital perpendicular to the plane of carbon atoms. The p-orbitals overlap sideways to form π-bonds. Due to the infinite planar structure of the graphite sheets, the electrons in the overlapping p-orbitals are delocalized along the hexagonal atomic sheets of carbon [49]. This results in the metal-like properties of graphite, e.g., electrical conductivity and broad-band light absorption. The broad-band light absorption of CPs, particularly CB and BC, also serves as a basis for their optical determination.

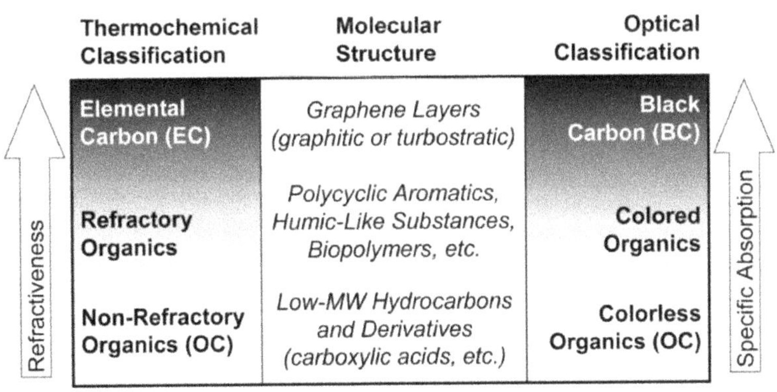

Figure 2. Classification of different components of CPs. Reproduced with permission from Ref. [40]. Copyright © 2002 Springer Nature.

Due to the sp^2 hybridization, the UV–Vis spectra of CB exhibit broad absorption in the UV band between 200 and 250 nm. It also merges with the long wavelength tail of the (σ-σ^*) band, located in the far UV end toward 100 nm, which is typical of sp^3 carbon sites [51]. With the shift of the (π-π^*) band's position toward the visible wavelength, the sp^2 character increases due to the extension of the sp^2 hybrid area [48,51]. Hence, this allows us to understand the graphitization process occurring in pyrolysis and combustion systems resulting in the production and CB and BC. Different structural parameters, i.e., the number of stacked graphitic layers and the curvature of aromatic layers, affect the (π-π^*) band position [52,53]. The increase in these structural parameters due to the graphitization process causes the shift toward the UV region of the band position, which is opposite to the shift toward the visible region caused by the growth of graphene.

CB and BC exhibit a very complex morphology spanning from the macrostructure to micro- and nano-structures. Even at the nanometer scale, the arrangement of aromatic units of different sizes (>2 rings) in non-stacked and stacked units from two to five stacks can give rise to a different degree of crystallinity and different classes of absorbers [48,54]. This structural complexity leads to changes in the electronic interactions and hence affects the correct interpretation of UV–Vis spectral features of CPs [48].

The bottom of the chart in Figures 1 and 2 shows the volatile organic compounds with their characteristic absorption in the UV range [39]. Between two extremes, the middle of the chart corresponds to the moderately volatile (refractory) compounds with poorly characterized molecular structures. Among these, a few compounds (e.g., HULIS and PAHs) correspond to the colored compounds with optical properties relevant to the BrC particles [47]. Compared to CB and BC, the optical absorption of BrC has strong wavelength dependence in the visible wavelength with the degree of absorption decreasing sharply from the UV to the visible region [39,55]. This is due to the presence of BrC chromophores, and the exact molecular identities of these chromophores are highly variable; however, they are expected to have a high degree of conjugation across the molecular skeleton and high absorption cross-sections. As fluorescent compounds often have the same characteristics, the compounds found in BrC particles can also act as efficient fluorophores [47,56]. In addition, studies have reported that nitroaromatic compounds are major BrC species resulting from the burning of biomass and are also responsible for 50% to 80% of the total light absorption by BrC (at 400 nm) [43,57,58]. Furthermore, aromatic volatile organic compounds, i.e., benzene homologs and derivatives, are also considered important precursors of BrC [57,59]. Specifically, nitrophenols and nitrocatechols are dominant chromophore species (>50% of the concentration) in BrC. In addition, Nitrophenols and nitrocatechols contribute more than 50% of the optical absorption of BrC between 300 and 400 nm [57]. Additionally, anthropogenic volatile organic compounds (e.g., benzene and toluene) and the oxidation of the biomass burning-related products (e.g., pyrocatechol and methylcatechols) can also generate similar BrC chromophores, indicating that these functionalized aromatic compounds play an important role in the optical absorption properties of BrC particles [57]. Furthermore, biogenic materials, their low-temperature oxidation, and polymerization products, e.g., fulvic substances and tannin/lignin compounds, also contribute to the light-absorbing properties of BrC particles [39].

The classification of CPs only based on their optical properties is not sufficient; hence, any potentially available chemical information can also be useful. For example, the H/C molar ratio can be a good indicator of the presence of organic substances in CPs [39]. The H/C molar ratio in BC (soot) is approximately 0.15, and it is well below the values for OC. In highly condensed materials (i.e., PAHs and lignin) it tends to be in the range of 0.5 to 1.5 [39,60]. Atmospheric HULIS have H/C ratios of about 1.4–1.6 [39,61]. Individual CP components are discussed below in detail.

2.1.1. Carbon Black (CB)

CB is an industrially manufactured product consisting of fine black powder obtained through the partial combustion or thermal decomposition of hydrocarbons [42,62]. Based on its manufacturing process, CB can be divided into acetylene black, channel black, furnace black, and thermal black [63]. Approximately ten million tons of CB are produced every year globally, making it one of the top 50 industrially manufactured chemicals [42,64]. Almost 90% of commercially produced CB is used in rubber applications, namely, tire-related automotive uses and other automotive and non-automotive uses of rubber [64,65]. The remaining 10% is used for various applications, e.g., as a black pigment or a conducting agent in plastics, inks, paints, and even in food (E153) [64–67]. The size of primary CB particles ranges from 15 to 300 nm [42]. There is a huge risk of occupational exposure to CB during different stages of its manufacturing process, production, collection, and handling [42,65]. The risk of occupational exposure to CB is also possible in downstream CB user industries such as the manufacture of rubber, paint, and ink; printing; in the plastics, paper, and ceramics industries; and in carbon electrode production [42,68].

2.1.2. Black Carbon (BC)

BC is a collective term used for different carbonaceous substances ranging from partly charred plant residues to highly graphitized soot particles resulting from incomplete combustion [69]. BC particles are a heterogeneous mixture of different species from various sources consisting of large aromatics and few functional groups [70]. Compared to CB, there is no universally defined chemical definition of BC. However, in a Report to Congress on Black Carbon, the United States (US) Environmental Protection Agency (EPA) defined BC as a "solid form of carbonaceous component of PM that absorbs solar radiation at all wavelengths" [71]. BC mostly originates from the incomplete combustion of biomass and fossil fuels linked to human activities [39,42]. Among these, diesel exhaust engines are a major contributor to the emissions of BC in the environment in the proximity of heavy-traffic areas [72]. In addition, the natural sources of BC emissions are volcanic eruptions and wildfires [73]. The level of global BC emissions from both natural and anthropogenic sources is estimated to be approximately nine million tons per year [42]. The size of BC particles ranges from a few nm to a few hundred nm for atmospheric BC usually sourced from diesel exhaust processes [74]. Atmospheric exposure to BC depends on the daily activities of individuals, whereas persons associated with the transportation sector are prone to higher exposure to BC [72].

2.1.3. Brown Carbon (BrC)

In contrast to CB and BC, BrC is a fraction of OC that does not absorb all visible light and is characterized by its (light) brown appearance [75]. Recently, BrC attracted the attention of the scientific community due to its strong light-absorbing abilities in the ultraviolet (UV) and visible (Vis) regions and for playing a role in climate change [76]. BrC is chemically very complex as it originates from the incomplete combustion of various materials but can also stem from non-combustion processes either through primary or secondary sources [75]. The major primary BrC sources include low-temperature biomass and coal burning through human activities, biogenic emissions from plant residue, and humic matter [77,78]. Secondary BrC is produced through atmospheric transformation reactions [79]. Compounds of high molecular weight and a light-absorbing nature can result from atmospheric multiphase transformations in the presence of gas-phase or cloud micro-droplets [80–82]. BrC is a dynamic mixture of organic compounds and very little information is available about the relationship between its chemical composition and emission sources. Atmospheric humic-like substances have been considered major components of BrC in addition to polycyclic aromatics and biopolymers such as lignin [39,43,61]. Atmospheric BrC can also be classified based on its soluble and insoluble fractions both in water and methanol [83,84]. The global level of BrC emissions is estimated to be approximately seven million tons [75,85]. Although the role of BrC in climate change is already known, the adverse health effects of BrC are also receiving more attention from scientists [86,87]. The characteristics of CB, BC, and BrC are provided in Table 1.

Table 1. Characteristics of CB, BC, and BrC.

Characteristics	Carbon Black (CB)	Black Carbon (BC)	Brown Carbon (BrC)
Sources/Origin	Large-scale commercial production processes, abrasion of materials containing CB.	Biomass burning, diesel exhaust, forest fires, volcanic eruptions.	Biomass burning, coal burning, forest fires, biogenic sources (dust, humic matter, etc.).
Production/Emission estimates	~9.8 million tons per year.	~8.5 million tons per year.	~6.9 million tons per year (carbon), variable.
Composition/Molecular structure	Elemental carbon > 97%	Source-dependent, often elemental carbon > 50%	Humic-like substances, PAHs, Biopolymers.
Primary particle sizes	~15–300 nm	Smaller than primary CB particles: diesel exhaust around ~15–40 nm.	Various sizes, depending on the molecular structure.
General morphology (shape, form)	Aciniform aggregates and agglomerates.	Complex chains, aciniform aggregates, agglomerated spherical particles.	Individual particles, aggregates with PAHs, tarballs.
Exposure sources	Environmental and Occupational.	Environmental: anthropogenic and natural sources.	Environmental: anthropogenic and natural sources.
Environmental/Health effects	Cytotoxicity, inflammation and oxidative stress, cardiovascular and respiratory diseases.	Global warming, cytotoxicity, inflammation and oxidative stress, cardiovascular and respiratory diseases, carrier for harmful substances.	Global warming, cell apoptosis, carrier for carcinogenic NPs, persistent organic pollutants (POPs).
Optical properties	Absorbs light at all wavelengths from UV–Vis-nIR.	Absorbs light at all wavelengths from UV–Vis-nIR.	Absorbs light strongly in the UV region.
References	[42,62–65,67]	[39,42,69–74]	[39,43,61,80–86]

3. Environmental Impact of CPs

Carbonaceous aerosols have a significant impact on the weather and climate through the absorption and scattering of sunlight [88]. When suspended in the atmosphere, BC contributes to the warming of the atmosphere by absorbing solar radiation at all wavelengths from UV–Vis to nIR and converting it into heat [89]. BC is the second most important chemical contributing to climate change after carbon dioxide [90]. The BC emitted through human activities and natural sources can also be deposited on the ice and snow, and the consequent decreased albedo effect results in an increased temperature and hence the melting the snow [91]. This also results in the enhanced melting of the ice caps in the arctic and other glaciated regions [92]. In addition, BC also has consequences on the health of the global ecosystem by changing rainfall patterns through its increased absorption of sunlight and by changing the number of liquid cloud droplets [93,94]. This can in turn affect both ecosystems and human living conditions, for example, by affecting agricultural productivity [93,95]. BC can also deposit on the leaves of trees, which consequently increases their temperature [96]. In the past few years, BrC has also received the attention of scientists, particularly in atmospheric research, due to its ability to strongly absorb light in the UV–blue region and hence contribute to climate change [97]. In contrast to BC, the light absorption efficiency of BrC is highly variable as it depends on the chemical composition and source [98]. BrC from the burning of biomass can cause strong circum-arctic warming and indirectly affect the ecosystem and the economy [99,100].

4. Health Effects of CPs

The toxicity of CPs is known to strongly depend on various factors such as their physical characteristics (size, shape, etc.) and chemical composition [101]. Environmental and occupational exposure to ambient CPs is associated with a multitude of diseases as well as a higher mortality rate [102–105]. CPs from the burning of biomass cause toxicity through the stimulation of oxidative stress, inflammation, and genotoxicity through localized or systemic toxicity [106]. Being a major fraction of ultrafine PM, very fine CPs are highly toxic due to their translocation to different organs via the bloodstream, and a large surface area can result in an enhanced inflammatory response [29]. The toxic effects of CPs on human health are discussed in detail in this section.

4.1. Pulmonary/Respiratory Effects

The lungs are the prime organs exposed to atmospheric CPs after inhalation [107]. The deposition of CPs in the lungs depends on the particle size, whereas the composition of the particles determines their toxicity [108]. CB particles consist mainly of EC whereas BC and BrC also contain organic (polyaromatic) components with chemical toxicity towards the lungs [39,42]. Exposure to CPs can cause two pulmonary diseases: COPD and asthma [109–111]. The pathophysiology of COPD and asthma includes inflammation of the airways, tissue remodeling and fibrosis, mucociliary dysfunctions, and structural changes [107,112]. Exposure to CPs can also cause alterations in the lungs through the interruption of different lung functions [107]. For example, the inflammation of the airways after the inhalation of CPs causes serious damage to lung function, as diesel exhaust particles (DEPs) alter the production of cytokines [113]. Furthermore, tissue remodeling and fibrosis are observed after inflammatory conditions leading to the accumulation of collagen fibers [107]. Susceptible individuals that already have COPD or asthma are more prone to CP-mediated oxidative damage. In addition, through the generation of reactive oxygen species (ROS), ultrafine CPs cause adverse effects in susceptible individuals with COPD or asthma [114]. The effects of CPs on various biochemical and molecular mediators can cause respiratory dysfunction. In vivo studies have shown the dose-dependent toxicity of synergized CB and Cd resulting in the inflammation of lungs [115]. In an inhalation study, rats were exposed to CB at 7 mg/m^3 and 50 mg/m^3 for 6 h per day and 5 days per week. After 13 weeks, inflammation and oxidative stress were observed in the bronchioalveolar lavage fluid from rats [116,117].

4.2. Cardiovascular Effects

Exposure to CPs from the atmosphere also affects cardiovascular functions, and several direct or indirect pathways explain the link between CPs and cardiovascular effects [118–120]. Through direct pathways, fine, and specifically ultrafine CPs, can translocate from the lungs into the bloodstream and hence target remote organs leading to potential cumulative toxicity [119]. Exposure to ultrafine particles causes cardiac depression effects leading to the deterioration of cardiac function [121]. Indirect pathways are mediated by pulmonary oxidative stress and the inflammatory response, as well as interaction with the autonomic nervous system through specific lung receptors [122]. After the deposition of particles in the lungs, they trigger an inflammation-related cascade resulting in an increased circulating level of pro-inflammatory cytokines, thereby contributing to the risks of atherosclerosis progression [26,123]. The dose-dependent exposure of rats to CB enhanced their cardiovascular risk by inducing hyperhomocysteinemia and platelet hyperactivity [124]. Furthermore, the ROS-dependent mechanism also involves the pro-inflammatory pathway triggered by CPs, which is linked to vascular dysfunction, cardiac arrhythmias, and myocardial infarction [125,126]. In addition, CPs can also stimulate the autonomic nervous system, resulting in impaired autonomic balance and a hyper-activated sympathetic tone that is related to increased cardiovascular risk [127]. Exposure to CPs is also linked with myocardial infarction and accelerated cardiovascular changes [128–130].

4.3. Reproductive and Developmental Toxicity

Maternal exposure to CPs can result in developmental toxicity and affect three major organ systems in offspring, namely, the central nervous system, the male reproductive system, and the immune system [131]. The dose-dependent maternal exposure to CPs (Printex 90) was linked with changes in the histology of different cell populations in the central nervous system (CNS) and the altered open-field behavior of the offspring in a murine model [132]. Maternal exposure to CPs resulted in the enlargement of lysosomal granules in brain perivascular macrophages (PVMs), as well as increased glial fibrillary acidic protein (GFAP) expression levels in astrocytes, indicating reactive astrogliosis in six-week-old offspring [132–134]. The observed changes in the CNS can cause early brain aging, and the offspring have an increased susceptibility to age-related brain disorders [132]. Additionally, the exposure to CPs, particularly CB particles, during gestation has effects on sperm counts and causes structural testicular changes in male offspring [135,136]. The exposure of mice to CB for 10 weeks decreased their daily sperm count and testosterone levels [131]. Maternal exposure to diesel exhaust and tobacco smoke particles can cause allergic immune responses in offspring [137]. Prenatal exposure to CPs can cause immune system impairment, resulting in the frequent onset of allergies during childhood [138]. Maternal exposure to CPs also affects the fetus, resulting in low birth weight [139].

4.4. Neurotoxic Effects

Epidemiological research along with in vivo and in vitro studies are further clarifying that exposure to fine particulates damages the nervous system and the brain [31]. Ambient CPs have toxic effects on the brain after potentially translocating through the olfactory tract/olfactory nerve, gastro-intestinal tract/vagus nerve, or blood–brain barrier [140]. The different potential pathways damaging the brain and CNS include direct toxic effects, neuroinflammation, and oxidative stress [141,142]. Ultrafine CPs can have a direct toxic effect through their deposition on the olfactory mucosa of the respiratory tract and subsequent translocation to the brain through the olfactory nerve [31]. Chronic exposure to particles from diesel exhaust can induce oxidative stress, neuroinflammation, and impaired neurogenesis in different brain regions, subsequently leading to brain cell death [143,144]. The dose-dependent exposure of mice to ultrafine soot-iron particles through inhalation was associated with indicators of neural inflammation [145]. Long-term exposure to fine particulates during adulthood accelerates the effects of aging in the brain, thereby increasing the risks of developing dementia or neurodegenerative diseases such as Alzheimer's disease or Parkinson's disease [146,147]. Long-term exposure to traffic exhaust CPs is associated with ischemic stroke, resulting in elevated stroke incidences [148].

4.5. Genotoxic and Carcinogenic Effects

The carcinogenic effects of CPs in humans are either caused at the exposure site such as the respiratory tract or at a distal location after the translocation of particles through the bloodstream [107,149]. Occupational exposure to PAHs, accounting for a significant fraction of BrC and BC, is linked to respiratory, urinary tract, and prostate cancers [150,151]. The regular inhalation of DEPs and CB particles results in pulmonary carcinogenic effects [152,153]. Exposure to CPs, particularly soot, can cause DNA mutations, while higher concentrations of PAHs are responsible for genotoxic effects by damaging DNA [154,155]. Exposure to CPs causes DNA modifications due to DNA adduct formation after the reaction of PAHs with DNA molecules [156,157]. CPs—particularly wood smoke particles—cause oxidative stress, which is associated with a failure of the DNA repair mechanism [158]. Exposure to fine and ultrafine particulates causes DNA methylation changes at the molecular level resulting in changing the expression profiles of genes such that they cause cancer [159]. Soot particles from the 1991 oil fires in Kuwait have shown that a dose-dependent increase results in the induction of genetic effects under in vitro conditions [107]. PAHs also cause epigenetic effects through histone modification, whereas CB can potentiate single- and double-stranded DNA breaks and hence cause genotoxic effects [160,161]. The presence of metal ions in CPs brings about chemical toxicity through the enhanced ROS formation capacity of CPs, resulting in genotoxic effects [162,163]. In addition, BrC particles attract persistent organic pollutants (POPs) and act as carriers for carcinogenic materials such as benzo[a]pyrene (Bap) [75,76,164].

4.6. Dermal Toxicity

The skin is the second most important route—after the respiratory tract—for the interaction and penetration of nanoparticle pollutants in the body [165]. Therefore, epidermal cells, like cells in other organs, are exposed to carbonaceous pollutants from the atmosphere, resulting in the production of pro-inflammatory cytokines by human dermal keratinocytes [166,167]. Diesel exhaust CPs and those from cigarette smoke are responsible for damaging effects on skin tissues and premature skin aging [167,168]. For example, DEPs are responsible for the development of allergic and non-allergic skin inflammation after the generation of ROS through redox reactions [169]. Organic extracts from CPs caused dose-dependent cytotoxicity in murine epidermal cells in model studies [170]. Although CPs have cytotoxic effects on the skin, no study was found regarding the toxic effects of CPs in other organs after full penetration through the skin. The toxic effects of CPs on human health are shown in the schematic in Figure 3.

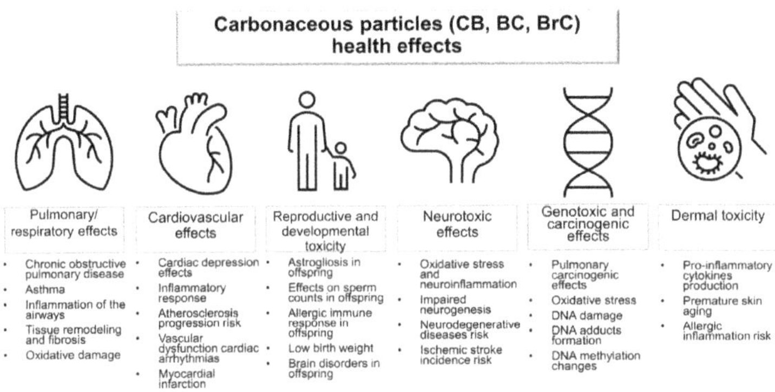

Figure 3. Schematic showing the toxic effects of exposure to CPs in humans.

5. Characterization Techniques for Carbonaceous Particles (CPs)

Adequate sampling and qualitative and quantitative characterization are essential to evaluate the adverse effects of CPs on the environment and humans. However, CPs' identification is not a simple task, as they are a heterogeneous group of particulates with a diverse chemical and structural nature and large variability in size [42,47]. In addition, various sampling conditions, such as those presented by polluted air or biological samples, pose many challenges for the qualitative and quantitative identification of CPs [39,171]. For example, elevated background signals from biological settings such as fluids or tissue samples are very intense compared to signals from CPs [171]. Due to the variability in different sampling conditions, it is difficult to develop generally applicable techniques for CPs' identification [171,172]. The generally used techniques for the characterization of CPs in atmospheric and biological samples are discussed in this section together with their advantages and limitations.

5.1. Characterization Techniques for Atmospheric CPs

The identification and quantification of CPs in atmospheric samples is not straightforward and the currently available techniques cannot accurately perform the sensitive detection of the different constituents of PM [37,38,173]. The detection of CPs in atmospheric samples is generally performed using light absorption and thermal radiation techniques; however, these two classes of techniques do not necessarily give similar results [40,94,174,175]. Light absorption techniques correlate the light absorption or attenuation from the samples to the mass (per volume) of the absorbing material using a light absorption coefficient and assuming that the absorbed or attenuated light is proportional to the mass of the CPs [55,176]. Using light absorption techniques, the bulk analysis of CPs is performed using a filter-based approach in which a gas stream (polluted air) is passed through a filter to concentrate the samples. Aethalometers use a filter-based technique to measure the light absorption of CPs and can estimate the contribution from different CPs based on wavelength-dependent light absorption measurements [176,177]. For example, emissions from the burning of fossil fuel absorb light dominantly from the Vis–nIR region, indicating the presence of BC, whereas emissions from the burning of biomass have more BrC, and hence an enhanced absorption in the UV and blue wavelengths [178,179]. In addition, based on the attenuation coefficient, the concentration of the CPs can be estimated [180]. Although this technique can identify and quantify CPs in environmental samples, there are several limitations to this technique such as: (i) the shadowing effect, (ii) the scattering of the light beam at the filter fibers, (iii) and a false response from non-CPs, which can lead to the false estimation of these measurements [176,181].

Thermal radiation techniques work on the principle of measuring the mass concentrations of CPs based on heating the light-absorbing carbonaceous aerosols followed by the analysis of emitted radiation [182]. Laser-induced incandescence (LII) is a powerful thermal radiation technique in which the emission signal from the particles is generated by the absorption of high-intensity laser radiation, and the thermal emission from the heated particles is detected in the selected detection window [175,183]. Depending on the laser's wavelength, LII can identify different fractions of CP in gaseous samples [182]. It can also quantify the mass concentrations (based on the mass per volume) of CPs, as well as measure the particle size based on the temporal decay of the LII signal [184]. It is used extensively for environmental applications such as ambient air quality or source monitoring [182]. There are several limitations of this technique; for example: (i) the instrument response depends on the type of CP, (ii) proper calibration is required to convert the intensity of the thermal radiation to the CP mass, and (iii) no established reference materials are available for calibration [174,181].

Furthermore, the combination of LII and light scattering from particles has also been employed to identify CP, as well as to measure their mass concentrations and particle size. This method is used in Single Particle Soot Photometers (SP2) for single-particle analysis [185,186]. In SP2, LII measures CPs such as BC, whereas single-particle light scattering is employed to measure the particle size and mass concentrations [186]. This instrument can also measure the content of CPs without interference from semi-volatile materials or mineral dust particles [175,186]. As SP2 works based on LII and light scattering, the limitations of this method in addition to LII's drawbacks include (i) the requirement for prior information about the refractive index and shape of the particles, and (ii) the necessity of calibrating the instrument for the accurate determination of the mass concentrations and particle size [175].

Other common optical techniques used for atmospheric CP measurements include cavity ring-down spectroscopy (CRDS) and multi-angle absorption photometer (MAAP) [187–189]. In addition, high-performance liquid chromatography (HPLC), as well as mass spectrometry (MS), are used for the characterization of chromophores in BrC samples from different emission sources [58,190]. The above-mentioned techniques are briefly discussed for atmospheric samples; however, the detailed discussion of atmospheric CP characterization exceeds the scope of this work. There is an evident need for the characterization of CPs in biological samples in order to gain insights into the toxicity of CPs towards humans.

5.2. Characterization Techniques for CPs in Biological Samples

Currently, the epidemiological research studying the toxic effects of CPs in biological samples is hampered by the scarcity of analytical tools that function in biological samples [171]. Most of the available techniques for the detection of CPs cannot be used because of their limited biocompatibility and the interference of background signals from biological samples [171]. Indirect measurements can make use of radiolabeled CPs to study the toxic effects of CPs in model organisms after deliberate exposure. Further mass spectrometry, electron microscopy, and optical techniques can—to some extent—also be employed for studying CPs from natural exposure.

5.2.1. Isotope Tracing/Radiolabeling Detection of CPs

Isotope tracing is a useful technique for tracking carbonaceous nanomaterials in different biological systems [191]. This approach has been used to trace CPs in lab animals as well as in clinical studies [191–193]. In animal studies on mice, the translocation of ultrafine carbon particles (elemental ^{13}C) from the respiratory tract, via the bloodstream, to the liver was observed [191,192]. Human respiratory detection and clearance studies use scintigraphy methods such as CPs labeled with gamma-ray-emitting radioisotopes, and gamma-ray emission can be used for the localization and quantification of CPs in the lungs as well as the clearance and translocation of these particles [194,195].

Among the different clinically employed radiolabeling techniques, the most predominant is the "Technegas approach", which uses ultrafine CP suspensions labeled with Technetium-99m (^{99m}Tc) [196]. The preparation of Technegas is carried out in a special machine at 2550 °C in an atmosphere of 100% argon gas and carbon, resulting in a thin layer of technetium encapsulating the carbon nanoparticles with a typical size of 30 to 60 nm [196,197]. The limitations of the employed isotopes in the Technegas technique are as follows: (i) ^{99m}Tc has a short physical half-life (~6 h); (ii) ^{99m}Tc leaches from the CPs, which limits this technique's use for clearance studies; and (iii) the hygroscopic properties of the particles as well as the presence of free pertechnetate can cause chemical instability of the generated particles [193,196,198]. Therefore, many modifications to this technique are used such as labeling with an indium isotope (^{111}In) or $^{68}GaCl_3$, which can extend the investigation time from a few hours to a few days [198–200]. Although these labeling methods offer sensitive localization and quantification of carbon nanoparticles in human studies, they have limitations, such as exposing the subject to radiation; the use of labeled carbon nanoparticles, which can modify the properties of the nanoparticles; and the limited post-administration follow-up, as radionuclides decay after few days [171,196].

5.2.2. Mass Spectrometry Detection of CPs

Laser desorption/ionization mass spectrometry (LDI-MS) is widely used for the characterization of CPs in environmental samples, and has been recently employed for the identification of CPs in biological samples such as cells and those in animal studies [86,201]. LDI-MS works on the principle of the ionization of the analyte using a laser beam, thereby removing the molecules from the surface of the material and ionizing them [202]. Afterwards, the ionized molecules are characterized based on their mass-to-charge ratio (m/z) in the analyzer through the time-of-flight (TOF) measurements [203,204]. Lin et al. used LDI-MS to study the biodistribution of soot particles in mice after their inhalation of $PM_{2.5}$ [201]. The samples with CPs were ionized under laser excitation in a vacuum atmosphere resulting in the formation of carbon cluster structures and characteristics peaks (C_n^-) with repeated mass units [201]. The identification and quantification in biological samples were performed using murine lung samples and consistent anionic carbon cluster peaks were observed, making this technique feasible for measurements in biological samples [201]. The use of mass spectroscopy to identify and quantify CPs (soot) in biological samples is very promising; however, there are limitations of this approach, for instance: (i) the removal of the impurities can make sample preparation very complicated, and (ii) mostly biological samples contain carbon; hence, mass spectrometry can result in the detection of carbon cluster peaks from biological samples instead of soot particles [205,206].

5.2.3. Electron Microscopy Detection of CPs

Electron microscopy (EM) is a conventional technique that can provide good spatial resolution to analyze the cellular uptake of nanoparticles [207–209]. By using EM, CPs are characterized based on their typically black aggregates in the cells or tissue samples [210,211]. Scanning electron microscopy (SEM) scans the surface of a sample using a focused electron beam and can provide images of the sample in 3D with a very high resolution (3–20 nm) [212,213]. The detailed visualization and internalization of the nanoparticles by a cellular organism can be performed by transmission electron microscopy (TEM), in which a beam of electrons is transmitted through a thin specimen [213]. TEM analysis also provides information about different parameters of the nano-particulates such as their size distribution, shape, and aggregation [213]. Owing to its sub-nanometer resolution, it can reveal the fine relationships between the nano-particulates and the cellular/tissue components [210,213]. For example, Jiang et al. visualized the cellular uptake of CB in the cytoplasm of the BEAS-2 cells. Komatsu et al. also observed CB and DEPs in the form of randomly dispersed aggregates in the cytoplasm of TM3 cells, whereas Zhang et al. found the particles trapped inside the alveolar macrophages from measurements of the lung sections of mice exposed to CB [209,210,214]. In addition, Belade et al. observed the distribution of CB NPs in MRC-5 and 16HBE cells using TEM and found aggregates in the cytosol and cytoplasmic vesicles (Figure 4) [211].

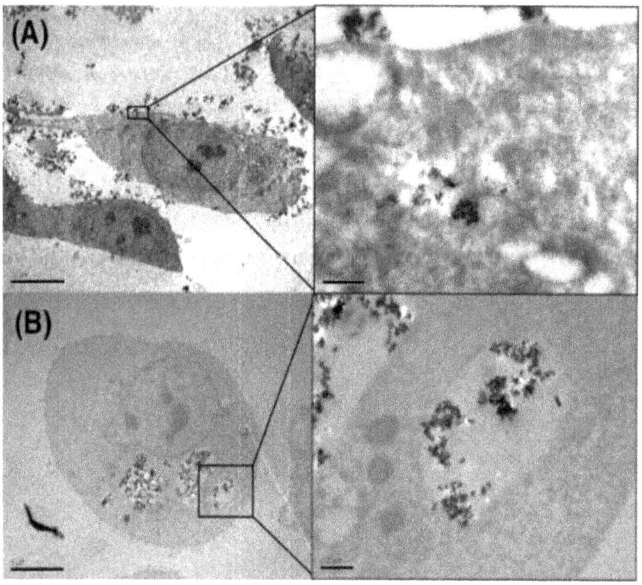

Figure 4. TEM images of 16HBE cells incubated with CB micro- and nanoparticles (MNPs). (**A**) 16HBE cell with CB13 MNPs in a vesicle (scale bars: left (10 μm); right (0.5 μm)) (**B**) 16HBE cell containing CB21 MNPs in a vesicle (scale bars: left (5 μm); right (0.5 μm)). Reproduced with permission from Ref. [211]. Copyright © 2011 Elsevier.

TEM can provide a very high resolution; however, some of the limitations of this technique are as follows. (i) CPs can only be identified based on their dark appearance in a bio-context, hence, it is difficult to discriminate different CPs. (ii) Biological samples need to be prepared carefully by skilled persons and sliced into very thin sections for analysis using TEM. (iii) The small field of view of TEM can hamper the imaging of large biological specimens such as entire organisms. (iv) Finally, TEM can only provide a static snapshot of the sample; thus, the investigation of dynamic cellular processes is limited [215,216]. Although conventional EM is widely used in NP research in biological systems, it cannot be used as the only tool for the qualitative or quantitative evaluation of the cellular uptake of NPs [217]. Hence, to obtain more abundant and accurate information, it is mostly combined with other methods such as optical microscopy techniques, electron energy loss spectroscopy (EELS), or energy dispersive X-ray (EDX) [218]. Another drawback of EM is that it involves the use of a very expensive instrument compared to most optical microscopes [213].

5.2.4. Optical Techniques for the Detection of CPs

Optical techniques can offer the capabilities of the non-destructive and label-free detection of CPs in their biological context. These techniques work based on different phenomena, including the optical absorption or scattering of light, molecular vibrational fingerprinting, or exploiting the non-linear optical behavior of nanoparticles.

- Absorption- and scattering-based microscopy

Absorption-based microscopy is one of the most straightforward forms of optical microscopy techniques and can be implemented in a bright-field microscopy system [219,220]. It works on the principle of the transmission of light through the sample and the generation of contrast based on the changes in the absorption of light in the denser regions of the sample [219]. Therefore, CB/BC NPs will appear darker compared to BrC NPs against the transparent biological media in the bright-field images [39,219]. In comparison, scattering-based microscopy works on the principle of the detection of Rayleigh scattering from the specimen, in which contrast between different materials is generated based on the scattering intensity [219,221]. The scattering intensity from the particles in a biological context depends not only on their size but also on their refractive index, the refractive index of the surroundings, and the optical configuration of the setup in which the scattered light is detected [222]. Scattering-based microscopy images can be obtained using a conventional dark-field microscope [219]. Hence, both of these optical techniques can be applied for the label-free observation of CPs based on their light absorption or scattering in biological fluids, as well as their uptake by cells or tissues [171].

You et al. observed the nanoparticulate CB in the lung $CD11c^+$ cells from mice deliberately exposed to CB using bright-field and dark-field microscopy [223]. In addition, Modrzynska et al. observed the black aggregates in the liver sections of mice exposed to CB NPs using bright-field microscopy [224]. Absorption- and scattering-based microscopy are simple yet effective approaches to the visualization of CPs in biological samples, but they are limited by: (i) low resolution; hence, the detection of very small CPs below ~200 nm is challenging and only large aggregates can be detected; (ii) visualization at a low concentration is difficult; and (iii) the identification of different types of CPs is not possible [171,225].

- Raman spectroscopy

Raman spectroscopy is a valuable technique for the label-free quantitative and qualitative characterization of CPs in biological samples [226,227]. It works on the principle of detecting inelastically scattered light to study the vibrational fingerprinting of different materials for their characterization [228]. Raman spectroscopy is used to obtain detailed information on different types of carbonaceous materials and their degrees of structural disorder [229]. Different types of CPs are distinguished based on their degree of graphitization using Raman spectroscopy [226,230,231]. The features of the first-order Raman spectra of carbonaceous materials are in the region from 1200 cm^{-1} to 1700 cm^{-1}, whereas second-order features are around 2700 cm^{-1} and can range up to 3500 cm^{-1} [226,232]. The first-order Raman transitions provide information about carbon material with a long-range order (graphitic carbon) or without a long-range order (amorphous carbon) [226]. Amorphous carbon is mostly an unknown mixture of sp^2- and sp^3- bonded carbon [226]. The most important Raman features observed for carbonaceous materials are around 1580 cm^{-1}, representing an ideal graphitic lattice (G-band), and at ~1350 cm^{-1}, representing a disordered graphitic lattice from graphene layer edges (D1-band) (Figure 5) [232,233]. Some smaller features are also observed at ~1620 cm^{-1} (D2-band), ~1500 cm^{-1} (D3-band), and ~1200 cm^{-1} (D4-band) [232,233]. The D2-, D3-, and D4-bands represent the disordered graphitic lattice at the surface graphene layers, amorphous carbon, and a disordered graphitic lattice due to polyenes or ionic impurities, respectively [233–235].

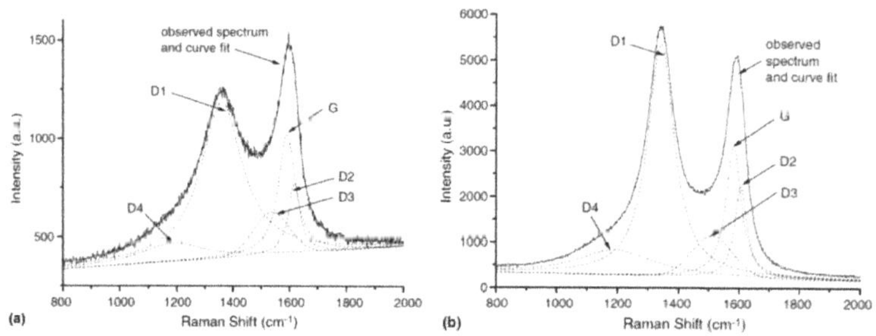

Figure 5. Peak fitting of first-order Raman spectra obtained using 514 nm laser. (**a**) Diesel soot; (**b**) Printex XE2. Reproduced with permission from Ref. [232]. Copyright © 2005 Elsevier.

Although Raman spectroscopy is mostly used for the characterization of CPs in environmental samples, few studies have reported the use of Raman spectroscopy for characterizing carbon materials in biological samples [223,236–238]. For example, Knief et al. studied the suitability of Raman spectroscopy to determining the toxicity of carbon nanotubes in human epithelial cells (A549) [236]. In addition, the Raman fingerprints of CPs in the lungs of mice exposed to cigarette smoke for four months were observed [223]. Furthermore, Čabanová et al. detected the Raman fingerprints of amorphous carbon in the mucosa and hypertrophic tissue samples of patients with chronic rhinitis, which indicates the potential of Raman spectroscopy to identify CPs in real exposure measurements [237,238].

Raman spectroscopy has the potential to characterize CPs based on their vibrational fingerprint; however, the application of Raman spectroscopy in most biological samples is limited by the strong autofluorescence background signal (from PAHs in BrC) [227,228]. In addition, to achieve a good signal-to-noise ratio (SNR), long acquisition times are needed, which greatly slows the associated measurements [227]. Hence, acquiring Raman measurements of CPs in (bio)liquid samples is not a simple task due to the long measurement times. A very weak Raman-scattering signal can be compensated with higher laser power densities, which can be detrimental to biological samples [171,228].

- Photothermal pump–probe microscopy

Pump–probe microscopy is a label-free optical technique used for the characterization of different types of nanoparticles in biological samples [239,240]. The basic idea behind the pump–probe approach is that the pump field excites the sample to a higher energy state or perturbs the electronic states of the materials, whereas the probe field determines the changes in the electronic states [240]. The probe field's absorption is either transiently enhanced or reduced [240,241]. In transient absorption microscopy, the absorption of the probe beam is transiently altered by the pump field, resulting in short-term changes in the probe field's intensity [240,242]. Ground-state depletion microscopy is another type of pump–probe microscopy approach, in which the population of the ground state of a molecule is altered and hence results in the reduced absorption of the probe field [243]. The photothermal approach to pump–probe microscopy also uses a pump field to heat the specimen in focus and induce changes in the refractive index of the surrounding medium, and modifications in the probe beam are observed due to the induced heat [244].

Steuwe et al. used photothermal pump–probe microscopy to identify CPs in different biological samples based on the contrast generation due to changes in the probe field intensity [245]. In addition to visualizing CB NPs in an automated fashion in a flow cell, CB NPs were also detected inside the human lung fibroblasts and spiked urine samples. The nonlinear nature of the signal provides 3D sectioning as well as large imaging depths due to the use of longer laser wavelengths.

This technique successfully detected CPs in different biological samples; however, some of the limitations of this approach are as follows: (i) the non-specific nature of the signal makes it difficult to distinguish different types of CPs; (ii) typically, the pump field can excite several molecular states simultaneously and result in the generation of spurious signals from non-CPs; and (iii) pump–probe microscopy requires the use of two ultrafast light sources mainly comprising a laser and an Optical Parametric Oscillator (OPO), hence making it an expensive approach [239,246]. Pump–probe approaches have also been proposed using cheaper diode laser; however, this can compromise the resolution [247,248].

- Femtosecond-pulsed laser microscopy (non-incandescence-related white light generation) for CP detection

Femtosecond- (fs) pulsed laser microscopy (FPLM) is an innovative approach for the label-free and non-invasive characterization of CPs in fluids, cells, and tissue samples [249]. This approach can be implemented using a multiphoton microscopy setup available in most biomedical research labs [250]. Some of the non-linear optical processes observed using a multiphoton microscope include two-photon-excited autofluorescence (TPAF), second-harmonic generation (SHG), third-harmonic generation (THG), etc. [250–253]. These non-linear processes work based on the simultaneous absorption of two or more photons by the specimen [250,254]. The multiphoton approach has the advantages of imaging living tissue samples with greater penetration depth and lower radiation damage due to the use of nIR laser wavelengths [250,255].

Recently, Bové et al. developed a novel biocompatible approach based on fs-pulsed laser microscopy for the label-free identification of CPs in biological media (Figure 6a) [249]. This approach works on the principle of non-incandescence-related white light (WL) generation from CB NPs when illuminated using a fs-pulsed laser at nIR wavelengths. The light emitted by the CB NPs covers the whole visible spectrum, and based on this spectral information, the emitted light can be detected in different spectral windows compatible with most multiphoton microscopes [256]. By using this approach, BC particles were detected in urine samples, human placenta samples (Figure 6b), and on plant leaves [257–259]. Further, the relationship between the WL emission intensity and particle size was also reported in aqueous suspensions [260]. Using FPLM, Bongaerts et al. recently revealed the presence of ambient BC particles in fetal blood and organs [261]. Table 2 provides a detailed overview of the techniques for the detection of CPs.

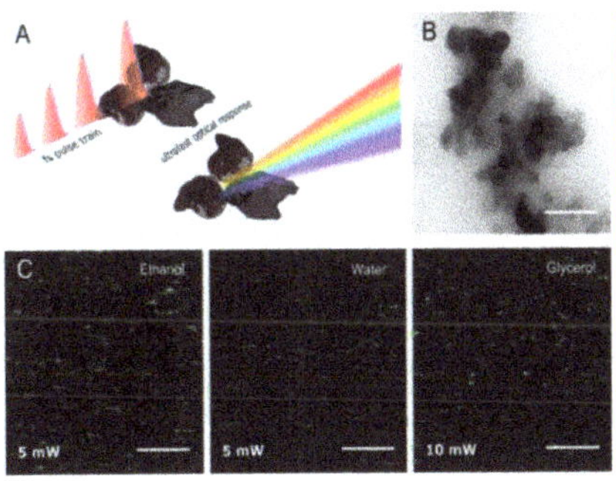

(a)

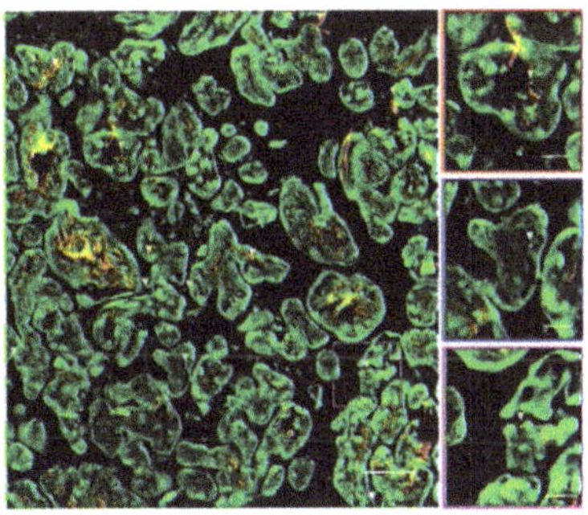

(b)

Figure 6. (**a**) The detection method of CB NPs based on WL emission: (**A**) Schematic showing the emission and excitation process. (**B**) TEM image of CB NP aggregates. Scale bar: 300 nm (CB). (**C**) CB detection using the FPLM detection method in different sampling conditions. Reproduced with permission from Ref. [249]. Copyright © 2016, American Chemical Society (ACS). Available from: https://pubs.acs.org/doi/10.1021/acs.nanolett.6b00502 (accessed on 14 April 2022). Further permissions related to the material excerpted should be directed to the ACS. (**b**) Detection of BC particles at the fetal side of human placenta based on WL emission originating from the BC particles under illumination with fs-pulsed laser. Reused from Ref. [258], originally published under Creative Commons Attribution 4.0 International License, Copyright © Authors 2019.

Table 2. Techniques for detection of CPs in biological samples.

Detection Techniques	Detection Mechanism	Sample Types	Label-Free Approach	Model Studies	Detection in Real Samples	Limitations	References
Radiolabeling detection	Half-life probe	Cells, tissues, clinical studies	No	Possible	Not Possible	Individual gets exposed to ionizing, limited post-administration follow-up, limited to deliberate exposure.	[171,194–196,198]
Mass spectrometry	Laser desorption/ionization mass to charge ratio	Solid/liquid samples	Yes	Possible	No studies reported	Removal of impurities, complicated sample preparation, interference from the biological samples.	[86,201,203–205]
Electron microscopy	Using a focused electron beam	Dried samples	Yes	Possible	No studies reported	Complicated sample preparation, small field-of-view, investigation of dynamic cellular processes are not possible.	[207,208,210–213,215]
Absorption- and scattering-based microscopy	Light absorption and scattering	Solid/liquid samples	Yes	Possible	No studies reported	Low resolution, detection of very large aggregates, challenging to detect low concentrations.	[171,219–224]
Raman spectroscopy	Molecular vibrational fingerprinting	Solid/liquid samples	Yes	Possible	Possible, detected in human nasal mucus and hypertrophic tissues	Strong background signal from cells or tissues, long acquisition times.	[171,226–235]
Photothermal pump-probe microscopy	Contrast generation due to changes in the probe field intensity	Solid/liquid samples	Yes	Possible	No studies reported	Generation of spurious signal from non-CPs, cannot distinguish among different CPs, expensive.	[239–242,244–248]
Femtosecond pulsed laser microscopy (non-incandescence related WL generation)	White light emission, spectral information	Solid/liquid samples	Yes	Possible	Possible, detected in human urine and placenta samples	Possible interference from non-CP components of PM, size determination is challenging, difficult to distinguish among different CPs.	[249–253,255–259,261]

6. Conclusions and Outlook

6.1. Conclusions

CPs, being a major fraction of ultrafine PM, are abundant in the atmosphere and originate from different anthropogenic and natural sources. Their abundant presence in the atmosphere not only contributes to environmental impacts such as global warming (by their absorption of solar radiation) but also presents a likely pathway for their inhalation by humans. The inhalation of CPs by humans results in a wide range of adverse effects, ranging from respiratory to carcinogenic effects, as ultrafine CPs can translocate through the bloodstream to different organs. The qualitative and quantitative detection of CPs in atmospheric and biological samples is very important to understand their adverse effects. Different analytical techniques such as laser-induced incandescence and optical absorption are used for CPs' detection in gaseous samples. However, CPs' detection in biological samples is very challenging due to their heterogeneous nature and complexity associated with biological samples. The different analytical techniques used for CPs' detection in model studies in biological samples include radiolabeling detection, electron microscopy, mass spectrometry, etc. Whereas Raman spectroscopy and FPLM are employed for CP detection in models and real-exposure samples.

6.2. Challenges and Outlook

Different techniques such as FPLM or Raman spectroscopy are very sensitive for the qualitative and quantitative detection of CPs in real-exposure biological samples; however, these microspectroscopy techniques are very labor-intensive and time-consuming due to the related sample preparation, microscopy measurements, and data analysis. Therefore, to detect CPs in biological samples for clinical applications, there is still a need for the development of robust, sensitive, and time-effective analytical techniques. There is also growing evidence that the harmful effects of PM pollutants are size-dependent, in wherein ultrafine particles are the most harmful [262]. Hence, the size determination of CPs in biological samples can provide additional information about the toxicity of CPs related to their size. The techniques discussed in this review, e.g., FPLM and Raman spectroscopy, cannot detect very small particles (<100 nm) in biological samples; hence, high-resolution techniques are needed to detect particles below 100 nm.

Although BC and CB constitute a major fraction of CPs, recently, BrC has also received significant attention from scientists. The complexity and heterogeneity of CPs, the variability of their molecular properties, and the nature of their emission sources pose great challenges to their detection and discrimination in biological samples. Hence, there is a need for the development of advanced and sensitive analytical tools for the qualitative and quantitative detection and discrimination of different components of CPs to understand their adverse effects on human health. The currently available approaches will require optimization to not only detect very small particles (<100 nm) but to also discriminate between different CPs for diagnostic purposes.

In addition to understanding the adverse effects of CPs on human health, their atmospheric concentrations from different anthropogenic sources need to be controlled by better policy making (and implementation) at the regional and national levels. Recently, the WHO released new Global Air Quality Guidelines intending to save millions of lives from the harmful effects of air pollution. Among the different constituents of air pollution, particulate matter (PM) is on the top of the agenda of the WHO guidelines. In addition, the EU key directive (2008/50/EC) on ambient air quality and cleaner air for Europe also stressed reducing air pollution levels to minimize the adverse effects on human health.

Author Contributions: Conceptualization, I.A. and M.B.J.R.; Writing—original draft preparation, I.A.; Writing—review and editing, I.A. and M.B.J.R.; Supervision, M.B.J.R.; Project administration, M.B.J.R.; Funding acquisition, M.B.J.R. All authors have read and agreed to the published version of the manuscript.

Funding: This work was supported by the Flemish Research Foundation Project (G082317N).

Data Availability Statement: Not Applicable.

Acknowledgments: We would like to thank Marcel Ameloot, Tim Nawrot, Hannelore Bové, and Eva Bongaerts for their collaboration.

Conflicts of Interest: The authors declare no conflict of interest.

References

1. OCDE. *The Economic Consequences of Outdoor Air Pollution*, Policy Highlights; OCDE: Paris, France, 2016; 116p. [CrossRef]
2. Sicard, P.; Agathokleous, E.; De Marco, A.; Paoletti, E.; Calatayud, V. Urban population exposure to air pollution in Europe over the last decades. *Environ. Sci. Eur.* **2021**, *33*, 28. [CrossRef] [PubMed]
3. Gakidou, E.; Afshin, A.; Abajobir, A.A.; Abate, K.H.; Abbafati, C.; Abbas, K.M.; Abd-Allah, F.; Abdulle, A.M.; Abera, S.F.; Aboyans, V.; et al. Global, regional, and national comparative risk assessment of 84 behavioural, environmental and occupational, and metabolic risks or clusters of risks, 1990–2016: A systematic analysis for the Global Burden of Disease Study 2016. *Lancet* **2017**, *390*, 1345–1422. [CrossRef]
4. Neira, M.; Prüss-Ustün, A. Preventing disease through healthy environments: A global assessment of the environmental burden of disease. *Toxicol. Lett.* **2016**, *259*, S1. [CrossRef]
5. Hamanaka, R.B.; Mutlu, G.M. Particulate Matter Air Pollution: Effects on the Cardiovascular System. *Front. Endocrinol.* **2018**, *9*, 680. [CrossRef] [PubMed]
6. Newby, D.E.; Mannucci, P.M.; Tell, G.S.; Baccarelli, A.A.; Brook, R.D.; Donaldson, K.; Forastiere, F.; Franchini, M.; Franco, O.H.; Graham, I.; et al. Expert position paper on air pollution and cardiovascular disease. *Eur. Heart J.* **2015**, *36*, 83–93. [CrossRef] [PubMed]
7. Brook, R.D.; Newby, D.E.; Rajagopalan, S. The Global Threat of Outdoor Ambient Air Pollution to Cardiovascular Health. *JAMA Cardiol.* **2017**, *2*, 353. [CrossRef]
8. Schwarze, P.E.; Øvrevik, J.; Låg, M.; Refsnes, M.; Nafstad, P.; Hetland, R.B.; Dybing, E. Particulate matter properties and health effects: Consistency of epidemiological and toxicological studies. *Hum. Exp. Toxicol.* **2006**, *25*, 559–579. [CrossRef]
9. Adams, K.; Greenbaum, D.S.; Shaikh, R.; van Erp, A.M.; Russell, A.G. Particulate matter components, sources, and health: Systematic approaches to testing effects. *J. Air Waste Manag. Assoc.* **2015**, *65*, 544–558. [CrossRef]
10. Okubo, M.; Kuwahara, T. Emission regulations. In *New Technologies for Emission Control in Marine Diesel Engines*; Elsevier: Amsterdam, The Netherlands, 2020.
11. Thorpe, A.; Harrison, R.M. Sources and properties of non-exhaust particulate matter from road traffic: A review. *Sci. Total Environ.* **2008**, *400*, 270–282. [CrossRef]
12. Kelly, F.J.; Fussell, J.C. Size, source and chemical composition as determinants of toxicity attributable to ambient particulate matter. *Atmos. Environ.* **2012**, *60*, 504–526. [CrossRef]
13. Louie, S.M.; Ma, R.; Lowry, G.V. Transformations of Nanomaterials in the Environment. In *Frontiers of Nanoscience*; Elsevier: Amsterdam, The Netherlands, 2014; ISBN 0013-936X.
14. Keller, A.A.; McFerran, S.; Lazareva, A.; Suh, S. Global life cycle releases of engineered nanomaterials. *J. Nanopart. Res.* **2013**, *15*, 1692. [CrossRef]
15. Kabir, E.; Ray, S.; Kim, K.-H.; Yoon, H.-O.; Jeon, E.-C.; Kim, Y.S.; Cho, Y.-S.; Yun, S.-T.; Brown, R.J.C. Current Status of Trace Metal Pollution in Soils Affected by Industrial Activities. *Sci. World J.* **2012**, *2012*, 1–18. [CrossRef]
16. Hime, N.J.; Marks, G.B.; Cowie, C.T. A comparison of the health effects of ambient particulate matter air pollution from five emission sources. *Int. J. Environ. Res. Public Health* **2018**, *15*, 1206. [CrossRef]
17. Reşitoğlu, I.A.; Altinişik, K.; Keskin, A. The pollutant emissions from diesel-engine vehicles and exhaust aftertreatment systems. *Clean Technol. Environ. Policy* **2015**, *17*, 15–27. [CrossRef]
18. Power, M.C.; Weisskopf, M.G.; Alexeeff, S.E.; Coull, B.A.; Spiro, A.; Schwartz, J. Traffic-Related Air Pollution and Cognitive Function in a Cohort of Older Men. *Environ. Health Perspect.* **2011**, *119*, 682–687. [CrossRef]
19. Dumka, U.C.; Kaskaoutis, D.G.; Devara, P.C.S.; Kumar, R.; Kumar, S.; Tiwari, S.; Gerasopoulos, E.; Mihalopoulos, N. Year-long variability of the fossil fuel and wood burning black carbon components at a rural site in southern Delhi outskirts. *Atmos. Res.* **2019**, *216*, 11–25. [CrossRef]
20. Fussell, J.C.; Kelly, F.J. Mechanisms underlying the health effects of desert sand dust. *Environ. Int.* **2021**, *157*, 106790. [CrossRef]
21. *Health Effects of Particulate Matter*; World Health Organization: Copenhagen, Denmark, 2013. Available online: https://www.euro.who.int/__data/assets/pdf_file/0006/189051/Health-effects-of-particulate-matter-final-Eng.pdf (accessed on 20 October 2022).
22. Manisalidis, I.; Stavropoulou, E.; Stavropoulos, A.; Bezirtzoglou, E. Environmental and Health Impacts of Air Pollution: A Review. *Front. Public Health* **2020**, *8*, 14. [CrossRef]
23. *Health Relevance of Particulate Matter from Various Sources*; Report on a WHO Workshop. World Health Organization: Copenhagen, Denmark, 2007. Available online: https://www.euro.who.int/__data/assets/pdf_file/0007/78658/E90672.pdf (accessed on 20 October 2022).
24. Lee, Y.M.; Lee, J.H.; Kim, H.C.; Ha, E. Effects of PM10 on mortality in pure COPD and asthma-COPD overlap: Difference in exposure duration, gender, and smoking status. *Sci. Rep.* **2020**, *10*, 2402. [CrossRef]

25. Lewis, T.C.; Robins, T.G.; Dvonch, J.T.; Keeler, G.J.; Yip, F.Y.; Mentz, G.B.; Lin, X.; Parker, E.A.; Israel, B.A.; Gonzalez, L.; et al. Air pollution-associated changes in lung function among asthmatic children in Detroit. *Environ. Health Perspect.* **2005**, *113*, 1068–1075. [CrossRef]
26. Du, Y.; Xu, X.; Chu, M.; Guo, Y.; Wang, J. Air particulate matter and cardiovascular disease: The epidemiological, biomedical and clinical evidence. *J. Thorac. Dis.* **2016**, *8*, E8–E19. [CrossRef] [PubMed]
27. Li, R.; Zhou, R.; Zhang, J. Function of PM2.5 in the pathogenesis of lung cancer and chronic airway inflammatory diseases. *Oncol. Lett.* **2018**, *15*, 7506–7514. [CrossRef]
28. Harrison, R.M. What is responsible for the carcinogenicity of PM2.5? *Occup. Environ. Med.* **2004**, *61*, 799–805. [CrossRef] [PubMed]
29. Schraufnagel, D.E. The health effects of ultrafine particles. *Exp. Mol. Med.* **2020**, *52*, 311–317. [CrossRef] [PubMed]
30. Oberdörster, G.; Celein, R.M.; Ferin, J.; Weiss, B. Association of particulate air pollution and acute mortality: Involvement of ultrafine particles? *Inhal. Toxicol.* **1995**, *7*, 111–124. [CrossRef]
31. Oberdörster, G.; Sharp, Z.; Atudorei, V.; Elder, A.; Gelein, R.; Kreyling, W.; Cox, C. Translocation of inhaled ultrafine particles to the brain. *Inhal. Toxicol.* **2004**, *16*, 437–445. [CrossRef]
32. Hong, G.; Jee, Y.K. Special issue on ultrafine particles: Where are they from and how do they affect us? *Exp. Mol. Med.* **2020**, *52*, 309–310. [CrossRef]
33. Bai, L.; Chen, H.; Hatzopoulou, M.; Jerrett, M.; Kwong, J.C.; Burnett, R.T.; Van Donkelaar, A.; Copes, R.; Martin, R.V.; Van Ryswyk, K.; et al. Exposure to ambient ultrafine particles and nitrogen dioxide and incident hypertension and diabetes. *Epidemiology* **2018**, *29*, 323–332. [CrossRef]
34. Schlesinger, R.B.; Kunzli, N.; Hidy, G.M.; Gotschi, T.; Jerrett, M. The health relevance of ambient particulate matter characteristics: Coherence of toxicological and epidemiological inferences. *Inhal. Toxicol.* **2006**, *18*, 95–125. [CrossRef]
35. Stanek, L.W.; Sacks, J.D.; Dutton, S.J.; Dubois, J.J.B. Attributing health effects to apportioned components and sources of particulate matter: An evaluation of collective results. *Atmos. Environ.* **2011**, *45*, 5655–5663. [CrossRef]
36. Krzyzanowski, M.; Dibbert, B.K.; Schneider, J. *Health Effects of Transport Related Air Pollution*; World Health Organization: Copenhagen, Denmark, 2005. Available online: https://www.euro.who.int/__data/assets/pdf_file/0006/74715/E86650.pdf (accessed on 20 October 2022).
37. Donaldson, K.; Tran, L.; Jimenez, L.A.; Duffin, R.; Newby, D.E.; Mills, N.; MacNee, W.; Stone, V. Combustion-derived nanoparticles: A review of their toxicology following inhalation exposure. *Part. Fibre Toxicol.* **2005**, *2*, 10. [CrossRef]
38. Janssen, N.A.H.; Hoek, G.; Simic-Lawson, M.; Fischer, P.; van Bree, L.; Ten Brink, H.; Keuken, M.; Atkinson, R.W.; Ross Anderson, H.; Brunekreef, B.; et al. Black carbon as an additional indicator of the adverse health effects of airborne particles compared with pm10 and pm2.5. *Environ. Health Perspect.* **2011**, *119*, 1691–1699. [CrossRef]
39. Andreae, M.O.; Gelencsér, A. Black carbon or brown carbon? the nature of light-absorbing carbonaceous aerosols. *Atmos. Chem. Phys.* **2006**, *6*, 3131–3148. [CrossRef]
40. Pöschl, U. Aerosol particle analysis: Challenges and progress. *Anal. Bioanal. Chem.* **2003**, *375*, 30–32. [CrossRef]
41. Lyu, R.; Zhang, J.; Wu, J.; Feng, Y. Primary Carbonaceous Particle Emission from Four Power Plants with Ultralow Emission in China. *ACS Omega* **2021**, *6*, 1309–1315. [CrossRef]
42. Long, C.M.; Nascarella, M.A.; Valberg, P.A. Carbon black vs. black carbon and other airborne materials containing elemental carbon: Physical and chemical distinctions. *Environ. Pollut.* **2013**, *181*, 271–286. [CrossRef]
43. Lin, P.; Bluvshtein, N.; Rudich, Y.; Nizkorodov, S.A.; Laskin, J.; Laskin, A. Molecular Chemistry of Atmospheric Brown Carbon Inferred from a Nationwide Biomass Burning Event. *Environ. Sci. Technol.* **2017**, *51*, 11561–11570. [CrossRef]
44. Medalia, A.I.; Rivin, D.; Sanders, D.R. A comparison of carbon black with soot. *Sci. Total Environ.* **1983**, *31*, 1–22. [CrossRef]
45. Schnaiter, M.; Horvath, H.; Möhler, O.; Naumann, K.-H.; Saathoff, H.; Schöck, O.W. UV-VIS-NIR spectral optical properties of soot and soot-containing aerosols. *J. Aerosol Sci.* **2003**, *34*, 1421–1444. [CrossRef]
46. Romshoo, B.; Müller, T.; Pfeifer, S.; Saturno, J.; Nowak, A.; Ciupek, K.; Quincey, P.; Wiedensohler, A. Optical properties of coated black carbon aggregates: Numerical simulations, radiative forcing estimates, and size-resolved parameterization scheme. *Atmos. Chem. Phys.* **2021**, *21*, 12989–13010. [CrossRef]
47. Laskin, A.; Laskin, J.; Nizkorodov, S.A. Chemistry of Atmospheric Brown Carbon. *Chem. Rev.* **2015**, *115*, 4335–4382. [CrossRef] [PubMed]
48. Russo, C.; Stanzione, F.; Alfè, M.; Ciajolo, A.; Tregrossi, A. Spectral Analysis in the UV-Visible Range for Revealing the Molecular Form of Combustion-Generated Carbonaceous Species. *Combust. Sci. Technol.* **2012**, *184*, 1219–1231. [CrossRef]
49. Robertson, J. Hard amorphous (diamond-like) carbons. *Prog. Solid State Chem.* **1991**, *21*, 199–333. [CrossRef]
50. Robertson, J.; O'Reilly, E.P. Electronic and atomic structure of amorphous carbon. *Phys. Rev. B* **1987**, *35*, 2946–2957. [CrossRef] [PubMed]
51. Llamas-Jansa, I.; Jäger, C.; Mutschke, H.; Henning, T. Far-ultraviolet to near-infrared optical properties of carbon nanoparticles produced by pulsed-laser pyrolysis of hydrocarbons and their relation with structural variations. *Carbon N. Y.* **2007**, *45*, 1542–1557. [CrossRef]
52. Duley, W.W.; Seahra, S. Graphite, Polycyclic Aromatic Hydrocarbons, and the 2175 Å Extinction Feature. *Astrophys. J.* **1998**, *507*, 874–888. [CrossRef]
53. Henning, T.; Jäger, C.; Mutschke, H. Laboratory Studies of Carbonaceous Dust Analogs. *Astrophys. Dust ASP Conf. Ser.* **2004**, *309*, 603–628.

54. Galvez, A.; Herlin-Boime, N.; Reynaud, C.; Clinard, C.; Rouzaud, J.-N. Carbon nanoparticles from laser pyrolysis. *Carbon N. Y.* **2002**, *40*, 2775–2789. [CrossRef]
55. Wang, X.; Heald, C.L.; Sedlacek, A.J.; de Sá, S.S.; Martin, S.T.; Alexander, M.L.; Watson, T.B.; Aiken, A.C.; Springston, S.R.; Artaxo, P. Deriving brown carbon from multiwavelength absorption measurements: Method and application to AERONET and Aethalometer observations. *Atmos. Chem. Phys.* **2016**, *16*, 12733–12752. [CrossRef]
56. Hegglin, M.I.; Krieger, U.K.; Koop, T.; Peter, T. Technical Note: Organics-Induced Fluorescence in Raman Studies of Sulfuric Acid Aerosols. *Aerosol Sci. Technol.* **2002**, *36*, 510–512. [CrossRef]
57. Li, X.; Hu, M.; Wang, Y.; Xu, N.; Fan, H.; Zong, T.; Wu, Z.; Guo, S.; Zhu, W.; Chen, S.; et al. Links between the optical properties and chemical compositions of brown carbon chromophores in different environments: Contributions and formation of functionalized aromatic compounds. *Sci. Total Environ.* **2021**, *786*, 147418. [CrossRef]
58. Lin, P.; Aiona, P.K.; Li, Y.; Shiraiwa, M.; Laskin, J.; Nizkorodov, S.A.; Laskin, A. Molecular Characterization of Brown Carbon in Biomass Burning Aerosol Particles. *Environ. Sci. Technol.* **2016**, *50*, 11815–11824. [CrossRef]
59. Soleimanian, E.; Mousavi, A.; Taghvaee, S.; Shafer, M.M.; Sioutas, C. Impact of secondary and primary particulate matter (PM) sources on the enhanced light absorption by brown carbon (BrC) particles in central Los Angeles. *Sci. Total Environ.* **2020**, *705*, 135902. [CrossRef]
60. Cachier, H.; Brémond, M.-P.; Buat-Ménard, P. Carbonaceous aerosols from different tropical biomass burning sources. *Nature* **1989**, *340*, 371–373. [CrossRef]
61. Graber, E.R.; Rudich, Y. Atmospheric HULIS: How humic-like are they? A comprehensive and critical review. *Atmos. Chem. Phys.* **2006**, *6*, 729–753. [CrossRef]
62. Bernauer, U.; Chaudhry, Q.; Coenraads, P.-J.; Degen, G.; Dusinska, M.; Lilienblum, W.; Nielsen, E.; Platzek, T.; Rousselle, C.; van Benthem, J. *Opinion on Carbon Black (Nano-Form)*; Scientific Committee on Consumer Safety: Brussels, Belgium, 2013; ISBN 9789279301209.
63. Singh, M.; Vander Wal, R. Nanostructure Quantification of Carbon Blacks. *C* **2018**, *5*, 2. [CrossRef]
64. Association, I.C.B. Carbon Black User's Guide, Safety, Health, & Environmental Information. *Int. Carbon Black Assoc.* 2016. Available online: https://cancarb.com/wp-content/uploads/2020/05/carbon-black-user-guide.pdf (accessed on 20 October 2022).
65. WHO. *IARC Monographs on the Evaluation of Carcinogenic Risks to Humans*; World Health Organization: Geneva, Switzerland, 1994. [CrossRef]
66. *Kirk-Othmer, Encyclopedia of Chemical Technology*; John Wiley and Sons: Hoboken, NJ, USA, 1993.
67. Scientific Opinion on the re-evaluation of vegetable carbon (E 153) as a food additive. *EFSA J.* **2012**, *10*, 2592. [CrossRef]
68. Sids Initial Assessment Profile for Carbon Black, SIAM 22, 18–21 April. 2006. Available online: https://hpvchemicals.oecd.org/ui/handler.axd?id=b2394717-5545-46c3-9194-ae15f987df25 (accessed on 22 April 2022).
69. Shrestha, G.; Traina, S.J.; Swanston, C.W. Black carbon's properties and role in the environment: A comprehensive review. *Sustainability* **2010**, *2*, 294–320. [CrossRef]
70. Dai, X.; Boutton, T.W.; Glaser, B.; Ansley, R.J.; Zech, W. Black carbon in a temperate mixed-grass savanna. *Soil Biol. Biochem.* **2005**, *37*, 1879–1881. [CrossRef]
71. Epa, U. *Report to Congress on Black Carbon: Executive Summary*; United States Environmental Protection Agency: Washington, DC, USA, 2012. Available online: https://19january2017snapshot.epa.gov/www3/airquality/blackcarbon/2012report/fullreport.pdf (accessed on 20 October 2022).
72. Dons, E.; Int Panis, L.; Van Poppel, M.; Theunis, J.; Wets, G. Personal exposure to Black Carbon in transport microenvironments. *Atmos. Environ.* **2012**, *55*, 392–398. [CrossRef]
73. Byčenkiene, S.; Ulevicius, V.; Dudoitis, V.; Pauraite, J. Identification and characterization of black carbon aerosol sources in the East Baltic Region. *Adv. Meteorol.* **2013**, *2013*, 380614. [CrossRef]
74. Ning, Z.; Chan, K.L.; Wong, K.C.; Westerdahl, D.; Močnik, G.; Zhou, J.H.; Cheung, C.S. Black carbon mass size distributions of diesel exhaust and urban aerosols measured using differential mobility analyzer in tandem with Aethalometer. *Atmos. Environ.* **2013**, *80*, 31–40. [CrossRef]
75. Sonwani, S.; Yadav, A.; Saxena, P. Atmospheric Brown Carbon: A Global Emerging Concern for Climate and Environmental Health. *Manag. Contam. Emerg. Concern Environ.* **2021**, *1*, 225–247. [CrossRef]
76. Yan, J.; Wang, X.; Gong, P.; Wang, C.; Cong, Z. Review of brown carbon aerosols: Recent progress and perspectives. *Sci. Total Environ.* **2018**, *634*, 1475–1485. [CrossRef] [PubMed]
77. Bond, T.C. Spectral dependence of visible light absorption by carbonaceous particles emitted from coal combustion. *Geophys. Res. Lett.* **2001**, *28*, 4075–4078. [CrossRef]
78. Lack, D.A.; Langridge, J.M.; Bahreini, R.; Cappa, C.D.; Middlebrook, A.M.; Schwarz, J.P. Brown carbon and internal mixing in biomass burning particles. *Proc. Natl. Acad. Sci. USA* **2012**, *109*, 14802–14807. [CrossRef]
79. Alexander, D.T.L.; Crozier, P.A.; Anderson, J.R. Brown carbon spheres in East Asian outflow and their optical properties. *Science* **2008**, *321*, 833–836. [CrossRef]
80. Lack, D.A.; Cappa, C.D. Impact of brown and clear carbon on light absorption enhancement, single scatter albedo and absorption wavelength dependence of black carbon. *Atmos. Chem. Phys.* **2010**, *10*, 4207–4220. [CrossRef]

81. De Haan, D.O.; Corrigan, A.L.; Tolbert, M.A.; Jimenez, J.L.; Wood, S.E.; Turley, J.J. Secondary organic aerosol formation by self-reactions of methylglyoxal and glyoxal in evaporating droplets. *Environ. Sci. Technol.* **2009**, *43*, 8184–8190. [CrossRef]
82. Powelson, M.H.; Espelien, B.M.; Hawkins, L.N.; Galloway, M.M.; De Haan, D.O. Brown carbon formation by aqueous-phase carbonyl compound reactions with amines and ammonium sulfate. *Environ. Sci. Technol.* **2014**, *48*, 985–993. [CrossRef]
83. Bai, Z.; Zhang, L.; Cheng, Y.; Zhang, W.; Mao, J.; Chen, H.; Li, L.; Wang, L.; Chen, J. Water/Methanol-Insoluble Brown Carbon Can Dominate Aerosol-Enhanced Light Absorption in Port Cities. *Environ. Sci. Technol.* **2020**, *54*, 14889–14898. [CrossRef]
84. Satish, R.; Rastogi, N.; Singh, A.; Singh, D. Change in characteristics of water-soluble and water-insoluble brown carbon aerosols during a large-scale biomass burning. *Environ. Sci. Pollut. Res.* **2020**, *27*, 33339–33350. [CrossRef]
85. Jo, D.S.; Park, R.J.; Lee, S.; Kim, S.-W.; Zhang, X. A global simulation of brown carbon: Implications for photochemistry and direct radiative effect. *Atmos. Chem. Phys.* **2016**, *16*, 3413–3432. [CrossRef]
86. Atwi, K.; Mondal, A.; Pant, J.; Cheng, Z.; El Hajj, O.; Ijeli, I.; Handa, H.; Saleh, R. Physicochemical properties and cytotoxicity of brown carbon produced under different combustion conditions. *Atmos. Environ.* **2021**, *244*, 117881. [CrossRef]
87. Lima de Albuquerque, Y.; Berger, E.; Li, C.; Pardo, M.; George, C.; Rudich, Y.; Géloën, A. The Toxic Effect of Water-Soluble Particulate Pollutants from Biomass Burning on Alveolar Lung Cells. *Atmosphere* **2021**, *12*, 1023. [CrossRef]
88. Marelle, L.; Thomas, J.L.; Ahmed, S.; Tuite, K.; Stutz, J.; Dommergue, A.; Simpson, W.R.; Frey, M.M.; Baladima, F. Implementation and Impacts of Surface and Blowing Snow Sources of Arctic Bromine Activation Within WRF-Chem 4.1.1. *J. Adv. Model. Earth Syst.* **2021**, *13*, e2020MS002391. [CrossRef]
89. Jacobson, M.Z. Control of fossil-fuel particulate black carbon and organic matter, possibly the most effective method of slowing global warming. *J. Geophys. Res. Atmos.* **2002**, *107*, ACH 16-1–ACH 16-22. [CrossRef]
90. Tollefson, J. Soot a major contributor to climate change. *Nature* **2013**, *15*, 15. [CrossRef]
91. Zhang, Y.; Gao, T.; Kang, S.; Sprenger, M.; Tao, S.; Du, W.; Yang, J.; Wang, F.; Meng, W. Effects of black carbon and mineral dust on glacial melting on the Muz Taw glacier, Central Asia. *Sci. Total Environ.* **2020**, *740*, 140056. [CrossRef]
92. Kang, S.; Zhang, Y.; Qian, Y.; Wang, H. A review of black carbon in snow and ice and its impact on the cryosphere. *Earth-Sci. Rev.* **2020**, *210*, 103346. [CrossRef]
93. Ming, Y.; Ramaswamy, V.; Persad, G. Two opposing effects of absorbing aerosols on global-mean precipitation. *Geophys. Res. Lett.* **2010**, *37*, 1–4. [CrossRef]
94. Bond, T.C.; Doherty, S.J.; Fahey, D.W.; Forster, P.M.; Berntsen, T.; Deangelo, B.J.; Flanner, M.G.; Ghan, S.; Kärcher, B.; Koch, D.; et al. Bounding the role of black carbon in the climate system: A scientific assessment. *J. Geophys. Res. Atmos.* **2013**, *118*, 5380–5552. [CrossRef]
95. Brown, M.E.; Antle, J.M.; Backlund, P.; Carr, E.R.; Easterling, W.E.; Walsh, M.K.; Ammann, C.; Attavanich, W.; Barrett, C.B.; Bellemare, M.F.; et al. *Climate Change, Global Food Security, and the U.S. Food System*; International Food Policy Research Institute: Washington, DC, USA, 2015. [CrossRef]
96. Gao, C.; He, J.; Cong, J.; Zhang, S.; Wang, G. Impact of forest fires generated black carbon deposition fluxes in Great Hinggan Mountains (China). *L. Degrad. Dev.* **2018**, *29*, 2073–2081. [CrossRef]
97. Lu, Z.; Streets, D.G.; Winijkul, E.; Yan, F.; Chen, Y.; Bond, T.C.; Feng, Y.; Dubey, M.K.; Liu, S.; Pinto, J.P.; et al. Light absorption properties and radiative effects of primary organic aerosol emissions. *Environ. Sci. Technol.* **2015**, *49*, 4868–4877. [CrossRef] [PubMed]
98. Saleh, R.; Hennigan, C.J.; McMeeking, G.R.; Chuang, W.K.; Robinson, E.S.; Coe, H.; Donahue, N.M.; Robinson, A.L. Absorptivity of brown carbon in fresh and photo-chemically aged biomass-burning emissions. *Atmos. Chem. Phys.* **2013**, *13*, 7683–7693. [CrossRef]
99. Yue, S.; Zhu, J.; Chen, S.; Xie, Q.; Li, W.; Li, L.; Ren, H.; Su, S.; Li, P.; Ma, H.; et al. Brown carbon from biomass burning imposes strong circum-Arctic warming. *One Earth* **2022**, *5*, 293–304. [CrossRef]
100. Myers-Smith, I.H.; Kerby, J.T.; Phoenix, G.K.; Bjerke, J.W.; Epstein, H.E.; Assmann, J.J.; John, C.; Andreu-Hayles, L.; Angers-Blondin, S.; Beck, P.S.A.; et al. Complexity revealed in the greening of the Arctic. *Nat. Clim. Chang.* **2020**, *10*, 106–117. [CrossRef]
101. Riediker, M.; Zink, D.; Kreyling, W.; Oberdörster, G.; Elder, A.; Graham, U.; Lynch, I.; Duschl, A.; Ichihara, G.; Ichihara, S.; et al. Particle toxicology and health—where are we? *Part. Fibre Toxicol.* **2019**, *16*, 19. [CrossRef]
102. Lelieveld, J.; Evans, J.S.; Fnais, M.; Giannadaki, D.; Pozzer, A. The contribution of outdoor air pollution sources to premature mortality on a global scale. *Nature* **2015**, *525*, 367–371. [CrossRef]
103. Chen, C.; Xu, D.; He, M.Z.; Wang, Y.; Du, Z.; Du, Y.; Qian, Y.; Ji, D.; Li, T. Fine Particle Constituents and Mortality: A Time-Series Study in Beijing, China. *Environ. Sci. Technol.* **2018**, *52*, 11378–11386. [CrossRef]
104. Olstrup, H.; Johansson, C.; Forsberg, B. The use of carbonaceous particle exposure metrics in health impact calculations. *Int. J. Environ. Res. Public Health* **2016**, *13*, 249. [CrossRef]
105. Hvidtfeldt, U.A.; Geels, C.; Sørensen, M.; Ketzel, M.; Khan, J.; Tjønneland, A.; Christensen, J.H.; Brandt, J.; Raaschou-Nielsen, O. Long-term residential exposure to PM2.5 constituents and mortality in a Danish cohort. *Environ. Int.* **2019**, *133*, 105268. [CrossRef]
106. Johnston, H.J.; Mueller, W.; Steinle, S.; Vardoulakis, S.; Tantrakarnapa, K.; Loh, M.; Cherrie, J.W. How Harmful Is Particulate Matter Emitted from Biomass Burning? A Thailand Perspective. *Curr. Pollut. Rep.* **2019**, *5*, 353–377. [CrossRef]
107. Niranjan, R.; Thakur, A.K. The toxicological mechanisms of environmental soot (black carbon) and carbon black: Focus on Oxidative stress and inflammatory pathways. *Front. Immunol.* **2017**, *8*, 763. [CrossRef]

108. Möller, W.; Felten, K.; Sommerer, K.; Scheuch, G.; Meyer, G.; Meyer, P.; Häussinger, K.; Kreyling, W.G. Deposition, Retention, and Translocation of Ultrafine Particles from the Central Airways and Lung Periphery. *Am. J. Respir. Crit. Care Med.* **2008**, *177*, 426–432. [CrossRef]
109. Nitta, H.; Sato, T.; Nakai, S.; Maeda, K.; Aoki, S.; Ono, M. Respiratory Health Associated with Exposure to Automobile Exhaust. I. Results of Cross-sectional Studies in 1979, 1982, and 1983. *Arch. Environ. Health Int. J.* **1993**, *48*, 53–58. [CrossRef]
110. English, P.; Neutra, R.; Scalf, R.; Sullivan, M.; Waller, L.; Zhu, L. Examining associations between childhood asthma and traffic flow using a geographic information system. *Environ. Health Perspect.* **1999**, *107*, 761–767. [CrossRef]
111. Brauer, M.; Hoek, G.; Van Vliet, P.; Meliefste, K.; Fischer, P.H.; Wijga, A., Koopman, L.P.; Neijens, H.J.; Gerritsen, J.; Kerkhof, M.; et al. Air Pollution from Traffic and the Development of Respiratory Infections and Asthmatic and Allergic Symptoms in Children. *Am. J. Respir. Crit. Care Med.* **2002**, *166*, 1092–1098. [CrossRef] [PubMed]
112. Ozier, A.; Bara, I.; Girodet, P.-O.; Marthan, R.; Berger, P. Pathophysiology of asthma. *Rev. Prat.* **2011**, *61*, 339–345.
113. Boland, S.; Baeza-Squiban, A.; Fournier, T.; Houcine, O.; Gendron, M.-C.; Chévrier, M.; Jouvenot, G.; Coste, A.; Aubier, M.; Marano, F. Diesel exhaust particles are taken up by human airway epithelial cells in vitro and alter cytokine production. *Am. J. Physiol. Cell. Mol. Physiol.* **1999**, *276*, L604–L613. [CrossRef]
114. Dick, C.A.J.; Brown, D.M.; Donaldson, K.; Stone, V. The Role of Free Radicals in the Toxic and Inflammatory Effects of Four Different Ultrafine Particle Types. *Inhal. Toxicol.* **2003**, *15*, 39–52. [CrossRef]
115. Wang, L.; Bao, S.; Liu, X.; Wang, F.; Zhang, J.; Dang, P.; Wang, F.; Li, B.; Lin, Y. Low-dose exposure to black carbon significantly increase lung injury of cadmium by promoting cellular apoptosis. *Ecotoxicol. Environ. Saf.* **2021**, *224*, 112703. [CrossRef]
116. Elder, A.; Gelein, R.; Finkelstein, J.N.; Driscoll, K.E.; Harkema, J.; Oberdörster, G. Effects of Subchronically Inhaled Carbon Black in Three Species. I. Retention Kinetics, Lung Inflammation, and Histopathology. *Toxicol. Sci.* **2005**, *88*, 614–629. [CrossRef] [PubMed]
117. *Environment Canada Screening Assessment for the Challenge: Carbon Black* Environment Canada, Health Canada: Ottawa, ON, Canada, 2013; pp. 1–48.
118. Nemmar, A.; Hoet, P.H.M.; Vanquickenborne, B.; Dinsdale, D.; Thomeer, M.; Hoylaerts, M.F.; Vanbilloen, H.; Mortelmans, L.; Nemery, B. Passage of inhaled particles into the blood circulation in humans. *Circulation* **2002**, *105*, 411–414. [CrossRef] [PubMed]
119. Shrey, K.; Suchit, A.; Deepika, D.; Shruti, K.; Vibha, R. Air pollutants: The key stages in the pathway towards the development of cardiovascular disorders. *Environ. Toxicol. Pharmacol.* **2011**, *31*, 1–9. [CrossRef]
120. Grahame, T.J.; Schlesinger, R.B. Cardiovascular health and particulate vehicular emissions: A critical evaluation of the evidence. *Air Qual. Atmos. Health* **2010**, *3*, 3–27. [CrossRef]
121. Simkhovich, B.Z.; Marjoram, P.; Kleinman, M.T.; Kloner, R.A. Direct and acute cardiotoxicity of ultrafine particles in young adult and old rat hearts. *Basic Res. Cardiol.* **2007**, *102*, 467–475. [CrossRef]
122. Dockery, D.W.; Stone, P.H. Cardiovascular Risks from Fine Particulate Air Pollution. *N. Engl. J. Med.* **2007**, *356*, 511–513. [CrossRef]
123. Meier, R.; Cascio, W.E.; Ghio, A.J.; Wild, P.; Danuser, B.; Riediker, M. Associations of Short-Term Particle and Noise Exposures with Markers of Cardiovascular and Respiratory Health among Highway Maintenance Workers. *Environ. Health Perspect.* **2014**, *122*, 726–732. [CrossRef]
124. Kim, H.; Oh, S.-J.; Kwak, H.-C.; Kim, J.-K.; Lim, C.-H.; Yang, J.-S.; Park, K.; Kim, S.-K.; Lee, M.-Y. The Impact of Intratracheally Instilled Carbon Black on the Cardiovascular System of Rats: Elevation of Blood Homocysteine and Hyperactivity of Platelets. *J. Toxicol. Environ. Health Part A* **2012**, *75*, 1471–1483. [CrossRef]
125. Gurgueira, S.A.; Lawrence, J.; Coull, B.; Krishna Murthy, G.G.; González-Flecha, B. Rapid increases in the steady-state concentration of reactive oxygen species in the lungs and heart after particulate air pollution inhalation. *Environ. Health Perspect.* **2002**, *110*, 749–755. [CrossRef]
126. Ying, Z.; Kampfrath, T.; Thurston, G.; Farrar, B.; Lippmann, M.; Wang, A.; Sun, Q.; Chen, L.C.; Rajagopalan, S. Ambient Particulates Alter Vascular Function through Induction of Reactive Oxygen and Nitrogen Species. *Toxicol. Sci.* **2009**, *111*, 80–88. [CrossRef]
127. Martinelli, N.; Olivieri, O.; Girelli, D. Air particulate matter and cardiovascular disease: A narrative review. *Eur. J. Intern. Med.* **2013**, *24*, 295–302. [CrossRef]
128. Von Klot, S.; Cyrys, J.; Hoek, G.; Kühnel, B.; Pitz, M.; Kuhn, U.; Kuch, B.; Meisinger, C.; Hörmann, A.; Wichmann, H.-E.; et al. Estimated Personal Soot Exposure Is Associated with Acute Myocardial Infarction Onset in a Case-Crossover Study. *Prog. Cardiovasc. Dis.* **2011**, *53*, 361–368. [CrossRef]
129. Bhaskaran, K.; Hajat, S.; Haines, A.; Herrett, E.; Wilkinson, P.; Smeeth, L. Effects of air pollution on the incidence of myocardial infarction. *Heart* **2009**, *95*, 1746–1759. [CrossRef]
130. Büchner, N.; Ale-Agha, N.; Jakob, S.; Sydlik, U.; Kunze, K.; Unfried, K.; Altschmied, J.; Haendeler, J. Unhealthy diet and ultrafine carbon black particles induce senescence and disease associated phenotypic changes. *Exp. Gerontol.* **2013**, *48*, 8–16. [CrossRef]
131. Larsen, P.B.; Mørck, T.A.; Nørgaard Andersen, D.; Hougaard, K.S. *A Critical Review of Studies on the Reproductive and Developmental Toxicity of Nanomaterials*; European Chemicals Agency: Helsinki, Finland, 2020; ISBN 9789294814234. [CrossRef]
132. Umezawa, M.; Onoda, A.; Korshunova, I.; Jensen, A.C.Ø.; Koponen, I.K.; Jensen, K.A.; Khodosevich, K.; Vogel, U.; Hougaard, K.S. Maternal inhalation of carbon black nanoparticles induces neurodevelopmental changes in mouse offspring. *Part. Fibre Toxicol.* **2018**, *15*, 36. [CrossRef]

133. Onoda, A.; Umezawa, M.; Takeda, K.; Ihara, T.; Sugamata, M. Effects of Maternal Exposure to Ultrafine Carbon Black on Brain Perivascular Macrophages and Surrounding Astrocytes in Offspring Mice. *PLoS ONE* **2014**, *9*, e94336. [CrossRef]
134. Onoda, A.; Takeda, K.; Umezawa, M. Dose-dependent induction of astrocyte activation and reactive astrogliosis in mouse brain following maternal exposure to carbon black nanoparticle. *Part. Fibre Toxicol.* **2017**, *14*, 4. [CrossRef]
135. Skovmand, A.; Jensen, A.C.Ø.; Maurice, C.; Marchetti, F.; Lauvås, A.J.; Koponen, I.K.; Jensen, K.A.; Goericke-Pesch, S.; Vogel, U.; Hougaard, K.S. Effects of maternal inhalation of carbon black nanoparticles on reproductive and fertility parameters in a four-generation study of male mice. *Part. Fibre Toxicol.* **2019**, *16*, 13. [CrossRef]
136. Yokota, S.; Takeda, K.; Oshio, S. Spatiotemporal Small Non-coding RNAs Expressed in the Germline as an Early Biomarker of Testicular Toxicity and Transgenerational Effects Caused by Prenatal Exposure to Nanosized Particles. *Front. Toxicol.* **2021**, *3*, 32. [CrossRef] [PubMed]
137. Watanabe, N.; Ohsawa, M. Elevated serum immunoglobulin E to Cryptomeria japonica pollen in rats exposed to diesel exhaust during fetal and neonatal periods. *BMC Pregnancy Childbirth* **2002**, *2*, 2. [CrossRef] [PubMed]
138. Dietert, R.R.; Holsapple, M.P. Methodologies for developmental immunotoxicity (DIT) testing. *Methods* **2007**, *41*, 123–131. [CrossRef] [PubMed]
139. Manangama, G.; Audignon-Durand, S.; Migault, L.; Gramond, C.; Zaros, C.; Teysseire, R.; Sentilhes, L.; Brochard, P.; Lacourt, A.; Delva, F. Maternal occupational exposure to carbonaceous nanoscale particles and small for gestational age and the evolution of head circumference in the French Longitudinal Study of Children—Elfe study. *Environ. Res.* **2020**, *185*, 109394. [CrossRef] [PubMed]
140. Kim, H.; Kim, W.-H.; Kim, Y.-Y.; Park, H.-Y. Air Pollution and Central Nervous System Disease: A Review of the Impact of Fine Particulate Matter on Neurological Disorders. *Front. Public Health* **2020**, *8*, 575330. [CrossRef]
141. Block, M.L.; Calderón-Garcidueñas, L. Air pollution: Mechanisms of neuroinflammation and CNS disease. *Trends Neurosci.* **2009**, *32*, 506–516. [CrossRef]
142. Bandyopadhyay, A. Neurological Disorders from Ambient (Urban) Air Pollution Emphasizing UFPM and PM2.5. *Curr. Pollut. Rep.* **2016**, *2*, 203–211. [CrossRef]
143. Costa, L.G.; Cole, T.B.; Coburn, J.; Chang, Y.-C.; Dao, K.; Roqué, P.J. Neurotoxicity of traffic-related air pollution. *Neurotoxicology* **2017**, *59*, 133–139. [CrossRef]
144. Swardfager, W.; Lanctôt, K.; Rothenburg, L.; Wong, A.; Cappell, J.; Herrmann, N. A Meta-Analysis of Cytokines in Alzheimer's Disease. *Biol. Psychiatry* **2010**, *68*, 930–941. [CrossRef]
145. Hopkins, L.E.; Laing, E.A.; Peake, J.L.; Uyeminami, D.; Mack, S.M.; Li, X.; Smiley-Jewell, S.; Pinkerton, K.E. Repeated Iron–Soot Exposure and Nose-to-brain Transport of Inhaled Ultrafine Particles. *Toxicol. Pathol.* **2018**, *46*, 75–84. [CrossRef]
146. Chen, J.-C.; Wang, X.; Wellenius, G.A.; Serre, M.L.; Driscoll, I.; Casanova, R.; McArdle, J.J.; Manson, J.E.; Chui, H.C.; Espeland, M.A. Ambient air pollution and neurotoxicity on brain structure: Evidence from women's health initiative memory study. *Ann. Neurol.* **2015**, *78*, 466–476. [CrossRef]
147. Jung, C.-R.; Lin, Y.-T.; Hwang, B.-F. Ozone, Particulate Matter, and Newly Diagnosed Alzheimer's Disease: A Population-Based Cohort Study in Taiwan. *J. Alzheimer's Dis.* **2015**, *44*, 573–584. [CrossRef]
148. Ljungman, P.L.S.; Andersson, N.; Stockfelt, L.; Andersson, E.M.; Nilsson Sommar, J.; Eneroth, K.; Gidhagen, L.; Johansson, C.; Lager, A.; Leander, K.; et al. Long-Term Exposure to Particulate Air Pollution, Black Carbon, and Their Source Components in Relation to Ischemic Heart Disease and Stroke. *Environ. Health Perspect.* **2019**, *127*, 107012. [CrossRef]
149. Tankersley, C.G.; Bierman, A.; Rabold, R. Variation in heart rate regulation and the effects of particle exposure in inbred mice. *Inhal. Toxicol.* **2007**, *19*, 621–629. [CrossRef]
150. Bosetti, C.; Boffetta, P.; La Vecchia, C. Occupational exposures to polycyclic aromatic hydrocarbons, and respiratory and urinary tract cancers: A quantitative review to 2005. *Ann. Oncol.* **2007**, *18*, 431–446. [CrossRef]
151. Seidler, A.; Heiskel, H.; Bickeböller, R.; Elsner, G. Association between diesel exposure at work and prostate cancer. *Scand. J. Work. Environ. Health* **1998**, *24*, 486–494. [CrossRef]
152. Mauderly, J. Diesel exhaust is a pulmonary carcinogen in rats exposed chronically by inhalation. *Fundam. Appl. Toxicol.* **1987**, *9*, 208–221. [CrossRef]
153. Mauderly, J.L.; Snipes, M.B.; Barr, E.B.; Belinsky, S.A.; Bond, J.A.; Brooks, A.L.; Chang, I.Y.; Cheng, Y.S.; Gillett, N.A.; Griffith, W.C. Pulmonary toxicity of inhaled diesel exhaust and carbon black in chronically exposed rats. Part I: Neoplastic and nonneoplastic lung lesions. *Res. Rep. Health Eff. Inst.* **1994**, 1–75. Available online: https://www.healtheffects.org/publication/pulmonary-toxicity-inhaled-diesel-exhaust-and-carbon-black-chronically-exposed-rats-1 (accessed on 20 October 2022).
154. Jacobsen, N.R.; Pojana, G.; White, P.; Møller, P.; Cohn, C.A.; Smith Korsholm, K.; Vogel, U.; Marcomini, A.; Loft, S.; Wallin, H. Genotoxicity, cytotoxicity, and reactive oxygen species induced by single-walled carbon nanotubes and C 60 fullerenes in the FE1-Muta^{TM}Mouse lung epithelial cells. *Environ. Mol. Mutagen.* **2008**, *49*, 476–487. [CrossRef]
155. Gminski, R.; Decker, K.; Heinz, C.; Seidel, A.; Könczöl, M.; Goldenberg, E.; Grobéty, B.; Ebner, W.; Gieré, R.; Mersch-Sundermann, V. Genotoxic effects of three selected black toner powders and their dimethyl sulfoxide extracts in cultured human epithelial A549 lung cells in vitro. *Environ. Mol. Mutagen.* **2011**, *52*, 296–309. [CrossRef]
156. Bond, J.A.; Johnson, N.F.; Snipes, M.B.; Mauderly, J.L.; Sega, G.A. DNA adduct formation in rat alveolar type II cells: Cells potentially at risk for inhaled diesel exhaust. *Environ. Mol. Mutagen.* **1990**, *16*, 64–69. [CrossRef] [PubMed]

157. Gallagher, J.; Heinrich, U.; George, M.; Hendee, L.; Phillips, D.H.; Lewtas, J. Formation of DNA adducts in rat lung following chronic inhalation of diesel emissions, carbon black and titanium dioxide particles. *Carcinogenesis* **1994**, *15*, 1291–1299. [CrossRef] [PubMed]
158. Danielsen, P.H.; Loft, S.; Jacobsen, N.R.; Jensen, K.A.; Autrup, H.; Ravanat, J.-L.; Wallin, H.; Møller, P. Oxidative Stress, Inflammation, and DNA Damage in Rats after Intratracheal Instillation or Oral Exposure to Ambient Air and Wood Smoke Particulate Matter. *Toxicol. Sci.* **2010**, *118*, 574–585. [CrossRef] [PubMed]
159. Baccarelli, A.; Wright, R.O.; Bollati, V.; Tarantini, L.; Litonjua, A.A.; Suh, H.H.; Zanobetti, A.; Sparrow, D.; Vokonas, P.S.; Schwartz, J. Rapid DNA Methylation Changes after Exposure to Traffic Particles. *Am. J. Respir. Crit. Care Med.* **2009**, *179*, 572–578. [CrossRef]
160. Mroz, R.M.; Schins, R.P.F.; Li, H.; Jimenez, L.A.; Drost, E.M.; Holownia, A.; MacNee, W.; Donaldson, K. Nanoparticle-driven DNA damage mimics irradiation-related carcinogenesis pathways. *Eur. Respir. J.* **2008**, *31*, 241–251. [CrossRef]
161. Baccarelli, A.; Bollati, V. Epigenetics and environmental chemicals. *Curr. Opin. Pediatr.* **2009**, *21*, 243–251. [CrossRef]
162. Wilson, M. Interactions between Ultrafine Particles and Transition Metals in Vivo and in Vitro. *Toxicol. Appl. Pharmacol.* **2002**, *184*, 172–179. [CrossRef]
163. Corrêa, A.X.R.; Cotelle, S.; Millet, M.; Somensi, C.A.; Wagner, T.M.; Radetski, C.M. Genotoxicity assessment of particulate matter emitted from heavy-duty diesel-powered vehicles using the in vivo Vicia faba L. micronucleus test. *Ecotoxicol. Environ. Saf.* **2016**, *127*, 199–204. [CrossRef]
164. Hardonnière, K.; Saunier, E.; Lemarié, A.; Fernier, M.; Gallais, I.; Héliès-Toussaint, C.; Mograbi, B.; Antonio, S.; Bénit, P.; Rustin, P.; et al. The environmental carcinogen benzo[a]pyrene induces a Warburg-like metabolic reprogramming dependent on NHE1 and associated with cell survival. *Sci. Rep.* **2016**, *6*, 30776. [CrossRef]
165. Baudouin, C.; Charveron, M.; Tarroux, R.; Gall, Y. Environmental pollutants and skin cancer. *Cell Biol. Toxicol.* **2002**, *18*, 341–348. [CrossRef]
166. Ushio, H.; Nohara, K.; Fujimaki, H. Effect of environmental pollutants on the production of pro-inflammatory cytokines by normal human dermal keratinocytes. *Toxicol. Lett.* **1999**, *105*, 17–24. [CrossRef]
167. Fiorito, S.; Mastrofrancesco, A.; Cardinali, G.; Rosato, E.; Salsano, F.; Su, D.S.; Serafino, A.; Picardo, M. Effects of carbonaceous nanoparticles from low-emission and older diesel engines on human skin cells. *Carbon N. Y.* **2011**, *49*, 5038–5048. [CrossRef]
168. Vierkötter, A.; Schikowski, T.; Ranft, U.; Sugiri, D.; Matsui, M.; Krämer, U.; Krutmann, J. Airborne Particle Exposure and Extrinsic Skin Aging. *J. Investig. Dermatol.* **2010**, *130*, 2719–2726. [CrossRef]
169. Aam, B.B.; Fonnum, F. ROS scavenging effects of organic extract of diesel exhaust particles on human neutrophil granulocytes and rat alveolar macrophages. *Toxicology* **2007**, *230*, 207–218. [CrossRef]
170. Ma, C.; Wang, J.; Luo, J. Activation of nuclear factor kappa B by diesel exhaust particles in mouse epidermal cells through phosphatidylinositol 3-kinase/Akt signaling pathway. *Biochem. Pharmacol.* **2004**, *67*, 1975–1983. [CrossRef]
171. Chen, C.; Wang, H. *Biomedical Applications and Toxicology of Carbon Nanomaterials*; John Wiley & Sons: Hoboken, NJ, USA, 2016.
172. Masiello, C.A. New directions in black carbon organic geochemistry. *Mar. Chem.* **2004**, *92*, 201–213. [CrossRef]
173. Su, X.; Sutarlie, L.; Loh, X.J. Sensors and Analytical Technologies for Air Quality: Particulate Matters and Bioaerosols. *Chem. – An Asian J.* **2020**, *15*, 4241–4255. [CrossRef]
174. Petzold, A.; Ogren, J.A.; Fiebig, M.; Laj, P.; Li, S.-M.; Baltensperger, U.; Holzer-Popp, T.; Kinne, S.; Pappalardo, G.; Sugimoto, N.; et al. Recommendations for reporting "black carbon" measurements. *Atmos. Chem. Phys.* **2013**, *13*, 8365–8379. [CrossRef]
175. Migliorini, F.; De Iuliis, S.; Maffi, S.; Zizak, G. Environmental application of pulsed laser-induced incandescence. *Appl. Phys. B* **2013**, *112*, 433–440. [CrossRef]
176. Weingartner, E.; Saathoff, H.; Schnaiter, M.; Streit, N.; Bitnar, B.; Baltensperger, U. Absorption of light by soot particles: Determination of the absorption coefficient by means of aethalometers. *J. Aerosol Sci.* **2003**, *34*, 1445–1463. [CrossRef]
177. Adam, M.G.; Chiang, A.W.J.; Balasubramanian, R. Insights into characteristics of light absorbing carbonaceous aerosols over an urban location in Southeast Asia. *Environ. Pollut.* **2020**, *257*, 113425. [CrossRef] [PubMed]
178. Sandradewi, J.; Prévôt, A.S.H.; Szidat, S.; Perron, N.; Alfarra, M.R.; Lanz, V.A.; Weingartner, E.; Baltensperger, U. Using Aerosol Light Absorption Measurements for the Quantitative Determination of Wood Burning and Traffic Emission Contributions to Particulate Matter. *Environ. Sci. Technol.* **2008**, *42*, 3316–3323. [CrossRef] [PubMed]
179. Kirchstetter, T.W.; Novakov, T.; Hobbs, P.V. Evidence that the spectral dependence of light absorption by aerosols is affected by organic carbon. *J. Geophys. Res. Atmos.* **2004**, *109*, 1–12. [CrossRef]
180. Good, N.; Mölter, A.; Peel, J.L.; Volckens, J. An accurate filter loading correction is essential for assessing personal exposure to black carbon using an Aethalometer. *J. Expo. Sci. Environ. Epidemiol.* **2017**, *27*, 409–416. [CrossRef] [PubMed]
181. Baumgardner, D.; Popovicheva, O.; Allan, J.; Bernardoni, V.; Cao, J.; Cavalli, F.; Cozic, J.; Diapouli, E.; Eleftheriadis, K.; Genberg, P.J.; et al. Soot reference materials for instrument calibration and intercomparisons: A workshop summary with recommendations. *Atmos. Meas. Tech.* **2012**, *5*, 1869–1887. [CrossRef]
182. Michelsen, H.A.; Schulz, C.; Smallwood, G.J.; Will, S. Laser-induced incandescence: Particulate diagnostics for combustion, atmospheric, and industrial applications. *Prog. Energy Combust. Sci.* **2015**, *51*, 2–48. [CrossRef]
183. Stephens, M.; Turner, N.; Sandberg, J. Particle identification by laser-induced incandescence in a solid-state laser cavity. *Appl. Opt.* **2003**, *42*, 3726. [CrossRef]
184. Axelsson, B.; Collin, R.; Bengtsson, P.-E. Laser-induced incandescence for soot particle size measurements in premixed flat flames. *Appl. Opt.* **2000**, *39*, 3683. [CrossRef]

185. Moteki, N.; Kondo, Y. Method to measure time-dependent scattering cross sections of particles evaporating in a laser beam. *J. Aerosol Sci.* **2008**, *39*, 348–364. [CrossRef]
186. Corbin, J.C.; Gysel-Beer, M. Detection of tar brown carbon with a single particle soot photometer (SP2). *Atmos. Chem. Phys.* **2019**, *19*, 15673–15690. [CrossRef]
187. Miller, J.L.; Orr-Ewing, A.J. Cavity ring-down spectroscopy measurement of single aerosol particle extinction. II. Extinction of light by an aerosol particle in an optical cavity excited by a cw laser. *J. Chem. Phys.* **2007**, *126*, 174303. [CrossRef]
188. Cignoli, F.; De Iuliis, S.; Manta, V.; Zizak, G. Two-dimensional two-wavelength emission technique for soot diagnostics. *Appl. Opt.* **2001**, *40*, 5370. [CrossRef]
189. Huang, Q.; Wang, F.; Yan, J.; Chi, Y. Simultaneous estimation of the 3-D soot temperature and volume fraction distributions in asymmetric flames using high-speed stereoscopic images. *Appl. Opt.* **2012**, *51*, 2968. [CrossRef]
190. Lin, P.; Liu, J.; Shilling, J.E.; Kathmann, S.M.; Laskin, J.; Laskin, A. Molecular characterization of brown carbon (BrC) chromophores in secondary organic aerosol generated from photo-oxidation of toluene. *Phys. Chem. Chem. Phys.* **2015**, *17*, 23312–23325. [CrossRef]
191. Kong, H.; Zhang, Y.; Li, Y.; Cui, Z.; Xia, K.; Sun, Y.; Zhao, Q.; Zhu, Y. Size-Dependent Cytotoxicity of Nanocarbon Blacks. *Int. J. Mol. Sci.* **2013**, *14*, 22529–22543. [CrossRef]
192. Oberdörster, G.; Sharp, Z.; Atudorei, V.; Elder, A.; Gelein, R.; Lunts, A.; Kreyling, W.; Cox, C. Extrapulmonary translocation of ultrafine carbon particles following whole-body inhalation exposure of rats. *J. Toxicol. Environ. Health Part A* **2002**, *65*, 1531–1543. [CrossRef]
193. Wiebert, P.; Sanchez-Crespo, A.; Falk, R.; Philipson, K.; Lundin, A.; Larsson, S.; Möller, W.; Kreyling, W.; Svartengren, M. No significant translocation of inhaled 35-nm carbon particles to the circulation in humans. *Inhal. Toxicol.* **2006**, *18*, 741–747. [CrossRef]
194. Brown, J.S.; Kim, C.S.; Reist, P.C.; Zeman, K.L.; Bennett, W.D. Generation of radiolabeled "soot-like" ultrafine aerosols suitable for use in human inhalation studies. *Aerosol Sci. Technol.* **2000**, *32*, 325–337. [CrossRef]
195. Pourchez, J.; Albuquerque-Silva, I.M.D.; Cottier, M.; Clotagatide, A.; Vecellio, L.; Durand, M.; Dubois, F. Generation and characterization of radiolabelled nanosized carbonaceous aerosols for human inhalation studies. *J. Aerosol Sci.* **2013**, *55*, 1–11. [CrossRef]
196. Mills, N.L.; Amin, N.; Robinson, S.D.; Anand, A.; Davies, J.; Patel, D.; De La Fuente, J.M.; Cassee, F.R.; Boon, N.A.; MacNee, W.; et al. Do inhaled carbon nanoparticles translocate directly into the circulation in humans? *Am. J. Respir. Crit. Care Med.* **2006**, *173*, 426–431. [CrossRef]
197. Thrall, J. *Nuclear Medicine and Molecular Imaging: The Requisites E-Book*; Elsevier Health Sciences: Amsterdam, The Netherlands, 2020; ISBN 9780323550741. Available online: https://www.elsevier.com/books/nuclear-medicine-and-molecular-imaging-the-requisites/978-0-323-53037-8 (accessed on 20 October 2022).
198. Sanchez-Crespo, A.; Klepczynska-Nyström, A.; Lundin, A.; Larsson, B.M.; Svartengren, M. 111 Indium-labeled ultrafine carbon particles; a novel aerosol for pulmonary deposition and retention studies. *Inhal. Toxicol.* **2011**, *23*, 121–128. [CrossRef] [PubMed]
199. Klepczyńska-Nyström, A.; Sanchez-Crespo, A.; Andersson, M.; Falk, R.; Lundin, A.; Larsson, B.-M.; Svartengren, M. The pulmonary deposition and retention of indium-111 labeled ultrafine carbon particles in healthy individuals. *Inhal. Toxicol.* **2012**, *24*, 645–651. [CrossRef] [PubMed]
200. Borges, J.B.; Velikyan, I.; Långström, B.; Sörensen, J.; Ulin, J.; Maripuu, E.; Sandström, M.; Widström, C.; Hedenstierna, G. Ventilation Distribution Studies Comparing Technegas and "Gallgas" Using 68 $GaCl_3$ as the Label. *J. Nucl. Med.* **2011**, *52*, 206–209. [CrossRef] [PubMed]
201. Lin, Y.; Huang, X.; Liu, Y.; Cao, D.; Lu, D.; Feng, Z.; Liu, Q.; Lin, Z.; Jiang, G. Identification, Quantification, and Imaging of the Biodistribution of Soot Particles by Mass Spectral Fingerprinting. *Anal. Chem.* **2021**, *93*, 6665–6672. [CrossRef] [PubMed]
202. Houck, M.M.; Siegel, J.A. *Fundamentals of Forensic Science*; Elsevier: Amsterdam, The Netherlands, 2015; ISBN 9780128002315. Available online: https://www.elsevier.com/books/fundamentals-of-forensic-science/houck/978-0-12-800037-3 (accessed on 20 October 2022).
203. Mandal, A.; Singha, M.; Addy, P.S.; Basak, A. Laser desorption ionization mass spectrometry: Recent progress in matrix-free and label-assisted techniques. *Mass Spectrom. Rev.* **2019**, *38*, 3–21. [CrossRef]
204. Guilhaus, M. Special feature: Tutorial. Principles and instrumentation in time-of-flight mass spectrometry. Physical and instrumental concepts. *J. Mass Spectrom.* **1995**, *30*, 1519–1532. [CrossRef]
205. Kehayias, J.J.; Heymsfield, S.B.; LoMonte, A.F.; Wang, J.; Pierson, R.N. In vivo determination of body fat by measuring total body carbon. *Am. J. Clin. Nutr.* **1991**, *53*, 1339–1344. [CrossRef]
206. Alberts, B.; Johnson, A.; Lewis, J.; Raff, M.; Roberts, K.; Walter, P. *Molecular Biology of the Cell*, 4th ed.; Garland Science: New York, NY, USA, 2002; ISBN 0815340729. Available online: https://www.ncbi.nlm.nih.gov/books/NBK21054/ (accessed on 20 October 2022).
207. Marquis, B.J.; Love, S.A.; Braun, K.L.; Haynes, C.L. Analytical methods to assess nanoparticle toxicity. *Analyst* **2009**, *134*, 425. [CrossRef]
208. Brown, A.; Hondow, N. *Electron Microscopy of Nanoparticles in Cells*; Elsevier: Amsterdam, The Netherlands, 2013; Volume 5, pp. 95–120. [CrossRef]

209. Jiang, S.; Shang, M.; Mu, K.; Jiang, N.; Wen, H.; Wang, R.; Wu, H.; Li, W. In vitro and in vivo toxic effects and inflammatory responses induced by carboxylated black carbon-lead complex exposure. *Ecotoxicol. Environ. Saf.* **2018**, *165*, 484–494. [CrossRef]
210. Komatsu, T.; Tabata, M.; Kubo-Irie, M.; Shimizu, T.; Suzuki, K.; Nihei, Y.; Takeda, K. The effects of nanoparticles on mouse testis Leydig cells in vitro. *Toxicol. Vitr.* **2008**, *22*, 1825–1831. [CrossRef]
211. Belade, E.; Armand, L.; Martinon, L.; Kheuang, L.; Fleury-Feith, J.; Baeza-Squiban, A.; Lanone, S.; Billon-Galland, M.-A.; Pairon, J.-C.; Boczkowski, J. A comparative transmission electron microscopy study of titanium dioxide and carbon black nanoparticles uptake in human lung epithelial and fibroblast cell lines. *Toxicol. Vitr.* **2012**, *26*, 57–66. [CrossRef]
212. Li, L.; Mak, K.Y.; Shi, J.; Koon, H.K.; Leung, C.H.; Wong, C.M.; Leung, C.W.; Mak, C.S.K.; Chan, N.M.M.; Zhong, W.; et al. Comparative In Vitro Cytotoxicity Study on Uncoated Magnetic Nanoparticles: Effects on Cell Viability, Cell Morphology, and Cellular Uptake. *J. Nanosci. Nanotechnol.* **2012**, *12*, 9010–9017. [CrossRef]
213. Malatesta, M. Transmission Electron Microscopy as a Powerful Tool to Investigate the Interaction of Nanoparticles with Subcellular Structures. *Int. J. Mol. Sci.* **2021**, *22*, 12789. [CrossRef]
214. Zhang, R.; Dai, Y.; Zhang, X.; Niu, Y.; Meng, T.; Li, Y.; Duan, H.; Bin, P.; Ye, M.; Jia, X.; et al. Reduced pulmonary function and increased pro-inflammatory cytokines in nanoscale carbon black-exposed workers. *Part. Fibre Toxicol.* **2014**, *11*, 73. [CrossRef]
215. Bozzola, J.J.; Russell, L.D. *Electron Microscopy: Principles and Techniques for Biologists*; Jones & Bartlett Learning: Burlington, MA, USA, 1999; ISBN 0763701920.
216. Hayat, M.A. *Principles and Techniques of Electron Microscopy: Biological Applications*; Macmillan Publishers: New York, NY, USA, 1974; pp. 229–231. [CrossRef]
217. Mühlfeld, C.; Rothen-Rutishauser, B.; Vanhecke, D.; Blank, F.; Gehr, P.; Ochs, M. Visualization and quantitative analysis of nanoparticles in the respiratory tract by transmission electron microscopy. *Part. Fibre Toxicol.* **2007**, *4*, 11. [CrossRef]
218. Parent, P.; Laffon, C.; Marhaba, I.; Ferry, D.; Regier, T.Z.; Ortega, I.K.; Chazallon, B.; Carpentier, Y.; Focsa, C. Nanoscale characterization of aircraft soot: A high-resolution transmission electron microscopy, Raman spectroscopy, X-ray photoelectron and near-edge X-ray absorption spectroscopy study. *Carbon N. Y.* **2016**, *101*, 86–100. [CrossRef]
219. Wang, G.; Fang, N. Detecting and Tracking Nonfluorescent Nanoparticle Probes in Live Cells. *Methods Enzymol.* **2012**, *504*, 83–108. [CrossRef]
220. Murphy, D.B. *Fundamentals of Light Microscopy and Electronic Imaging*; John Wiley & Sons: Hoboken, NJ, USA, 2002; ISBN 047123429X.
221. Priest, L.; Peters, J.S.; Kukura, P. Scattering-based Light Microscopy: From Metal Nanoparticles to Single Proteins. *Chem. Rev.* **2021**, *121*, 11937–11970. [CrossRef]
222. Wayne, R. *Light and Video Microscopy*; Elsevier: Amsterdam, The Netherlands, 2013; ISBN 9780124114845.
223. You, R.; Lu, W.; Shan, M.; Berlin, J.M.; Samuel, E.L.; Marcano, D.C.; Sun, Z.; Sikkema, W.K.; Yuan, X.; Song, L.; et al. Nanoparticulate carbon black in cigarette smoke induces DNA cleavage and Th17-mediated emphysema. *elife* **2015**, *4*, e09623. [CrossRef]
224. Modrzynska, J.; Berthing, T.; Ravn-Haren, G.; Jacobsen, N.R.; Weydahl, I.K.; Loeschner, K.; Mortensen, A.; Saber, A.T.; Vogel, U. Primary genotoxicity in the liver following pulmonary exposure to carbon black nanoparticles in mice. *Part. Fibre Toxicol.* **2018**, *15*, 2. [CrossRef]
225. Pedata, P.; Stoeger, T.; Zimmermann, R.; Peters, A.; Oberdörster, G.; D'Anna, A. Are we forgetting the smallest sub 10 nm combustion generated nanoparticles. *Part. Fibre Toxicol.* **2015**, *12*, 34. [CrossRef] [PubMed]
226. Sze, S.K.; Siddique, N.; Sloan, J.J.; Escribano, R. Raman spectroscopic characterization of carbonaceous aerosols. *Atmos. Environ.* **2001**, *35*, 561–568. [CrossRef]
227. Beyssac, O.; Goffé, B.; Petitet, J.-P.; Froigneux, E.; Moreau, M.; Rouzaud, J.-N. On the characterization of disordered and heterogeneous carbonaceous materials by Raman spectroscopy. *Spectrochim. Acta Part A Mol. Biomol. Spectrosc.* **2003**, *59*, 2267–2276. [CrossRef]
228. Smith, E.; Dent, G. *Modern Raman Spectroscopy—A Practical Approach*; John Wiley & Sons, Ltd: Chichester, UK, 2004; ISBN 9780470011836.
229. Bokobza, L.; Bruneel, J.-L.; Couzi, M. Raman spectroscopy as a tool for the analysis of carbon-based materials (highly oriented pyrolitic graphite, multilayer graphene and multiwall carbon nanotubes) and of some of their elastomeric composites. *Vib. Spectrosc.* **2014**, *74*, 57–63. [CrossRef]
230. Dippel, B.; Heintzenberg, J. Soot characterization in atmospheric particles from different sources by NIR FT Raman spectroscopy. *J. Aerosol Sci.* **1999**, *30*, S907–S908. [CrossRef]
231. Gruber, T.; Zerda, T.W.; Gerspacher, M. Raman studies of heat-treated carbon blacks. *Carbon N. Y.* **1994**, *32*, 1377–1382. [CrossRef]
232. Sadezky, A.; Muckenhuber, H.; Grothe, H.; Niessner, R.; Pöschl, U. Raman microspectroscopy of soot and related carbonaceous materials: Spectral analysis and structural information. *Carbon N. Y.* **2005**, *43*, 1731–1742. [CrossRef]
233. Wang, Y.; Alsmeyer, D.C.; McCreery, R.L. Raman spectroscopy of carbon materials: Structural basis of observed spectra. *Chem. Mater.* **1990**, *2*, 557–563. [CrossRef]
234. Dippel, B.; Jander, H.; Heintzenberg, J. NIR FT Raman spectroscopic study of flame soot. *Phys. Chem. Chem. Phys.* **1999**, *1*, 4707–4712. [CrossRef]
235. Cuesta, A.; Dhamelincourt, P.; Laureyns, J.; Martínez-Alonso, A.; Tascón, J.M.D. Raman microprobe studies on carbon materials. *Carbon N. Y.* **1994**, *32*, 1523–1532. [CrossRef]

236. Knief, P.; Clarke, C.; Herzog, E.; Davoren, M.; Lyng, F.M.; Meade, A.D.; Byrne, H.J. Raman spectroscopy—A potential platform for the rapid measurement of carbon nanotube-induced cytotoxicity. *Analyst* **2009**, *134*, 1182. [CrossRef]
237. Čabanová, K.; Motyka, O.; Čábalová, L.; Hrabovská, K.; Bielniková, H.; Kuzníková, L'.; Dvořáčková, J.; Zeleník, K.; Komínek, P.; Kukutschová, J. Metal particles in mucus and hypertrophic tissue of the inferior nasal turbinates from the human upper respiratory tract. *Environ. Sci. Pollut. Res.* **2020**, *27*, 28146–28154. [CrossRef]
238. Čabanová, K.; Motyka, O.; Bielniková, H.; Čábalová, L.; Handlos, P.; Zabiegaj, D.; Zeleník, K.; Dvořáčková, J.; Komínek, P.; Heviánková, S.; et al. Identification of the phase composition of solid microparticles in the nasal mucosa of patients with chronic hypertrophic rhinitis using Raman microspectroscopy. *Sci. Rep.* **2021**, *11*, 18989. [CrossRef]
239. Fischer, M.C.; Wilson, J.W.; Robles, F.E.; Warren, W.S. Invited Review Article: Pump-probe microscopy. *Rev. Sci. Instrum.* **2016**, *87*, 031101. [CrossRef]
240. Tong, L.; Liu, Y.; Dolash, B.D.; Jung, Y.; Slipchenko, M.N.; Bergstrom, D.E.; Cheng, J.-X. Label-free imaging of semiconducting and metallic carbon nanotubes in cells and mice using transient absorption microscopy. *Nat. Nanotechnol.* **2012**, *7*, 56–61. [CrossRef]
241. Frank, H.A.; Cogdell, R.J. 8.6 Light Capture in Photosynthesis. In *Comprehensive Biophysics*; Elsevier: Amsterdam, The Netherlands, 2012; pp. 94–114.
242. Hartland, G.V. Ultrafast studies of single semiconductor and metal nanostructures through transient absorption microscopy. *Chem. Sci.* **2010**, *1*, 303–309. [CrossRef]
243. Chong, S.; Min, W.; Xie, X.S. Ground-State Depletion Microscopy: Detection Sensitivity of Single-Molecule Optical Absorption at Room Temperature. *J. Phys. Chem. Lett.* **2010**, *1*, 3316–3322. [CrossRef]
244. Gaiduk, A.; Ruijgrok, P.V.; Yorulmaz, M.; Orrit, M. Detection limits in photothermal microscopy. *Chem. Sci.* **2010**, *1*, 343–350. [CrossRef]
245. Steuwe, C.; Bové, H.; Clerinx, J.; VandeVen, M.; Fron, E.; Nawrot, T.; Ameloot, M.; Roeffaers, M. Rapid and label-free optical detection of individual carbon air pollutant nanoparticulates in biomedical samples. *J. Biophotonics* **2018**, *11*, e201700233. [CrossRef]
246. Jiang, J.; Grass, D.; Zhou, Y.; Warren, W.S.; Fischer, M.C. Beyond intensity modulation: New approaches to pump-probe microscopy. *Opt. Lett.* **2021**, *46*, 1474. [CrossRef]
247. Miyazaki, J.; Tsurui, H.; Hayashi-Takagi, A.; Kasai, H.; Kobayashi, T. Sub-diffraction resolution pump-probe microscopy with shot-noise limited sensitivity using laser diodes. *Opt. Express* **2014**, *22*, 9024. [CrossRef] [PubMed]
248. Miyazaki, J.; Kawasumi, K.; Kobayashi, T. Resolution improvement in laser diode-based pump–probe microscopy with an annular pupil filter. *Opt. Lett.* **2014**, *39*, 4219. [CrossRef] [PubMed]
249. Bové, H.; Steuwe, C.; Fron, E.; Slenders, E.; D'Haen, J.; Fujita, Y.; Uji-I, H.; Vandeven, M.; Roeffaers, M.; Ameloot, M. Biocompatible Label-Free Detection of Carbon Black Particles by Femtosecond Pulsed Laser Microscopy. *Nano Lett.* **2016**, *16*, 3173–3178. [CrossRef] [PubMed]
250. Larson, A.M. Multiphoton microscopy. *Nat. Photonics* **2011**, *5*, 1. [CrossRef]
251. Lévêque-Fort, S.; Georges, P. MICROSCOPY | Nonlinear Microscopy. In *Encyclopedia of Modern Optics*; Elsevier: Amsterdam, The Netherlands, 2005; pp. 92–103.
252. Li, D.; Zheng, W.; Qu, J.Y. Two-photon autofluorescence microscopy of multicolor excitation. *Opt. Lett.* **2009**, *34*, 202–204. [CrossRef]
253. Chen, X.; Nadiarynkh, O.; Plotnikov, S.; Campagnola, P.J. Second harmonic generation microscopy for quantitative analysis of collagen fibrillar structure. *Nat. Protoc.* **2012**, *7*, 654–669. [CrossRef]
254. Borile, G.; Sandrin, D.; Filippi, A.; Anderson, K.I.; Romanato, F. Label-free multiphoton microscopy: Much more than fancy images. *Int. J. Mol. Sci.* **2021**, *22*, 2657. [CrossRef]
255. Kobat, D.; Durst, M.E.; Nishimura, N.; Wong, A.W.; Schaffer, C.B.; Xu, C. Deep tissue multiphoton microscopy using longer wavelength excitation. *Opt. Express* **2009**, *17*, 13354–13364. [CrossRef]
256. Aslam, I.; Roeffaers, M.B.J. Unique emissive behavior of combustion-derived particles under illumination with femtosecond pulsed near-infrared laser light. *Nanoscale Adv.* **2021**, *3*, 5355–5362. [CrossRef]
257. Saenen, N.D.; Bové, H.; Steuwe, C.; Roeffaers, M.B.J.; Provost, E.B.; Lefebvre, W.; Vanpoucke, C.; Ameloot, M.; Nawrot, T.S. Children's urinary environmental carbon load: A novel marker reflecting residential ambient air pollution exposure? *Am. J. Respir. Crit. Care Med.* **2017**, *196*, 873–881. [CrossRef]
258. Bové, H.; Bongaerts, E.; Slenders, E.; Bijnens, E.M.; Saenen, N.D.; Gyselaers, W.; Van Eyken, P.; Plusquin, M.; Roeffaers, M.B.J.; Ameloot, M.; et al. Ambient black carbon particles reach the fetal side of human placenta. *Nat. Commun.* **2019**, *10*, 3866. [CrossRef]
259. Witters, K.; Plusquin, M.; Slenders, E.; Aslam, I.; Ameloot, M.; Roeffaers, M.B.J.; Vangronsveld, J.; Nawrot, T.S.; Bové, H. Monitoring indoor exposure to combustion-derived particles using plants. *Environ. Pollut.* **2020**, *266*, 115261. [CrossRef]
260. Aslam, I.; Fron, E.; Roeffaers, M.B.J. Label-free detection and size estimation of combustion-derived carbonaceous particles in a microfluidic approach. *Nanoscale Adv.* **2022**, *4*, 3272–3281. [CrossRef]
261. Bongaerts, E.; Lecante, L.L.; Bové, H.; Roeffaers, M.B.J.; Ameloot, M.; Fowler, P.A.; Nawrot, T.S. Maternal exposure to ambient black carbon particles and their presence in maternal and fetal circulation and organs: An analysis of two independent population-based observational studies. *Lancet Planet. Health* **2022**, *6*, e804–e811. [CrossRef]
262. Kim, K.-H.; Kabir, E.; Kabir, S. A review on the human health impact of airborne particulate matter. *Environ. Int.* **2015**, *74*, 136–143. [CrossRef]

nanomaterials

Article

Comparative Study on Nanotoxicity in Human Primary and Cancer Cells

In Young Kim, Minjeong Kwak, Jaeseok Kim, Tae Geol Lee and Min Beom Heo *

Nano-Safety Team, Safety Measurement Institute, Korea Research Institute of Standards and Science (KRISS), Yuseong-gu, Daejeon 34113, Korea; inyoungkim@kriss.re.kr (I.Y.K.); kwakmj@kriss.re.kr (M.K.); jaeseok.kim@kriss.re.kr (J.K.); tglee@kriss.re.kr (T.G.L.)
* Correspondence: mbheo@kriss.re.kr; Tel.: +82-42-6041-1052

Abstract: Nanomaterial toxicity tests using normal and cancer cells may yield markedly different results. Here, nanomaterial toxicity between cancer and primary human cells was compared to determine the basic cell line selection criteria for nanomaterial toxicity analyses. Specifically, we exposed two cancer (A549 and HepG2) and two normal cell lines (NHBE and HH) cell lines to SiO_2 nanoparticles (NPs) and evaluated the cytotoxicity (MTS assay), cell death mode, and intracellular NP retention. MTS assay results revealed higher sensitivity of HH cells to SiO_2 NPs than HepG2 cells, while no difference was observed between NHBE and A549 cells. In addition, SiO_2 NPs primarily induced necrosis in all the cell lines. Moreover, we evaluated NP accumulation by treating the cell lines with fluorescein-isothiocyanate-labeled SiO_2 NPs. After 48 h of treatment, less than 10% of A549 and HepG2 cells and more than 30% of NHBE and HH cells contained the labeled NPs. Collectively, our results suggest that cell viability, death mode, and intracellular compound accumulation could be assessed using cancer cells. However, the outcomes of certain investigations, such as intracellular NP retention, may differ between cancer and normal cells.

Keywords: nanotoxicity; nanomaterial; cancer cells; silica nanoparticles

1. Introduction

Nanotechnology has become a valuable and effective essential technology across several fields in recent years. Various nanomaterials, with sizes smaller than those of human cells, are widely used in cosmetics [1,2], sunscreens [3], food packaging [4], and pharmaceuticals [5]. However, their small size can make them toxic, posing risks to human health and safety [6–8].

As one of the most-produced nanoparticles (NPs) globally, silica NPs (SiNPs) are used in all aspects of life, including agriculture, food, and consumer goods [9]. In addition, the biocompatibility and stability of SiNPs make them promising candidates in various biomedical fields, such as gene carriers, drug delivery, and molecular imaging [10,11]. However, previous studies have reported on the toxic effects of SiNPs on essential organs such as the liver, lungs, brain, and kidneys [12,13]. Thus, extensive use has increased the risk of SiNPs exposure in humans.

Therefore, it is essential to investigate the safety of nanomaterials. Traditionally, toxicity studies using animals have been conducted to estimate the adverse effects of chemicals and nanomaterials on humans. However, many countries, particularly those in Europe, have recently begun regulating and limiting animal studies for ethical reasons; therefore, other test methods that estimate chemical or nanomaterial toxicity reliably are required [14]. Accordingly, human cell models that reduce unnecessary animal sacrifice and lower costs have been established. Presently, nanomaterial toxicity is being investigated using human cancer cells or immortalized cells [15,16]. Owing to their low cost, ease of handling, and excellent reproducibility, human cancer cells have considerable advantages in the rapid toxicity screening of nanomaterials or chemicals over other cells. However, unlike normal cells, cancer cells have unique characteristics, such as proliferating indefinitely due to excessive cell division from lack of cell cycle control and the apoptosis function [17]. In addition, cancer cells acquire energy by metabolizing glucose. The glycoprotein structure expressed on their membrane surface significantly differs from those on normal cells [18,19], resulting in studies raising concerns regarding the misjudgment of toxicity outcomes. Studies have associated nanomaterial toxicity with the cell lines used [20,21]. Therefore, results can be entirely different from those of normal cells when analyzing cell viability and the mechanism underlying the toxicity of nanomaterials using cancer cells [22].

Here, we compared the differences in nanomaterial toxicity between cancer and primary human cells. The liver and lungs, most affected by toxic substances, were the target organs in the current study. We used A549 (human lung cancer cell line) and NHBE cells (normal human bronchial epithelial cell line) as the lung model cells, and HepG2 (human liver cancer cell line) and HH cells (normal human hepatocyte line) as the liver model cells. Our goal was to aid the selection of cell lines for nanomaterial toxicity studies by comparing the toxicity test results and confirming differences in modes of cell death caused by SiO_2 NPs using different target organ cell lines.

2. Materials and Methods

2.1. Materials and Reagents

For the experiments, we used 20 nm SiO_2 NPs, which are certified reference materials (CRM, 301-01-002) produced by the Korea Research Institute of Standards and Science (Daejeon, Korea). Cadmium sulfate ($CdSO_4$, #383082) and stauroporine (STS, #S5921) were purchased from Sigma-Aldrich (St. Louis, MO, USA).

2.2. Preparation and Characterization of SiO_2 NPs

The SiO_2 NPs were dispersed in deionized water or the Dulbecco's modified Eagle medium (DMEM; #LM001-05, Welgene, Gyeongsan, South Korea) via vortexing for 1 min. The morphology and size of the primary SiO_2 NPs were analyzed using transmission electron microscopy (TEM; JEM-ARM200F, JEOL Ltd., Tokyo, Japan) and scanning electron microscopy (SEM; Gemini SEM 500, Carl Zeiss, Oberkochen, Germany). The size distribution of the SiO_2 NPs was measured using dynamic light scattering (DLS; Nano ZS90, Malvern Panalytical, Worcestershire, UK) and scanning mobility particle sizer (SMPS; TSI Incorporated, Shoreview, MN, USA). The zeta-potential was measured using the Zetasizer instrument (Nano ZS90, Malvern, UK).

2.3. Cell Culture

A549 and HepG2 cells, purchased from the American Type Culture Collection (Manassas, VA, USA), were cultured in DMEM supplemented with 10% fetal bovine serum (FBS; #SH30084.03, HyClone™, Marlborough, MA, USA) and 1% penicillin-streptomycin (#LS202-02, Welgene™). NHBE cells were purchased from Lonza™ (Alps, Switzerland). These cells were cultured in bronchial epithelial cell growth basal medium (#CC-3171, Lonza™) containing a supplement pack (#CC-4175, Lonza™). HH cells purchased from ScienCell™ (Carlsbad, CA, USA) were cultured in hepatocyte medium (#5201, ScienCell™) containing 5% FBS (#0025, ScienCell™), 1% penicillin-streptomycin (#0503, ScienCell™), and hepatocyte growth supplement (#5252, ScienCell™). All cells were incubated at 37 °C in a humidified atmosphere containing 5% CO_2. All cells were expanded after passage number three and then frozen in liquid nitrogen. After thawing, cells having passage numbers between 4 and 7 were used for the toxicity test.

2.4. Cytotoxicity Assay and Statistical Analysis

All cells were seeded at a density of 1.0×10^4 cells/well onto a 96-well plate (#3596, Corning®, Glendale, AZ, USA) and incubated for 24 h. After washing the cells with phosphate-buffered saline (PBS; LB001-02, Welgene™), they were treated with the prepared SiO_2 NPs, dispersed in serum-free DMEM, for 4 h. Cytotoxicity was quantified using

MTS assay (CellTiter96® AQueous One Solution Cell Proliferation Assay kit, Promega, Madison, WI, USA) according to the manufacturer's instructions. The 96-well plates were then incubated at 37 °C in a humidified atmosphere containing 5% CO_2 for 1 h. Cell viability was calculated using Equation (1). IC_{50} values of the NPs were calculated using the SoftMax Pro software (Molecular Devices, San Jose, CA, USA).

$$\text{Cell viability } (\%) = \frac{\text{Absorbance value of test cells}}{\text{Absorvance value of control cells}} \times 100 \quad (1)$$

The results of all experiments were statistically analyzed using GraphPad Prism software (version 7.0; GraphPad Software Inc., San Diego, CA, USA) and represented as the mean and standard error of independent experiments (n = 12). The normality of the data was assessed using the Kolmogorov–Smirnov test, and equal variance using Bartlett's test. For normally distributed data, statistical differences were determined using variance analysis followed by Bonferroni's multiple comparison test; else, the Kruskal–Wallis test was performed, followed by Dunn's test.

2.5. *Investigation of Cell Death Mode*

An Annexin V–Fluorescein Isothiocyanate (FITC) Apoptosis Detection kit (#130-092-052, Miltenyi Biotec, Bergisch Gladbach, Germany) was used to determine apoptosis according to the manufacturer's instructions. Using this kit, we washed the cells twice with 1 × binding buffer and then resuspended in 100 µL of 1 × binding buffer per 1.0 × 10^6 cells. Next, the cells were incubated with 10 µL of Annexin V-FITC in the dark at room temperature for 15 min and washed with 1 mL of 1 × binding buffer. Thereafter, the cells were stained with propidium iodide (PI), before resuspending in 500 µL of 1 × binding buffer per 1.0 × 10^6 cells. We added 5 µL of PI solution prior to analysis by fluorescence microscopy or flow cytometry. Fluorescence microscopy was performed using a confocal laser scanning microscope (Olympus, Tokyo, Japan), and flow cytometry was performed using Eclipse™ (Sony Biotechnology, San Jose, CA, USA).

2.6. *Visualization of FITC-Labeled 20 nm SiO_2 NPs Using Fluorescence Microscopy*

Cells (5.0 × 10^4 cells/well), seeded on a coverslip and incubated for 48 h, were treated with FITC-labeled 20 nm SiO_2 NPs for 4 h and fixed with 4% paraformaldehyde for 20 min. Next, the cells were washed twice with PBS and observed using fluorescence microscopy (Leica Microsystems, Wetzlar, Germany).

2.7. *FITC-Labeled 20 nm SiO_2 NPs Uptake Analysis Using Flow Cytometry*

For uptake analysis, cells (2.5 × 10^5 cells/well) were first seeded on 6 well plates and incubated. The next day, FITC-labeled 20 nm SiO_2 NPs were co-cultured with the cells for 4 h. The cells were washed twice with PBS and analyzed using flow cytometry. The live single-cell population was gated in a plot of FSC versus SSC after excluding cell FITC-negative cell population was gated in a plot of FSC versus FITC in the untreated sample. In all samples, the percentage of FITC-positive cells in a single-cell population decided the intracellular localization of FITC-SiO_2 NPs.

3. Results

3.1. *Characterization of SiO_2 NPs*

For characterization of SiO_2 NPs in deionized water, their morphology and size were investigated using electron microscopy. SiO_2 NPs exhibit uniform spherical morphology and size distribution in SEM (Figure 1a) and TEM (Figure 1b) images. DLS (Figure 1c) and SMPS (Figure 1d) determined the hydrodynamic sizes of SiO_2 NPs, which were 19.2 and 21.7 nm, respectively. Since NPs can have significant variations in size distribution in biological environments, we compared SiO_2 NPs in deionized water and serum-free DMEM. Unexpectedly, z-average and polydispersity index are similar in the two dispersions (Figure 1e). In addition, the zeta-potential show negative charges in both conditions (Figure 1e). These results indicate that SiO_2 NPs are monodispersed without aggregation.

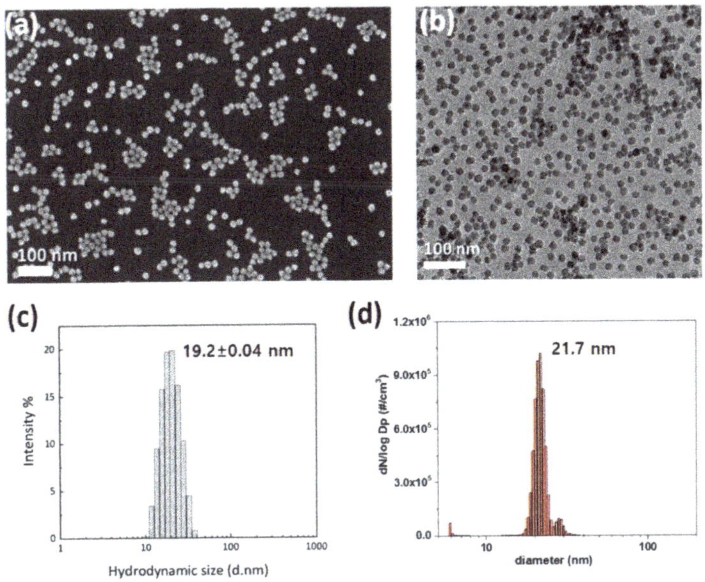

(e)

	Water	Serum-free DMEM
z-average (d, nm)	19.2 ± 0.04	22.2 ± 0.1
Polydispersity index	0.17	0.14
Z-potential (mV)	- 41.1 ± 0.67	- 21.1 ± 1.25

Figure 1. Characterization of 20 nm SiO_2 NPs: (**a**) Scanning electron microscopy image; (**b**) Transmission electron microscopy image; (**c**) Dynamic light scattering size analysis; (**d**) Scanning mobility particle sizer analysis; (**e**) Performance comparison in deionized water and serum-free DMEM.

3.2. Cytotoxicity of SiO_2 NPs

Our study compared SiO2 NPs toxicity between cancer (A549 and HepG2) and normal cell lines (NHBE and HH), MTS assay results revealed that the viability of all four cell lines significantly decreased depending on the concentration of SiO_2 NPs in the cells (Figure 2). Notably, IC_{50} values of SiO_2 NPs were almost similar between the lung-derived A549 and NHBE cells (Figure 2a), suggesting that there was no difference in toxicity caused by SiO_2 NPs between the two cell lines. However, in the liver-derived cell lines, the IC_{50} values of SiO_2 NPs in normal HH cells were approximately 1.5-fold higher than that in HepG2 cells, suggesting that the higher sensitivity of HH cells to SiO_2 NPs was higher than that of HepG2 cells (Figure 2b).

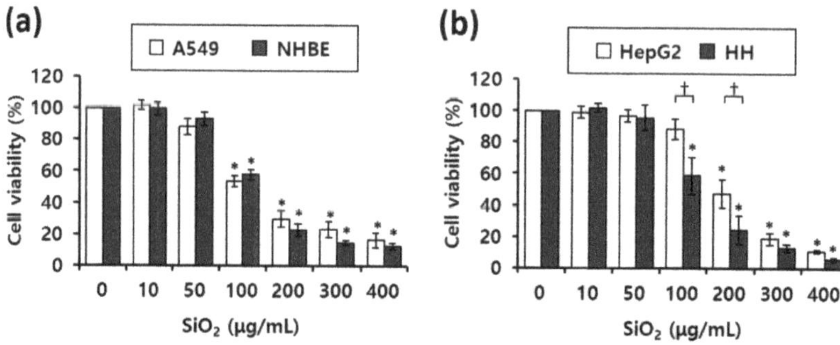

(μg/mL)	A549	NHBE
IC_{50}	111.0 ± 9.7	113.0 ± 6.8
IC_{20}	60.1 ± 6.6	69.9 ± 5.3

	HepG2	HH
IC_{50}	49.1 ± 6.2	29.5 ± 6.1
IC_{20}	30.6 ± 4.6	17.6 ± 4.2

Figure 2. MTS cytotoxicity assay results showing the different inhibitory concentrations of 20 nm SiO_2 NPs against cells subjected to NP treatment for 4 h (total number of replicates = 12); (**a**) Lung-derived cells. * $p < 0.005$, compared with untreated cells. (**b**) Liver-derived cells. * $p < 0.005$, compared with untreated cells; † $p < 0.05$, compared with HepG2 cells.

3.3. Cell Death Mode

Annexin V/PI staining was performed to determine differences in the mode of death caused by SiO_2 NPs between cancer and normal cells. STS was used as a positive control for apoptosis, and $CdSO_4$ for necrosis. In distinguishing the mode of cell death, Annexin V(+)/PI(−) cell populations were considered apoptotic, while Annexin V(−)/PI(+) or Annexin V(+)/PI(+) cell populations were considered necrotic. Fluorescence microscopy shows that the STS treatment increased green fluorescence intensity, while $CdSO_4$ treatment increased green and red fluorescence intensities in the lung-derived cell lines, A549 and NHBE (Figure 3a,d). Green and red fluorescence intensities were also significantly increased by SiO_2 NP treatment. Next, the mode of cell death was analyzed by sorting cells according to fluorescence using flow cytometry. STS primarily induced apoptosis, while $CdSO_4$ and SiO_2 NPs induced necrosis in A549 cells (Figure 3b,c). In NHBE cells, STS induced a mixture of apoptosis and necrosis, while $CdSO_4$ primarily induced necrosis. SiO_2 NPs induced necrosis in NHBE cells (Figure 3e,f). These results suggest that the mode of death induced by SiO_2 NPs was similar in A549 and NHBE cells. Similar to the lung-derived cell lines, STS treatment increased green fluorescence intensity, while $CdSO_4$ and SiO_2 NPs increased green/red fluorescence intensity in HepG2 and HH cell lines (Figure 4a,d). Flow cytometry analysis revealed that STS induced both apoptosis and necrosis, while $CdSO_4$ mostly induced necrosis (Figure 4b,c,e,f). In contrast, SiO_2 NP treatment mostly induced necrosis in HepG2 cells and apoptosis and necrosis in HH cells (Figure 4b,c,e,f). These results suggest that the mode of death induced by SiO_2 NPs in HepG2 and HH cells differ partially.

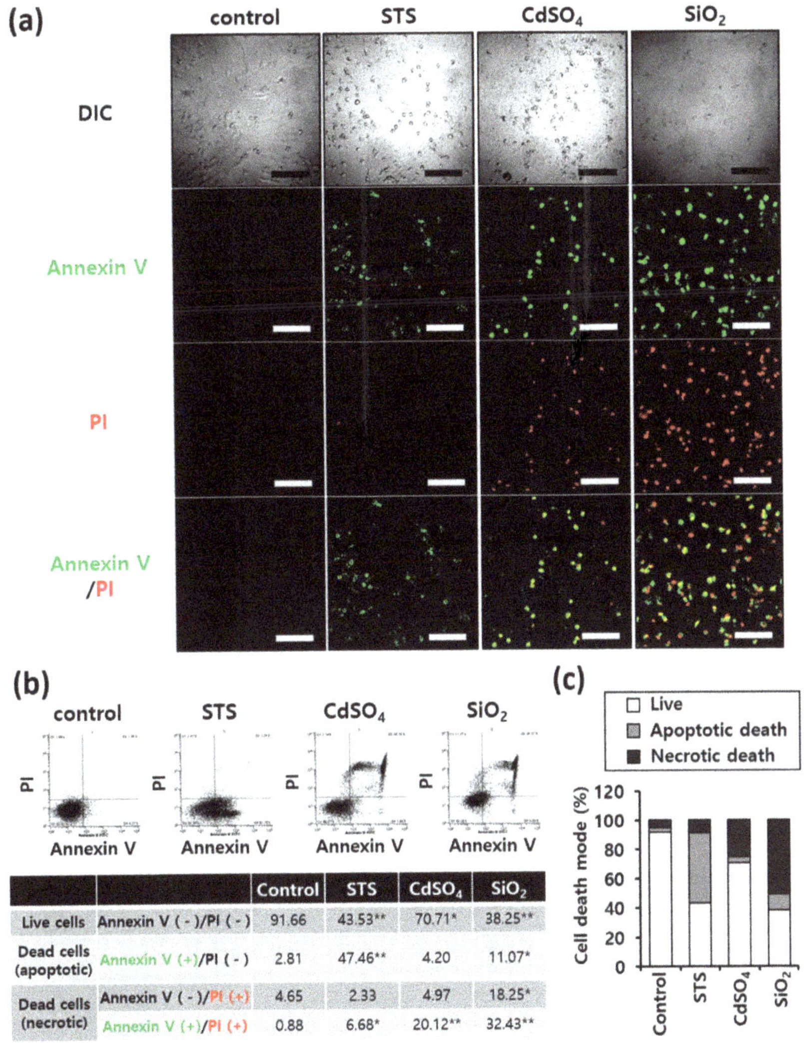

		Control	STS	$CdSO_4$	SiO_2
Live cells	Annexin V (-)/PI (-)	91.66	43.53**	70.71*	38.25**
Dead cells (apoptotic)	Annexin V (+)/PI (-)	2.81	47.46**	4.20	11.07*
Dead cells (necrotic)	Annexin V (-)/PI (+)	4.65	2.33	4.97	18.25*
	Annexin V (+)/PI (+)	0.88	6.68*	20.12**	32.43**

Figure 3. *Cont.*

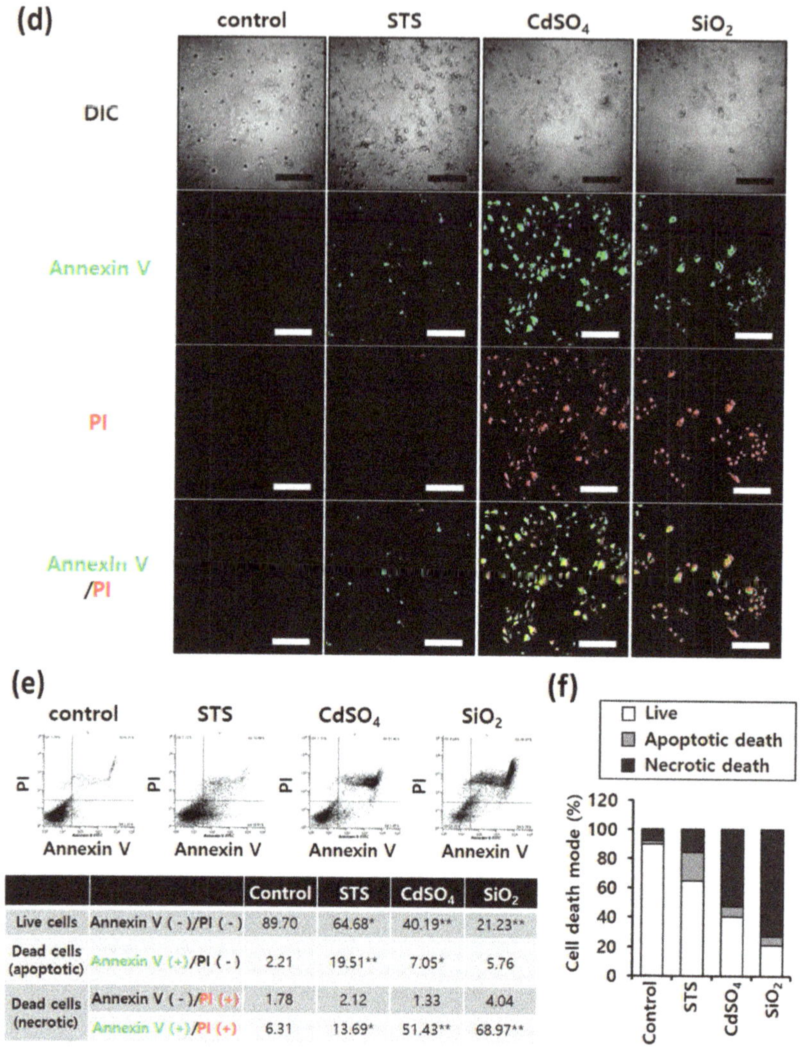

Figure 3. Annexin-V/PI double-staining assay of A549 and NHBE cells; After treating A549 and NHBE cells with SiO_2 NPs (at IC50) and two positive controls, the cells were stained with annexin V-fluorescein isothiocyanate and propidium iodide and analyzed by fluorescence microscopy and flow cytometry. $CdSO_4$ (1.0 mM) was used to induce necrosis, and staurosporine (STS; 1.0 μM) was used to induce apoptosis; replicate number = 3. (**a**,**d**) Confocal fluorescence microscopy images; scale bars represent 200 μm. (**b**,**e**) Flow cytometry analysis. * $p < 0.05$, compared with control; ** $p < 0.005$, compared with control. (**c**,**f**) Percentage distribution of necrotic, apoptotic, and viable cells.

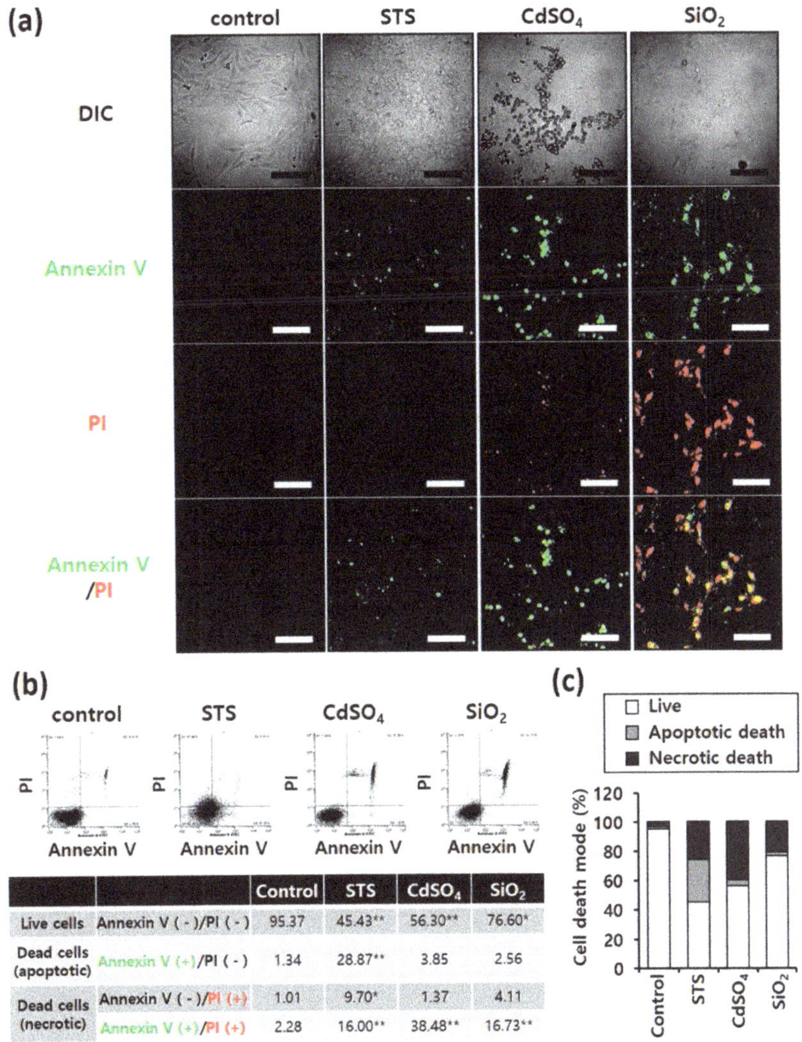

		Control	STS	$CdSO_4$	SiO_2
Live cells	Annexin V (-)/PI (-)	95.37	45.43**	56.30**	76.60*
Dead cells (apoptotic)	Annexin V (+)/PI (-)	1.34	28.87**	3.85	2.56
Dead cells (necrotic)	Annexin V (-)/PI (+)	1.01	9.70*	1.37	4.11
	Annexin V (+)/PI (+)	2.28	16.00**	38.48**	16.73**

Figure 4. *Cont.*

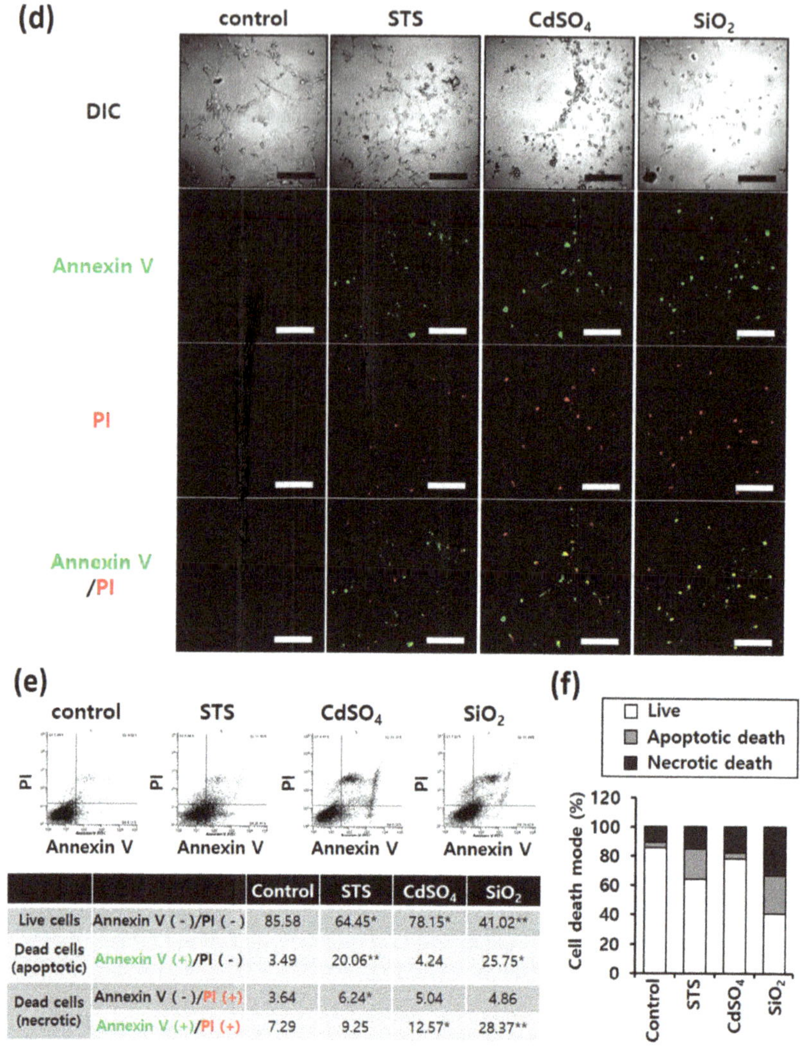

		Control	STS	$CdSO_4$	SiO_2
Live cells	Annexin V (-)/PI (-)	85.58	64.45*	78.15*	41.02**
Dead cells (apoptotic)	Annexin V (+)/PI (-)	3.49	20.06**	4.24	25.75*
Dead cells (necrotic)	Annexin V (-)/PI (+)	3.64	6.24*	5.04	4.86
	Annexin V (+)/PI (+)	7.29	9.25	12.57*	28.37**

Figure 4. Annexin-V/PI double-staining assay of HepG2 and HH cells; After treating HepG2 and HH cells with SiO_2 NPs (at IC50) and two positive controls, the cells were stained with annexin V-fluorescein isothiocyanate and propidium iodide and analyzed by fluorescence microscopy and flow cytometry. $CdSO_4$ (1.0 mM) was used to induce necrosis, and staurosporine (STS; 1.0 μM) was used to induce apoptosis; replicate number = 3. (**a**,**d**) Confocal fluorescence microscopy images; scale bars represent 200 μm. (**b**,**e**) Flow cytometry analysis. * $p < 0.05$, compared with control; ** $p < 0.005$, compared with control. (**c**,**f**) Percentage distribution of necrotic, apoptotic, and viable cells.

3.4. Examination of SiO_2 NPs Retention in Cells

Because intracellular nanomaterials can majorly cause cytotoxicity, the intracellular accumulation of SiO_2 NPs was confirmed. First, confocal microscopy revealed FITC-labeled SiO_2 NPs in the four cell lines when treated with these NPs (Figure 5a). In addition, the proportion of cells containing FITC-labeled SiO_2 NPs, determined by flow cytometry, revealed that more than 95% of all A549, NHBE, HepG2, and HH cells contained FITC-labeled SiO_2 NPs (Figure 5b). We conclude that SiO_2 NPs were internalized in the four cell lines without difficulty. For analysis of residual NPs, the culture medium was changed after 4 h of FITC-labeled SiO_2 NP treatment (Figure 5c). The number of cells containing FITC-labeled SiO_2 NPs was counted using flow cytometry at 0 and 48 h after treatment. After 48 h of treatment, FITC-labeled SiO_2 NPs were less than 10% in A549 and HepG2 cancer cells and more than 30% in NHBE and HH normal cells. Thus, the removal rate of NPs from cancer cells was markedly faster than that from normal cells. Therefore, regardless of their origin, cancer and normal cells differ.

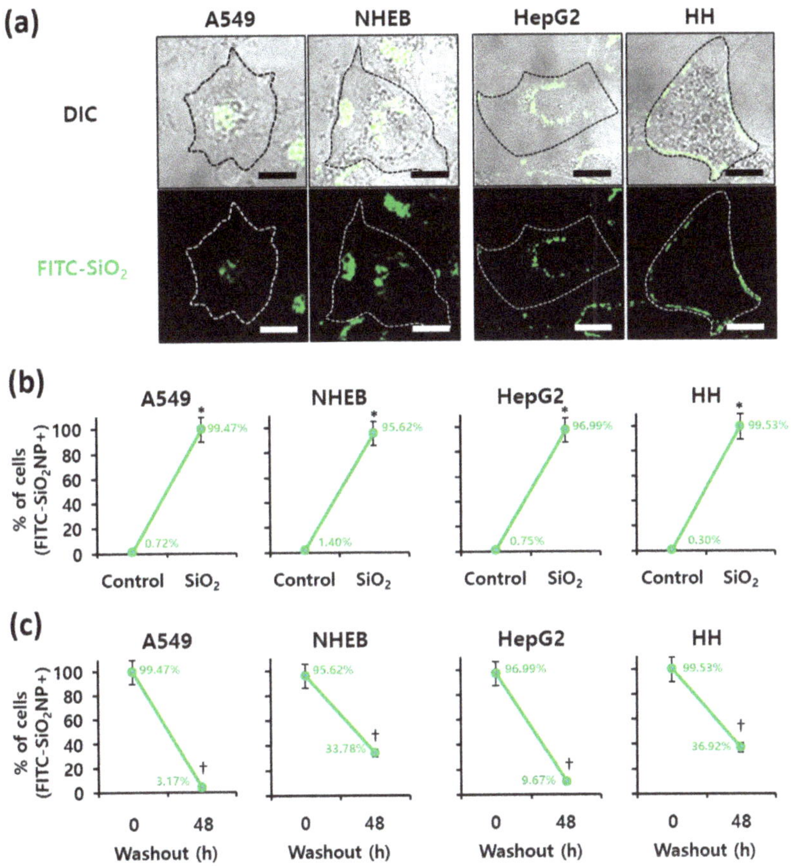

Figure 5. Examination of SiO_2 NPs retention in cells; Localization of fluorescein isothiocyanate (FITC)-labeled 20 nm SiO_2 NPs in each cell line, as determined by (**a**) confocal fluorescence microscopy (scale bars represent 10 μm) and (**b**) flow cytometry after 4 h of treatment. * $p < 0.005$, compared with control. (**c**) Retention of FITC-labeled 20 nm SiO_2 NPs in each cell line, as determined by flow cytometry at 0 and 48 h after 4 h of treatment. † $p < 0.005$, compared with washout for 0 h.

4. Discussion

When screening NPs, selecting the appropriate cell type is critical. Primary cells have several disadvantages, such as limited lifespan, specific culture conditions, and batch-to-batch differences [23,24]. It has been shown that NP-induced effects can vary according to species, tissue, and cell line in many groups. Therefore, cancer cell lines are used as a universal model in nanosafety evaluation. However, uncertainty can be introduced depending on the cell line type and characteristics when generalizing data [25–28]. A few studies have compared normal and cancer cells in the same species and tissue. Ekstrand-

Hammarstrom and colleagues compared the cellular uptake and response of TiO_2 NPs to NHBE (primary), BEAS-2B (human bronchial epithelial cell line, immortalized), and A549 (cancerous). NHBE, BEAS-2B, and A549 cells had similar cell viability and oxidative stress induction, but different proinflammatory responses [29]. Kermanizadeh and co-workers showed similar cellular responses to ZnO, MWCNT, Ag, and TiO_2 NPs in primary human hepatocytes (normal) and C3A cell lines (clonal derivative of HepG2, hepatocellular carcinoma) [30].

This study is the first to compare the cytotoxic effects and cellular localization of SiO_2 NPs in cancer cells (lung-A549, liver-HepG2) and corresponding primary human cells (lung-NHBE, liver-HH). While lung-origin cells had similar sensitivity to SiO_2 NPs (Figures 2 and 3), hepatic-origin had higher HH sensitivity to SiO_2 NPs than HepG2 (Figures 2 and 4). In addition, we confirmed that the degree of intracellular accumulation of SiO_2 NPs was similar in the four cell types (Figure 5). Therefore, we infer no difficulty using cancer cells for short-term cytotoxicity screening of NPs, while probably abandoning the notion of representing all cells.

We showed that the clearance of intracellular SiO_2 NPs was much faster in cancer cells (Figure 5c). The pathway of eliminating intracellular NPs compared to the cellular uptake mechanism is less well known. Intracellular NP concentration can be reduced through (1) apoptosis, (2) cell proliferation, (3) NP diffusion, (4) lysosomal degradation, and (5) exocytosis [31]. Since intracellular NPs fail to reach sufficient concentrations to cause cell death, cell death or proliferation need not be the main mechanism for cellular excretion of NPs [31]. Because cancer cells have a significantly higher proliferation rate than normal cells, the intracellular particle concentration is reduced more efficiently in our study. In addition, the difference in the ability of lysosomes between normal and cancer cells may affect the SiO_2 NP removal rate. Lysosomes remove intracellular NPs through direct degradation [32,33] and exocytosis [34–39]. Many researchers argue that lysosomal stability is necessary to clear intracellular NP [34–36,40]. During transformation into malignancy, cancer cells have enhanced lysosomal functions, such as biosynthesis, hydrolase activity, and exocytosis, to obtain substances necessary for assimilation and catabolism [41]. Therefore, we suggest that cancer cells with improved lysosomal function can reduce the intracellular particle concentration more efficiently than normal cells. In addition, cell viability and response between normal and cancer cells may differ upon long-term exposure to low SiO_2 NP concentration because the accumulated NPs can destroy organelles such as lysosomes and reach the cytoplasm.

Many studies on the toxic effects of SiNPs have been performed using in vivo models. However, awareness of interspecies differences [42–44] and restrictions on animal ethics [EU-Directive 2010/63] drive interest in developing human-based in vitro models. Despite their drawbacks in representing effects on humans or tissues due to their simplicity, they can be a valuable source in evaluating toxicity and mechanical data for drugs, chemicals, and NPs. Therefore, finding a cell line that is more similar to the tissue of origin is important in developing a standardized protocol for nanosafety evaluation. Although the current study cannot represent all cells, it provides a rationale for using cancer cells for short-term nanosafety screening. More research is needed for more accurate results on clinical trials, which requires data accumulation.

5. Conclusions

This study determines the possibility of using cancer cells instead of normal cells to evaluate the safety of nanomaterials in vitro. Viability, mode of death, and intracellular compound accumulation were compared between cancer and normal cells after treating them with SiO_2 NPs. The results suggested that cancer cells can be used to investigate the above parameters, having short evaluation periods. However, long-term evaluations, such as recovery after exposure to NPs, may differ between cancer and normal cells. In conclusion, the standardization of nanomaterial safety evaluation is required in future studies because the effect of NPs may vary depending on the origin of the test cells.

Author Contributions: Conceptualization, I.Y.K. and M.B.H.; Methodology, M.K., J.K. and M.B.H.; Validation, M.B.H.; Data curation, I.Y.K.; Writing—original draft preparation, I.Y.K. and M.B.H.; Writing—review and editing, I.Y.K.; Project administration, M.B.H.; Funding acquisition, T.G.L. and M.B.H. All authors have read and agreed to the published version of the manuscript.

Funding: This research was supported by the Nano Material Technology Development Program [grant number 2016M3A7B6908929, 2021M3C1C3097638] of the National Research Foundation (NRF) of Korea, funded by the Ministry of Science and ICT (MSIT); Development of Measurement Standards and Technology for Biomaterials and Medical Convergence, funded by the Korea Research Institute of Standards and Science [KRISS–2022–GP2022-0006]; and the Industrial Strategic Technology Development Program [20009773, Commercialization of 3D Multifunction Tissue Mimetics Based Drug Evaluation Platform], funded by the Ministry of Trade, Industry & Energy (MOTIE, Korea).

Institutional Review Board Statement: Not applicable.

Informed Consent Statement: Not applicable.

Data Availability Statement: The data presented in this study are available on request from the corresponding author.

Conflicts of Interest: The authors declare no conflict of interest.

References

1. Fytianos, G.; Rahdar, A.; Kyzas, G.Z. Nanomaterials in Cosmetics: Recent Updates. *Nanomaterials* **2020**, *10*, 979–995. [CrossRef]
2. Raj, S.; Jose, S.; Sumod, U.S.; Sabitha, M. Nanotechnology in Cosmetics: Opportunities and Challenges. *J. Pharm. Bioallied Sci.* **2012**, *4*, 186–193. [CrossRef] [PubMed]
3. Nasir, A.; Wang, S.; Friedman, A. The Emerging Role of Nanotechnology in Sunscreens: An Update. *Expert Rev. Dermatol.* **2011**, *6*, 437–439. [CrossRef]
4. Chaudhary, P.; Fatima, F.; Kumar, A. Relevance of Nanomaterials in Food Packaging and Its Advanced Future Prospects. *J. Inorg. Organomet. Polym. Mater.* **2020**, *30*, 5180–5192. [CrossRef] [PubMed]
5. Uddin, I.; Venkatachalam, S.; Mukhopadhyay, A.; Usmani, M.A. Nanomaterials in the Pharmaceuticals: Occurrence, Behaviour and Applications. *Curr. Pharm. Des.* **2016**, *22*, 1472–1484. [CrossRef] [PubMed]
6. Sharifi, S.; Behzadi, S.; Laurent, S.; Forrest, M.L.; Stroeve, P.; Mahmoudi, M. Toxicity of Nanomaterials. *Chem. Soc. Rev.* **2012**, *41*, 2323–2343. [CrossRef] [PubMed]
7. Sukhanova, A.; Bozrova, S.; Sokolov, P.; Berestovoy, M.; Karaulov, A.; Nabiev, I. Dependence of Nanoparticle Toxicity on Their Physical and Chemical Properties. *Nanoscale Res. Lett.* **2018**, *13*, 44. [CrossRef] [PubMed]
8. Hassanpour, P.; Panahi, Y.; Ebrahimi-Kalan, A.; Akbarzadeh, A.; Davaran, S.; Nasibova, A.N.; Khalilov, R.; Kavetskyy, T. Biomedical Applications of Aluminium Oxide Nanoparticles. *Micro Nano Lett.* **2018**, *13*, 1227–1231. [CrossRef]
9. Liu, Y.; Li, H.; Xiao, K. Distribution and biological effects of nanoparticles in the reproductive system. *Curr. Drug Metab.* **2016**, *17*, 478–496. [CrossRef]
10. Bitar, A.; Ahmad, N.M.; Fessi, H.; Elaissari, A. Silica-based nanoparticles for biomedical applications. *Drug Discov. Today* **2012**, *17*, 1147–1154. [CrossRef]
11. Tang, L.; Cheng, J.J. Nanoporous silica nanoparticle for nanomedicine application. *Nano Today* **2013**, *8*, 290–312. [CrossRef] [PubMed]
12. McAuliffe, M.E.; Perry, M.J. Are nanoparticles potential male reproductive toxicants? A literature review. *Nanotoxicology* **2007**, *1*, 204–210. [CrossRef]
13. Wang, R.; Song, B.; Wu, J.; Zhang, Y.; Chen, A.; Shao, L. Potential adverse effects of nanoparticles on the reproductive system. *Int. J. Nanomed.* **2018**, *13*, 8487. [CrossRef] [PubMed]
14. Valic, M.S.; Zheng, G. Research Tools for Extrapolating the Disposition and Pharmacokinetics of Nanomaterials from Preclinical Animals to Humans. *Theranostics* **2019**, *9*, 3365–3387. [CrossRef]
15. Bahadar, H.; Maqbool, F.; Niaz, K.; Abdollahi, M. Toxicity of Nanoparticles and an Overview of Current Experimental Models. *Iran. Biomed. J.* **2016**, *20*, 1–11. [CrossRef]
16. Kim, I.Y.; Lee, T.G.; Reipa, V.; Heo, M.B. Titanium Dioxide Induces Apoptosis Under UVA Irradiation via the Generation of Lysosomal Membrane Permeabilization-Dependent Reactive Oxygen Species in HaCat Cells. *Nanomaterials* **2021**, *11*, 1943. [CrossRef]
17. Feitelson, M.A.; Arzumanyan, A.; Kulathinal, R.J.; Blain, S.W.; Holcombe, R.F.; Mahajna, J.; Marino, M.; Martinez-Chantar, M.L.; Nawroth, R.; Sanchez-Garcia, I.; et al. Sustained Proliferation in Cancer: Mechanisms and Novel Therapeutic Targets. *Semin. Cancer Biol.* **2015**, *35*, S25–S54. [CrossRef]
18. San-Millán, I.; Brooks, G.A. Reexamining Cancer Metabolism: Lactate Production for Carcinogenesis Could Be the Purpose and Explanation of the Warburg Effect. *Carcinogenesis* **2017**, *38*, 119–133. [CrossRef]
19. Granchi, C.; Minutolo, F. Anti-Cancer Agents Counteracting Tumor Glycolysis. *ChemMedChem* **2012**, *7*, 1318–1350. [CrossRef]

20. Stefano, D.D.; Carnuccio, R.; Maiuri, M.C. Nanomaterials Toxicity and Cell Death Modalities. *J. Drug Deliv.* **2012**, *2012*, 1–14. [CrossRef]
21. Yildirimer, L.; Thanh, N.T.K.; Loizidou, M.; Seifalian, A.M. Toxicology and Clinical Potential of Nanoparticles. *Nano Today.* **2011**, *6*, 585–607. [CrossRef]
22. Barabadi, H.; Alizadeh, A.; Ovais, M.; Ahmadi, A.; Shinwari, Z.K.; Saravanan, M. Efficacy of Green Nanoparticles Against Cancerous and Normal Cell Lines: A Systematic Review and Meta-Analysis. *IET Nanobiotechnol.* **2018**, *12*, 377–391. [CrossRef]
23. Joris, F.; Manshian, B.B.; Peynshaert, K.; de Smedt, S.C.; Braeckmans, K.; Soenen, S.J. Assessing nanoparticle toxicity in cell-based assays: Influence of cell culture parameters and optimized models for bridging the in vitro-in vivo gap. *Chem. Soc. Rev.* **2013**, *42*, 8339–8359. [CrossRef]
24. Kermanizadeh, A.; Lohr, M.; Roursgaard, M.; Messner, S.; Gunness, P.; Kelm, J.M.; Moller, P.; Stone, V.; Loft, S. Hepatic toxicology following single and multiple exposure of engineered nanomaterials utilizing a novel primary human 3D liver microtissue model. *Part Fibre Toxicol.* **2014**, *11*, 56. [CrossRef]
25. Zhang, H.; Wang, X.; Wang, M.; Li, L.; Chang, C.H.; Ji, Z.; Xia, T.; Nel, A.E. Mammalian cells exhibit a range of sensitivities to silver nanoparticles that are partially explicable by variations in antioxidant defense and metallothionein expression. *Small* **2015**, *11*, 3797–3805. [CrossRef]
26. Luengo, Y.; Nardecchia, S.; Morales, M.P.; Serrano, M.C. Different cell responses induced by exposure to maghemite nanoparticles. *Nanoscale* **2013**, *5*, 11428–11437. [CrossRef]
27. Mukherjee, S.G.; O'Claonadh, N.; Casey, A.; Chambers, G. Comparative in vitro cytotoxicity study of silver nanoparticle on two mammalian cell lines. *Toxicol. In Vitro* **2012**, *26*, 238–251. [CrossRef]
28. Wang, Y.; Aker, W.G.; Hwang, H.M.; Yedjou, C.G.; Yu, H.; Tchounwou, P.B. A study of the mechanism of in vitro cytotoxicity of metal oxide nanoparticles using catfish primary hepatocytes and human HepG2 cells. *Sci. Total Environ.* **2011**, *409*, 4753–4762. [CrossRef]
29. Ekstrand Hammarstrom, B.; Akfur, C.M.; Andersson, P.O.; Lejon, C.; Osterlund, L.; Bucht, A. Human primary bronchial epithelial cells respond differently to titanium dioxide nanoparticles than the lung epithelial cell lines A549 and BEAS-2B. *Nanotoxicology* **2012**, *6*, 623–634. [CrossRef]
30. Kermanizadeh, A.; Gaiser, B.K.; Ward, M.B.; Stone, V. Primary human hepatocytes versus hepatic cell line: Assessing their suitability for in vitro nanotoxicology. *Nanotoxicology* **2013**, *7*, 1255–1271. [CrossRef]
31. Frohlich, E. Cellular elimination of nanoparticles. *Environ. Toxicol. Pharmacol.* **2016**, *46*, 90–94. [CrossRef] [PubMed]
32. Laskar, A.; Ghosh, M.; Khattak, S.I.; Li, W.; Yuan, X.M. Degradation of superparamagnetic iron oxide nanoparticle-induced ferritin by lysosomal cathepsins and related immune response. *Nanomedicine* **2012**, *7*, 705–717. [CrossRef] [PubMed]
33. Lunov, O.; Syrovets, T.; Rocker, C.; Tron, K.; Nienhaus, G.U.; Rasche, V.; Mailander, V.; Landfester, K.; Simmet, T. Lysosomal degradation of the carboxydextran shell of coated superparamagnetic iron oxide nanoparticles and the fate of professional phagocytes. *Biomaterials* **2010**, *31*, 9015–9022. [CrossRef] [PubMed]
34. Chithrani, B.D.; Chan, W.C. Elucidating the mechanism of cellular uptake and removal of protein-coated gold nanoparticles of different sizes and shapes. *Nano Lett.* **2007**, *7*, 1542–1550. [CrossRef] [PubMed]
35. Chu, Z.; Huang, Y.; Tao, Q.; Li, Q. Cellular uptake, evolution, and excretion of silica nanoparticles in human cells. *Nanoscale* **2011**, *3*, 3291–3299. [CrossRef]
36. Fang, C.Y.; Vaijayanthimala, V.; Cheng, C.A.; Yeh, S.H.; Chang, C.F.; Li, C.L.; Chang, H.C. The exocytosis of fluorescent nanodiamond and its use as a long-term cell tracker. *Small* **2011**, *7*, 3363–3370. [CrossRef]
37. Jiang, X.; Rocker, C.; Hafner, M.; Brandholt, S.; Dorlich, R.M.; Nienhaus, G.U. Endo- and exocytosis of zwitterionic quantum dot nanoparticles by live HeLa cells. *ACS Nano* **2010**, *4*, 6787–6797. [CrossRef]
38. Jin, H.; Heller, D.A.; Sharma, R.; Strano, M.S. Size-dependent cellular uptake and expulsion of single-walled carbon nanotubes: Single particle tracking and a generic uptake model for nanoparticles. *ACS Nano* **2009**, *3*, 149–158. [CrossRef] [PubMed]
39. Serda, R.E.; Mack, A.; van de Ven, A.L.; Ferrati, S.; Dunner, K., Jr.; Godin, B.; Chiappini, C.; Landry, M.; Brousseau, L.; Liu, X.; et al. Logic-embedded vectors for intracellular partitioning, endosomal escape, and exocytosis of nanoparticles. *Small* **2010**, *6*, 2691–2700. [CrossRef]
40. Yanes, R.E.; Tarn, D.; Hwang, A.A.; Ferris, D.P.; Sherman, S.P.; Thomas, C.R.; Lu, J.; Pyle, A.D.; Zink, J.I.; Tamanoi, F. Involvement of lysosomal exocytosis in the excretion of mesoporous silica nanoparticles and enhancement of the drug delivery effect by exocytosis inhibition. *Small* **2013**, *9*, 697–704. [CrossRef]
41. Kallukall, T.; Olsen, O.D.; Jaatella, M. Cancer-associated lysosomal changes: Friends or foes? *Oncogene* **2013**, *32*, 1995–2004.
42. Hartung, T. Toxicology for the twenty-first century. *Nature* **2009**, *460*, 208–212. [CrossRef]
43. Rangarajan, A.; Hong, S.J.; Gifford, A.; Weinberg, R.A. Species- and cell type-specific requirements for cellular transformation. *Cancer Cell* **2004**, *6*, 171–183. [CrossRef]
44. Yang, K.; Pfeifer, N.D.; Kock, K.; Brouwer, K.L. Species differences in hepatobiliary disposition of taurocholic acid in human and rat sandwich-cultured hepatocytes: Implications for drug-induced liver injury. *J. Pharmacol. Exp. Ther.* **2015**, *353*, 415–423. [CrossRef]

nanomaterials

MDPI

Article

Assessment of Systemic Toxicity, Genotoxicity, and Early Phase Hepatocarcinogenicity of Iron (III)-Tannic Acid Nanoparticles in Rats

Chi Be Hlaing [1], Arpamas Chariyakornkul [1], Chalermchai Pilapong [2], Charatda Punvittayagul [3], Somdet Srichairatanakool [1] and Rawiwan Wongpoomchai [1,*]

1 Department of Biochemistry, Faculty of Medicine, Chiang Mai University, Chiang Mai 50200, Thailand; hlaingchibe26o@gmail.com (C.B.H.); arpamas.ch@cmu.ac.th (A.C.); somdet.s@cmu.ac.th (S.S.)
2 Center of Excellence for Molecular Imaging (CEMI), Department of Radiologic Technology, Faculty of Associated Medical Sciences, Chiang Mai University, Chiang Mai 50200, Thailand; chalermchai.pilapong@cmu.ac.th
3 Research Affairs, Faculty of Veterinary Medicine, Chiang Mai University, Chiang Mai 50100, Thailand; charatda.pun@cmu.ac.th
* Correspondence: rawiwan.wong@cmu.ac.th; Tel.: +66-53-935325; Fax: +66-53-894031

Abstract: Iron-tannic acid nanoparticles (Fe-TA NPs) presented MRI contrast enhancement in both liver cancer cells and preneoplastic rat livers, while also exhibiting an anti-proliferative effect via enhanced autophagic death of liver cancer cells. Hence, a toxicity assessment of Fe-TA NPs was carried out in the present study. Acute and systemic toxicity of intraperitoneal Fe-TA NPs administration was investigated via a single dose of 55 mg/kg body weight (bw). Doses were then repeated 10 times within a range of 0.22 to 5.5 mg/kg bw every 3 days in rats. Furthermore, clastogenicity was assessed by rat liver micronucleus assay. Carcinogenicity was evaluated by medium-term carcinogenicity assay using glutathione *S*-transferase placental form positive foci as a preneoplastic marker, while three doses ranging from 0.55 to 17.5 mg/kg bw were administered 10 times weekly via intraperitoneum. Our study found that the LD_{50} value of Fe-TA NPs was greater than 55 mg/kg bw. Repeated dose administration of Fe-TA NPs over a period of 28 days and 10 weeks revealed no obvious signs of systemic toxicity, clastogenicity, and hepatocarcinogenicity. Furthermore, Fe-TA NPs did not alter liver function or serum iron status, however, increased liver iron content at certain dose in rats. Notably, antioxidant response was observed when a dose of 17.5 mg/kg bw was given to rats. Accordingly, our study found no signs of toxicity, genotoxicity, and early phase hepatocarcinogenicity of Fe-TA NPs in rats.

Keywords: acute toxicity; repeated dose toxicity; carcinogenicity; genotoxicity; nanoparticle

1. Introduction

The global cancer burden has rose to 19.3 million new cases and 9.9 million deaths in 2020. Primary liver cancer is the sixth most common cause of cancer and the third leading cause of cancer mortality worldwide [1]. About 80% of primary liver cancer cases involve hepatocellular carcinoma (HCC). The high rate of mortality observed in HCC patients can be attributed to late diagnosis, drug resistance, disease recurrence, and metastasis leading to poor clinical outcomes [2].

Imaging is one of the most important aspects of diagnosis, therapy, and follow-up for HCC patients [3]. Magnetic resonance imaging (MRI), or multiphase contrast-enhanced computed tomography (CT), is recommended for initial diagnostic evaluation of clinically suspicious HCC. The therapeutic options depend upon the stage of the disease and what curative and noncurative interventions are available. Curative forms of therapy include surgical resection, orthotopic liver transplant, and ablative techniques such as thermal ablation. Noncurative forms of therapy include transarterial chemoembolization (TACE), transarterial radioembolization (TARE), stereotactic body radiation therapy (SBRT), and systemic chemotherapy [4]. However, early diagnosis and effective treatment have remained challenging with regard to HCC management.

The use of nanomaterials in cancer management is one of the most promising and advanced approaches. Many studies have reported that nanoparticles (NPs) can play a role in the diagnosis and treatment of cancer [5]. Alshehri S et al. have recently reviewed current studies on theranostic applications of nanoparticles and reported that these studies reported favourable outcomes in both in vitro and in vivo cancer models [6]. In addition, many research studies are now focused on the application of nanodrug delivery systems for diagnostic and therapeutic agents in liver cancer patients in order to improve clinical efficacy [7]. Among the various types of NPs, metal-polyphenol nanoparticles exhibit a potential role in cancer theranostics. Polyphenols, such as tannic acid (TA), gallic acid, and epigallocatechin gallate, can assemble with certain metal ions, such as Fe^{3+}, Mn^{2+}, and Cu^{2+}, and then be loaded with chemotherapeutic agents or exist in their pure form. These nanoparticles have been studied in tumor imaging studies that involved magnetic resonance imaging (MRI), CT imaging, photothermal imaging, photoacoustic imaging, as well as tumor therapy photothermal therapy, photodynamic therapy, and chemotherapy [8]. In this context, metals and polyphenols can display their biological effects via individual or synergistic actions for both diagnostic and therapeutic purposes in cancer treatments [9]. Self assembled ferric and TA nanoparticles (Fe-TA NPs) can provide MRI signals in hepatocellular carcinoma cells [10] and rat liver preneoplasia models [11]. Furthermore, Fe-TA NPs also showed an antiproliferative effect via enhanced autophagic cell death in hepatocellular carcinoma cells [10]. This finding highlights the potential of Fe-TA NPs as a nanotheranostic agent for early diagnosis and treatment of HCC.

However, the potential toxicity of nanoparticles is of concern as there is limited available knowledge on the toxicity of self-assembled metal-polyphenol theranostic nanoparticles in vivo [6]. Furthermore, a dose–response relationship is an integral factor that is involved in evaluating the toxicity of a chemical substance. Some physical and chemical agents such as low-dose ionizing radiation and non-genotoxic carcinogens, including phenobarbital and α-benzene hexachloride, as well as certain phytochemicals such as resveratrol, quercetin, and sulforaphane, have exhibited a hormetic dose response [12–14]. To explore the further diagnostic and therapeutic potential of Fe-TA NPs, the present study assessed the toxicity of Fe-TA NPs through single and repeated dose systemic toxicity tests in rats, in vivo genotoxicity tests, and medium-term carcinogenicity test in rats.

2. Materials and Methods

2.1. Chemicals

Diethylnitrosamine (DEN), 3,3′-diaminobenzidine tetrahydrochloride hydrate (DAB), 5,6-diphenyl-1,2,4-triazine-p, p′-disulfonic acid disodium salt hydrate (ferrozine), and butylated hydroxytoluene (BHT) were purchased from Sigma-Aldrich (St. Louis, MO, USA). Collagenase type IV and 4′,6-diamidino-2-phenylindole (DAPI) were obtained from Invitrogen (Waltham, MA, USA). Assay kits for serum iron (RX SERIES SI 382) and serum TIBC (RX SERIES TI 3858) were purchased from Randox Laboratories Ltd. (Antrim, Northern Ireland, UK). Ferritin Rat ELISA Kit (ab157732) was purchased from Abcam (Cambridge, UK). Anti-rat GST-placental form was obtained from MBL (Nagoya, Japan) and mouse monoclonal purified anti-rat PCNA antibody was obtained from BioLegend (San Diego, CA, USA). ApopTag Peroxidase In Situ Apoptosis Detection Kit (S7100) for TUNEL assay was purchased from Merck KGaA (Darmstadt, Germany). EnvisionTM G/2 Doublestain System, Rabbit/Mouse (DAB+/Permanent Red) was obtained from Dako (Glostrup, Denmark). Vectastain ABC kit was obtained from Vector Laboratories, Inc. (Burlingame, CA, USA). β-NADPH was purchased from Oriental Yeast (Tokyo, Japan) and glucose-6-phosphate dehydrogenase was purchased from Worthington Biochemical Corporation (Lakewood, NJ, USA). Reduced and oxidized glutathione, glutathione reductase, 5,5′-

dithiobis (2-nitrobenzoic acid (DTNB), glucose-6-phosphate, and hemin were obtained from Sigma-Aldrich, (St. Louis, MO, USA). All other chemicals were analytical grade.

2.2. Iron (III)—Tannic Acid Nanoparticles

Fe-TA NPs were obtained from Dr. Chalermchai Pilapong, Center of Excellence for Molecular Imaging, Department of Radiologic Technology, Faculty of Associated Medical Sciences, Chiang Mai University, Chiang Mai, Thailand. Fe-TA NPs are known to exist in stable tris-coordinated complexes in PBS buffer (pH 7.4). Fe-TA NPs are spherical in shape with a hydrodynamic diameter of 3.14 $\pm$ 1.0 nm. Fe-TA NPs present a zeta potential of -23 ± 2.1 mV and exhibit good colloidal stability, water solubility, and high stability against both transchelation and transmetallation. The quantity of Fe-TA NPs was expressed as an equivalent concentration of iron spectrophotometrically [10].

2.3. Animal Studies

Wistar rats (5 to 7-week-olds) were obtained from Nomura Siam International Co., Ltd., Bangkok, Thailand. Rats were housed in stainless steel cages and acclimated for a week before beginning investigations. Housing conditions included a 12:12-h light/dark cycle at temperatures of 23 $\pm$ 2 °C and the level of humidity maintained within a range of 50 to 60%. Both a pellet diet and water were provided ad libitum. The animal protocols used in this study followed the relevant international and national guidelines for animal use and were approved of by The Animal Ethics Committee of the Faculty of Medicine, Chiang Mai University, Thailand (numbers: 39/2560 and 06/2563).

2.4. Single Dose Toxicity Test

Eight-week-old male Wistar rats were randomly divided into 2 groups, control and Fe-TA NPs treated groups, with 5 animals in each group. Fe-TA NPs were injected intraperitoneally at a dose of 55 mg/kg body weight a maximum dose for its solubility. Animals were observed for a total of 14 days for any sign of toxicity or mortality. Body weights of the rats were recorded twice a week. Animals were sacrificed under isoflurane inhalation on day 14 of the experiment. The livers, kidneys, and spleens of the animals were collected and weighed. Hematological parameters, including red blood cell count, leucocyte count, and platelet count, were determined with the use of an automatic analyzer at the Veterinary Diagnostic Centre Co., Ltd. (Chiang Mai, Thailand).

2.5. Repeated Dose Toxicity Test

Six-week-old male and female Wistar rats were randomly allocated into 6 groups with 10 animals in each group (5 rats of each gender). Group 1 served as a vehicle control group. Groups 2–4 were administered with Fe-TA NPs at doses of 0.22, 1.1, and 5.5 mg/kg bw, respectively. The remaining two groups were satellite groups, and were comprised of a control and a group that received 5.5 mg/kg bw of Fe-TA NPs. Fe-TA NPs were injected intraperitoneally every three days for a total of 10 times. Test doses of Fe-TA NPs were selected based on the single dose toxicity test and aimed to achieve a cumulative dose of 55 mg/kg bw. The dosage interval was based on an established liver clearance time of Fe-TA NPs [15]. Rats were monitored by measuring body weight every three days and food and water intake twice per week. After 24 h of the last injection and overnight fasting, rats were sacrificed under isoflurane inhalation. For the satellite groups, this duration was extended for a period of 14 days after the last injection, while body weight, food, and water intake were monitored according to the method that had been previously explained. Rats were observed for any delayed or persistent effect of, or recovery from, toxic effects during that period. Whole blood was collected for hematological and biochemical investigations in the Veterinary Diagnostic Laboratory, Faculty of Veterinary Medicine, Chiang Mai University and the Veterinary Diagnostic Centre Co. Ltd., Chiang Mai, Thailand. Urine samples were taken for urinalysis using a dipstick test (Multistix 10 SG, Siemens, Melbourne, Australia). Vital internal organs were removed for gross necropsy and organ weight measurement before being fixed in buffered 10% formalin for histopathological assessment. The remaining portion of the liver was kept in −20 °C until further analysis.

2.6. Clastogenicity Test

To determine the clastogenic activity of intraperitoneal Fe-TA NP administration, rat liver micronucleus assay was used in this study. Male Wistar rats (6-week-old, 180–190 g) were divided into five groups as is shown in Figure 1. Group 1 was established as a negative control receiving phosphate buffered saline, while groups 2 and 3 received a single dose of Fe-TA NPs at 55 and 5.5 mg/kg bw, respectively. Group 4 rats were injected with repeated doses of 5.5 mg/kg bw of Fe-TA NPs on days 1, 4 and 7 of the experiment. Group 5 served as a positive control receiving 10 mg/kg bw of diethylnitrosamine (DEN) on days 4 and 7 of the experiment. Twenty-four hours after the last injection, all rats were lightly anesthetized with isoflurane inhalation and underwent 2/3 partial hepatectomy. After the experiment was conducted for 4 days, rats were euthanized by overdose thiopental and their hepatocytes were isolated using the 2-step collagenase perfusion technique. Liver cells were stained with 4′,6-diamidino-2-phenylindole (DAPI) and analyzed under a fluorescent microscope. Micronucleated and mitotic hepatocytes were observed as 2000 hepatocytes/rat. The criteria for micronucleated hepatocyte scoring had been previously established by Chariyakornkul et al., 2019 [16].

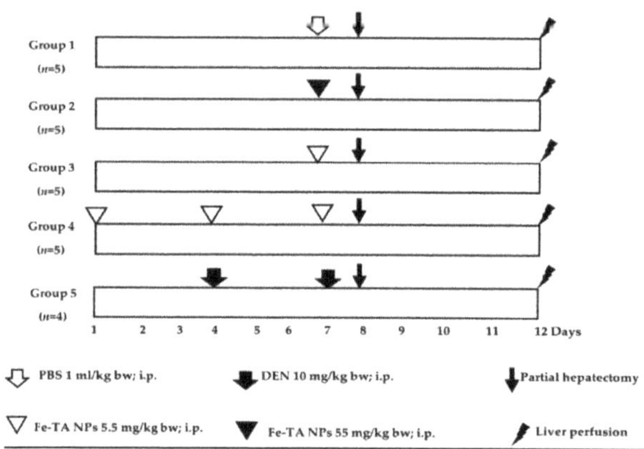

Figure 1. Experimental protocol for clastogenicity test of Fe-TA NPs in rats.

2.7. Medium-Term Liver Carcinogenicity Test

Six-week-old male Wistar rats were divided into 5 groups. Group 1 was a negative control receiving phosphate buffered saline and groups 2–4 (n = 4–6) served as treated groups that were intraperitoneally injected with Fe-TA NPs at doses of 0.55, 5.5 and 17.5 mg/kg bw once a week on 10 occasions, while group 5 (n = 3) was considered a positive control that received weekly injections of diethylnitrosamine at 100 mg/kg body weight on three occasions. One week before injections were given, all animals underwent two-third partial hepatectomies to induce regenerative cell proliferation. Twenty-four hours after the last injection was administered, all rats were sacrificed under isoflurane inhalation and blood sampling was collected. Serum alanine aminotransferase (ALT) was collected after the clotted blood was centrifuged and analyzed using an automatic analyzer at the Veterinary Diagnostic Centre Co., Ltd., Chiang Mai, Thailand. Subsequently, livers were perfused with 0.15 M KCl to remove blood samples. Livers, spleens, and kidneys of the rats were then collected and weighed. Liver portions were fixed in 10% buffered formalin and then embedded in paraffin for histopathological assessment of the formation of GST-P positive foci as a preneoplastic lesion in the livers, proliferating cell nuclear antigen (PCNA)-positive cells as cell proliferation markers and apoptotic cells by terminal deoxynucleotidyl transferase (TdT) dUTP nick end labelling (TUNEL) assay. These procedures were performed according to the method previously described [17].

2.8. Measurement of Serum and Liver Iron Status Parameters

The levels of serum iron and total iron-binding capacity were measured colorimetrically using commercial assay kits and a fully automatic biochemical analyzer (RX Daytona, Antrim, Ireland) according to the manufacturer's instructions. The percentage of transferrin saturation was calculated as a ratio of serum iron to TIBC.

Liver iron concentration was determined using a modified ferrozine assay as has been described elsewhere [18]. After dried liver tissues were cold homogenized in 50 mM phosphate buffer at a pH of 2.0 containing 0.005% methanolic BHT, they were further centrifuged and the supernatant was collected. The iron standards were prepared from a serial dilution of ferrous ammonium sulfate (0–800 μM). Standards and the supernatant were incubated with a chromogenic solution containing 0.508 mM ferrozine, 1.5 M sodium acetate, and 1.5% (*v*/*v*) thioglycolic acid at room temperature for 30 min. Absorbance was measured spectrophotometrically at 562 nm. The concentration of liver iron was calculated from the standard curve and expressed as ng of iron per mg of dry liver weight.

Liver ferritin values were determined using a Ferritin Rat ELISA Kit. Liver tissue samples were homogenized with ice cold NaCl and then centrifuged at 15,000 rpm at 4 °C for 20 min. The supernatant was collected and diluted with diluent buffer. A standard curve was prepared with a serial dilution of rat ferritin calibrator (0–100 ng/mL), and then ferritin levels were measured according to the manufacturer's instructions. Liver ferritin concentration values were calculated using a calibration curve that was normalized by protein concentrations determined by Lowry's method [19] and then expressed as a value of ng/mg liver protein.

2.9. Determination of Antioxidant Parameters in Rat Livers

Total glutathione (GSH) and glutathione disulfide (GSSG) levels were determined using the enzymatic recycling method based on DTNB with some modifications [20]. Liver homogenate in 0.01 M phosphate buffer was treated with 5% 5-sulfosalicylic acid. After cold centrifugation, supernatants were reacted with 4-vinylpyridine to protect GSH auto-oxidation for GSSG assay and then used directly for total GSH determination. The standards or samples were mixed with a reaction mixture containing glutathione reductase, DTNB, and β-NADPH. The degree of absorbance was measured kinetically at 405 nm for 5 min with a minute interval. The concentration was calculated with the use of a regression curve generated from the standards of GSH and GSSG and expressed as a value of nmol/mg protein. Reduced glutathione levels were calculated by subtracting values of GSSG from total GSH.

For the determination of antioxidant enzyme activity, the liver tissue was homogenized on ice with homogenizing buffer containing 1.15% (*w*/*v*) KCl and 0.25 mM PMSF. After medium-speed centrifugation, the supernatant was subsequently centrifuged at 30,000 rpm for 60 min at 4 °C in order to obtain cytosolic and microsomal fractions. The protein content of both fractions was then determined by applying the Lowry method using bovine serum albumin (BSA) as a standard [19].

Catalase (CAT) activity was determined using the method described by Aebi, 1984 [21]. In this assay, 30 mM H_2O_2 was used as a substrate and catalase activity was detected by a decrease in absorbance of H_2O_2 at 240 nm. CAT activity was calculated using molar extinction coefficient for H_2O_2 at 240 nm and expressed as μmol of H_2O_2 per minute per mg protein.

Glutathione reductase (GR) activity was determined according to the method previously described by Carlberg and Mannervik, 1985 [22]. Assay samples were reacted with a reaction mixture that contained phosphate buffer (pH 7.0) prepared from 0.1 M KH_2PO_4, 1 mM $MgCl_2.6H_2O$, 25 mM GSSG, and 0.1 mM NADPH. The oxidation of NADPH during the reaction was estimated spectrophotometrically at 340 nm for a single minute. GR activity was expressed in terms of μmol of NADPH consumed per minute per mg protein using the molar extinction coefficient for NADPH at 340 nm.

Glutathione peroxidase (GPx) activity was determined with slight modifications according to the method described by Nagalakshmi and Prasad (2001) [23]. Assay samples were added to the reaction mixture containing 0.1 M Tris-EDTA buffer (pH 8), 0.1 M GSH, 2 mM β-NADPH, 7 mM t-BHP, and 10 U GR. β-NADPH was oxidized during the reaction and was measured by estimating a decrease in the absorbance at 340 nm. GPx activity was calculated using the molar extinction coefficient for NADPH at 340 nm and was expressed in terms of μmol of NADPH consumed per minute per mg protein.

Heme oxygenase (HO) activity assay was performed using a slightly modified method [24]. The liver microsome was incubated with 0.1 M potassium phosphate buffer containing 2 mM $MgCl_2$ at a pH of 7.4 with 8 mM glucose-6-phosphate, 4 U/mL glucose-6-phosphate dehydrogenase, 6.4 mM NADPH, 400 μM hemin substrate, and 10 mg/mL of liver cytosol as a source of biliverdin reductase at 37 °C in the dark. The reaction was stopped using chloroform. The activity of HO was measured at wavelengths of 460 and 530 nm and expressed as nmol/min/mg protein using a molar extinction coefficient of 40.

2.10. Statistical Analysis

Data are presented as mean ± standard deviation values (S.D) for each variable. Statistical differences among three or more groups were assessed by one-way analysis of variance (ANOVA) followed by Bonferroni as a post-hoc test for toxicity tests and least significant differences for the relevant biochemical parameters. Statistical differences between the two groups were analyzed by Student's *t*-test. Data were considered statistically significant when a *p*-value of less than 0.05 was observed.

3. Results

3.1. Single and Repeated Dose Toxicity Levels of Iron (III)–Tannic Acid Nanoparticles in Rats

Single dose toxicity test aims to estimate the median lethal dose (LD_{50}) of Fe-TA NPs in rats. During the 14-day observation period, a single intraperitoneal injection of 55 mg/kg bw of Fe-TA NPs was not found to have caused any toxic symptoms or mortality. The gross necropsy indicated a non-disorder in organ morphology. Body weight, absolute and relative organ weights, and hematological parameters, including red blood cell, leucocyte, and platelet counts, were not found to be significantly different when compared to the control group (data not shown). It was indicated that the LD_{50} value of intraperitoneal Fe-TA NPs administration was more than 55 mg/kg body weight in male Wistar rats.

The repeated dose test aims to assess the most affected target organs and dosages of the test substance after a certain period of continuous exposure. The intraperitoneal treatment of Fe-TA NPs ranged from 0.22 to 5.5 mg/kg bw every 3 days for 10 times. This indicated no incidences of morbidity and mortality in both genders of rats. No significant changes in food and water intake and body weight were observed in both sexes of the experimental rats when compared to the control animals (Table 1). The absolute and relative organ weights of ovaries in 5.5 mg/kg bw dose groups and those of the seminal vesicles in 1.1 mg/kg bw dose groups were significantly increased when compared to those of the control group. However, no histopathological changes were observed. Apart from these findings, there were no significant changes in organ weight of Fe-TA NPs treated animals when compared to the control animals of both sexes (Table 1). In the satellite group of female rats, the brain weights of the rats in the Fe-TA NPs treated group were significantly reduced when compared to the control group (Table 1). No other significant changes were observed in terms of gross necropsy and organ weights of the members of both sexes of the satellite group (Table 1).

Table 1. Effect of 28-day administration of Fe-TA NPs on general observation and organ weights.

Parameters	Fe-TA NPs Treatment (mg/kg bw)					
	0	0.22	1.1	5.5	Satellite	
					0	5.5
Female						
Food intake (g/day)	14.3 ± 2.0	15.9 ± 2.3	14.5 ± 1.3	15.1 ± 1.8	15.4 ± 1.8	16.4 ± 2.1
Water intake (mL/day)	18.9 ± 3.2	19.7 ± 5.5	20.8 ± 5.1	21.0 ± 3.7	22.5 ± 1.9	19.0 ± 2.4
IBW (g)	120 ± 6.12	120 ± 8.66	121 ± 5.48	121 ± 4.18	120 ± 7.07	121 ± 2.24
FBW (g)	187 ± 4.47	193 ± 11.0	188 ± 17.2	193 ± 9.08	214 ± 15.5	204 ± 9.46
Liver (g)	6.36 ± 0.54	6.88 ± 0.30	6.53 ± 0.43	6.81 ± 0.47	6.94 ± 0.81	6.99 ± 0.48
Spleen (g)	0.37 ± 0.04	0.40 ± 0.03	0.36 ± 0.06	0.45 ± 0.04	0.39 ± 0.07	0.38 ± 0.03
Kidneys (g)	1.50 ± 0.06	1.57 ± 0.12	1.52 ± 0.14	1.49 ± 0.09	1.57 ± 0.17	1.51 ± 0.10
Lungs (g)	0.97 ± 0.12	0.89 ± 0.10	0.88 ± 0.08	0.90 ± 0.02	0.92 ± 0.14	0.83 ± 0.05
Heart (g)	0.60 ± 0.02	0.59 ± 0.05	0.56 ± 0.07	0.58 ± 0.07	0.62 ± 0.05	0.61 ± 0.03
Thymus (g)	0.38 ± 0.08	0.46 ± 0.03	0.43 ± 0.08	0.43 ± 0.10	0.46 ± 0.04	0.31 ± 0.04
Pancreas (g)	0.62 ± 0.16	0.68 ± 0.25	0.72 ± 0.29	0.92 ± 0.21	0.57 ± 0.07	0.69 ± 0.08
Brain (g)	1.60 ± 0.08	1.66 ± 0.06	1.64 ± 0.04	1.65 ± 0.06	1.36 ± 0.13	1.18 ± 0.03 #
Adrenal (mg)	64.0 ± 18.3	55.5 ± 9.06	60.8 ± 11.7	58.3 ± 3.61	64.9 ± 4.78	71.1 ± 13.8
Ovary (mg)	69.5 ± 15.4	91.1 ± 11.2	89.0 ± 12.5	110 ± 7.81 *	103 ± 31.2	80.7 ± 15.3
Male						
Food intake (g/day)	23.9 ± 2.6	26.9 ± 6.5	25.4 ± 5.2	23.4 ± 1.6	23.5 ± 1.7	23.8 ± 2.9
Water intake (ml/day)	27.9 ± 8.9	32.6 ± 6.8	32.4 ± 8.2	33.3 ± 5.1	30.0 ± 4.25	29.7 ± 4.22
IBW (g)	180 ± 12.3	180 ± 9.35	180 ± 6.12	180 ± 0.00	180 ± 14.1	180 ± 9.35
FBW (g)	321 ± 30.9	338 ± 23.9	314 ± 25.1	316 ± 9.62	365 ± 32.4	363 ± 18.6
Liver (g)	11.0 ± 1.33	12.0 ± 1.17	11.7 ± 1.56	11.0 ± 0.83	11.1 ± 1.91	11.0 ± 0.57
Spleen (g)	0.63 ± 0.14	0.64 ± 0.06	0.63 ± 0.07	0.67 ± 0.05	0.65 ± 0.14	0.66 ± 0.04
Kidneys (g)	2.46 ± 0.21	2.54 ± 0.15	2.62 ± 0.28	2.34 ± 0.08	2.38 ± 0.21	2.47 ± 0.31
Lungs (g)	1.18 ± 0.12	1.29 ± 0.18	1.38 ± 0.32	1.21 ± 0.15	1.13 ± 0.09	1.29 ± 0.20
Heart (g)	0.91 ± 0.15	0.93 ± 0.13	0.89 ± 0.07	0.88 ± 0.05	0.91 ± 0.05	0.96 ± 0.09
Thymus (g)	0.67 ± 0.14	0.67 ± 0.23	0.62 ± 0.09	0.62 ± 0.14	0.55 ± 0.14	0.48 ± 0.10
Pancreas (g)	0.79 ± 0.14	0.94 ± 0.22	0.87 ± 0.19	0.76 ± 0.22	0.73 ± 0.24	0.97 ± 0.17
Brain (g)	1.92 ± 0.09	1.85 ± 0.08	1.90 ± 0.09	1.82 ± 0.06	1.86 ± 0.10	1.84 ± 0.08
Adrenal (mg)	58.90 ± 6.32	55.00 ± 7.36	63.94 ± 17.90	64.36 ± 12.54	60.60 ± 10.06	66.58 ± 12.35
Testes (g)	3.34 ± 0.26	3.20 ± 0.23	3.30 ± 0.17	3.10 ± 0.19	3.65 ± 0.11	3.64 ± 0.17
Prostate (g)	0.39 ± 0.12	0.38 ± 0.09	0.41 ± 0.04	0.31 ± 0.06	0.44 ± 0.09	0.39 ± 0.05
Epididymis (g)	0.79 ± 0.12	0.78 ± 0.18	0.66 ± 0.12	0.63 ± 0.11	1.05 ± 0.07	1.03 ± 0.09
Seminal (g)	0.72 ± 0.11	0.76 ± 0.09	0.95 ± 0.09 *	0.84 ± 0.08	0.89 ± 0.19	0.81 ± 0.25

Values are expressed as Mean ± SD. (IBW: Initial body weight, FBW: Final body weight). * $p < 0.05$, significantly different when compared with the control. # $p < 0.05$, significantly different when compared with the control satellite.

The systemic toxic effects of Fe-TA NPs in rats were further studied. The hematological profile of female rats did not reveal any significant changes when compared to the control rats (Table 2). In male rats, the neutrophil count was significantly increased in groups receiving 1.1 and 5.5 mg/kg bw doses, while monocyte counts were significantly reduced in groups receiving 0.22 and 1.1 mg/kg bw doses when compared to the control group (Table 2). However, these values were found to be within the reference range. Although the biochemical profiles of female rats revealed significant reductions in BUN levels in both the medium- and high-dose treated groups, these levels were also determined to be within the reference range (Table 3). In male rats, serum potassium levels were significantly increased in the group receiving 5.5 mg/kg bw doses (Table 3). Moreover, significant changes in levels of creatinine, sodium, and TCO_2 in the satellite group of Fe-TA NPs treated male rats were also found to be within the reference range (Table 3). No significant differences were observed in liver function tests in terms of lipid profiles and blood glucose levels in rats of both sexes (Table 3). Urinalysis revealed no abnormalities apart from hematuria in some samples obtained from female rats. However, various levels of proteinuria in male rats were detected in most of the urine samples, while hematuria and ketonuria were detected in some of the other samples. Furthermore, the effects of Fe-TA NPs on serum and liver iron status were assessed in male rats. Serum iron, serum TIBC, and transferrin saturation values in the treated groups were not changed significantly when compared to the vehicle control groups. Fe-TA NPs treatment significantly increased liver iron content at 5.5 mg/kg bw dose; however, ferritin content between the treated groups and the vehicle control group were not determined to be different (Figure 2).

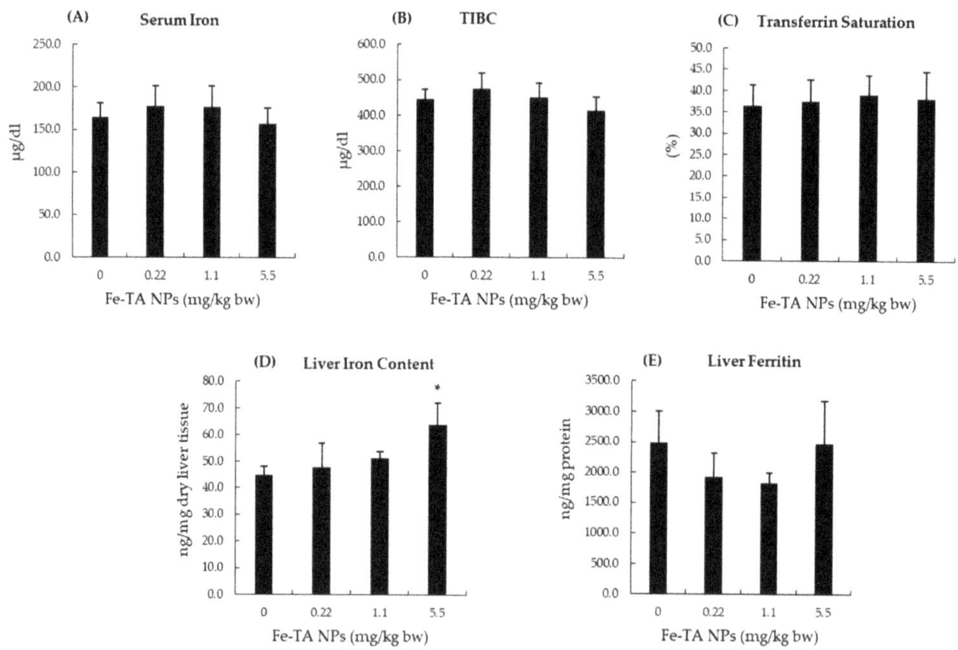

Figure 2. Effect of 28-day experiment of Fe-TA NPs on serum and liver iron parameters in male rats. (**A**): Serum iron, (**B**): TIBC (Total iron binding capacity), (**C**): Transferrin saturation, (**D**): Liver iron content, and (**E**) Liver ferritin. Data are expressed as Mean ± SD. * $p < 0.05$, significantly different when compared with a control group.

Table 2. Effect of Fe-TA NPs on hematological profile in rats exposed to repeated dose toxicity tests.

Parameters	Normal Range	Fe-TA NPs Treatment (mg/kg bw)					
		0	0.22	1.1	5.5	Satellite	
						0	5.5
Female							
RBC ($\times 10^6$ cell/μL)	4.6–9.2	7.3 ± 0.3	6.9 ± 0.6	7.6 ± 0.5	7.3 ± 0.4	8.5 ± 0.1	8.7 ± 0.6
Hemoglobin (g/dL)	11–19.2	15.7 ± 0.4	14.4 ± 1.2	16.1 ± 0.8	15.2 ± 0.8	18.1 ± 0.6	18.8 ± 1.1
Hematocrit (%)	36–53	48 ± 0.6	44 ± 3.2	49 ± 2.3	45.9 ± 2.0	52.9 ± 1.3	52.9 ± 2.7
MCV (fL)	48–70	66 ± 2.4	64 ± 1.3	65 ± 2.7	63 ± 1.1	62 ± 2.0	61 ± 2.0
MCH (pg)	16.0–23.1	21.7 ± 0.5	21.0 ± 0.8	21.3 ± 0.9	20.9 ± 0.8	21.3 ± 0.8	21.6 ± 0.6
MCHC (g/dL)	28.2–34.1	33.1 ± 0.7	32.7 ± 0.6	33.0 ± 0.3	33.2 ± 0.7	34.2 ± 0.7	35.5 ± 0.2
RDW (%)	10.0–16.0	12.2 ± 0.3	12.3 ± 0.8	12.2 ± 0.4	12.7 ± 0.5	12.6 ± 1.0	13.0 ± 0.9
Reticulocytes (%)	1.7–4.9	4.9 ± 1.6	3.8 ± 1.4	3.9 ± 0.8	3.7 ± 1.4	3.0 ± 1.0	2.9 ± 0.9
WBC ($\times 10^3$ cell/mm^3)	2–17	4 ± 1	3 ± 1	5 ± 1	5 ± 0	5 ± 1	4 ± 1
Neutrophil (%)	10–30	12 ± 3.9	12 ± 3.4	15 ± 10.4	12 ± 10.3	10 ± 1.0	16 ± 2.8
Lymphocyte (%)	65–85	83 ± 4.3	83 ± 4.30	79 ± 7.9	84 ± 8.5	85 ± 2.1	77.5 ± 2.4
Monocyte (%)	0–5	4 ± 2.5	4.8 ± 2.5	6 ± 2.95	4 ± 2.1	5 ± 0.1	7 ± 1.5
Eosinophil (%)	0–6	1 ± 1.2	0.4 ± 0.89	0.2 ± 0.45	0.0 ± 0.0	0.3 ± 0.5	0.0 ± 0.0
Basophil (%)	0–1	0.0 ± 0.0	0.0 ± 0.0	0.0 ± 0.0	0.0 ± 0.0	0.0 ± 0.0	0.0 ± 0.0
PLT ($\times 10^3$ cell/mm^3)	500–1300	787 ± 67	480 ± 337	762 ± 77	67.7 ± 206	793 ± 38	625 ± 315
Male							
RBC ($\times 10^6$ cell/μL)	4.6–9.2	5.8 ± 0.2	5.8 ± 0.3	6.0 ± 0.2	5.8 ± 0.2	6.7 ± 0.8	6.4 ± 0.9
Hemoglobin (g/dL)	11–19.2	16.7 ± 0.3	16.7 ± 0.3	16.7 ± 0.4	16.1 ± 0.5	15.1 ± 0.6	14.7 ± 0.5
Hematocrit (%)	36–53	49.1 ± 1.4	48.3 ± 1.6	49.3 ± 1.4	46.8 ± 1.1	45.4 ± 1.8	44.0 ± 1.6
MCV (fL)	48–70	84.0 ± 1.6	83.2 ± 2.6	82.4 ± 1.9	81.1 ± 1.8	68.1 ± 6.3	70.2 ± 8.4
MCH (pg)	16.0–23.1	28.6 ± 0.8	28.7 ± 1.0	28.0 ± 0.5	28.0 ± 0.5	22.6 ± 2.1	23.5 ± 2.8
MCHC (g/dL)	28.2–34.1	34.0 ± 0.5	34.5 ± 0.4	33.9 ± 0.5	34.5 ± 0.3	33.4 ± 0.3	33.4 ± 0.1
RDW (%	10.0–16.0	10.9 ± 0.3	11.1 ± 0.7	11.2 ± 0.5	11.3 ± 0.6	14.6 ± 3.6	12.8 ± 3.4
Reticulocytes (%)	1.7–4.9	3.2 ± 2.1	3.0 ± 1.2	5.0 ± 1.2	5.4 ± 2.0	2.1 ± 1.7	2.3 ± 1.0
WBC ($\times 10^3$ cell/mm^3)	2–17	6 ± 1	6 ± 2	7 ± 1	7 ± 2	5 ± 3	5 ± 2
Neutrophil (%)	10–30	7 ± 2.1	10 ± 0.5	14 ± 3.5 *	15 ± 5.2 *	17 ± 0.9	15 ± 6.0
Lymphocyte (%)	65–85	89 ± 2.2	87 ± 0.9	86 ± 3.5	85 ± 5.2	82 ± 1.1	83 ± 6.3
Monocyte (%)	0–5	5 ± 0.8	3 ± 1.1 *	0 ± 0.0 *	0 ± 0.0 *	1 ± 0.5	2 ± 0.5
Eosinophil (%)	0–6	0 ± 0	0 ± 0	0 ± 0	0 ± 0	0 ± 0	0 ± 0
Basophil (%)	0–1	0 ± 0	0 ± 0	0 ± 0	0 ± 0	0 ± 0	0 ± 0
PLT ($\times 10^3$ cell/mm^3)	500–1300	770 ± 177	802 ± 151	770 ± 103	915 ± 216	731 ± 65.2	590 ± 129

Values are expressed as Mean ± SD. * $p < 0.05$, significantly different when compared with the vehicle control. (RBC, red blood cell; MCH, mean corpuscular hemoglobin; MCHC, mean corpuscular hemoglobin concentration; MCV, mean corpuscular volume; RDW, red cell distribution width; PLT, platelet; WBC, white blood cell).

Table 3. Effect of Fe-TA NPs on biochemical profiles in rats exposed to repeated dose toxicity tests.

Parameters	Normal Range	Fe-TA NPs Treatment (mg/kg bw)					
		0	0.22	1.1	5.5	Satellite 0	Satellite 5.5
Female							
Glucose (mg/dL)	85–132	129 ± 14	144 ± 25	142 ± 13	167 ± 20	61 ± 21	173 ± 61
Cholesterol (mg/dL)	40–130	63 ± 10	70 ± 9	65 ± 8	67 ± 12	69 ± 4	68 ± 10
Triglyceride (mg/dL)	80–190	55 ± 12	49 ± 30	39± 8	41 ± 13	80 ± 11	55 ± 20
BUN (mg/dL)	10–21	24 ± 2	22 ± 3	18 ± 3 *	19 ± 2 *	25 ± 5	22 ± 2
Creatinine (mg/dL)	0.5–1.0	0.3 ± 0.1	0.2 ± 0.0	0.2 ± 0.0	0.2 ± 0.0	0.3± 0.0	0.3 ± 0.1
Total protein (g/dL)	6.3–8.6	7.1 ± 0.2	6.8 ± 0.6	7.0 ± 0.2	7.1 ± 0.3	8.5 ± 0.3	9.0 ± 0.5
Albumin (g/dL)	3.3–4.9	4.6 ± 0.1	4.3 ± 0.3	4.4 ± 0.2	4.4 ± 0.1	4.7 ± 0.2	4.7 ± 0.2
AST (U/L)	39–92	37 ± 17	42 ± 19	34 ± 7	28 ± 5.9	37 ± 12	42 ± 14
ALT (U/L)	17–50	117 ± 42	102 ± 16	122 ± 31	95 ± 26	96 ± 16	103 ± 20
ALP (U/L)	39–216	49 ± 7	55 ± 10	51 ± 10	54 ± 17	46 ± 6	33 ± 2
Total bil (mg/dL)	0.05–0.17	0.14 ± 0.02	0.09 ± 0.02	0.12 ± 0.06	0.12 ± 0.05	0.23 ± 0.03	0.22 ± 0.01
Direct bil (mg/dL)	0.03–0.07	0.08 ± 0.02	0.08 ± 0.01	0.09 ± 0.03	0.01 ± 0.03	0.09 ± 0.02	0.08 ± 0.01
Uric acid (mg/dL)	1.4–3.7	4.2 ± 0.6	3.5 ± 1.1	3.9 ± 0.7	4.8 ± 1.0	5.0 ± 2.0	4.5 ± 1.8
Sodium (mmol/L)	140–150	143 ± 2	141 ± 1	143 ± 1	142 ± 1	143 ± 1	144 ± 1
Potassium (mmol/L)	4–5.9	5.3 ± 0.7	5.9 ± 0.5	5.5 ± 0.4	5.6 ± 0.5	4.9 ± 0.2	5.2 ± 0.3
Chloride (mmol/L)	96–107	104 ± 2	105 ± 1	105 ± 1	105 ± 1	102 ± 2	104 ± 1
TCO_2 (mmol/L)	13–27.1	20.6 ± 1.5	22.6 ± 1.3	22.4 ± 1.6	22.9 ± 2.7	22.6 ± 2.5	19.5 ± 1.5
Male							
Glucose (mg/dL)	85–132	214 ± 30	227 ± 44	290 ± 39	211 ± 45	176 ± 38	164 ± 23
Cholesterol (mg/dL)	40–130	66 ± 9	73 ± 20	64 ± 15	61 ± 9	73 ± 6	72 ± 10
Triglyceride (mg/dL)	80–190	40 ± 18	75 ± 24	47 ± 26	38 ± 12	75 ± 45	48 ± 23
BUN (mg/dL)	10–21	19 ± 2	18 ± 2	19 ± 2	17 ± 3	19 ± 3	17 ± 1
Creatinine (mg/dL)	0.5–1.0	0.8 ± 0.1	0.8 ± 0.1	0.8 ± 0.1	0.7 ± 0.0	0.8 ± 0.1	0.7 ± 0.0 #
Total protein (g/dL)	6.3–8.6	6.7 ± 0.2	6.7 ± 0.2	6.9 ± 0.2	6.6 ± 0.1	6.8 ± 0.3	6.5 ± 0.1
Albumin (g/dL)	3.3–4.9	3.7 ± 0.1	3.6 ± 0.1	3.7 ± 0.1	3.6 ± 0.1	3.6 ± 0.1	3.5 ± 0.1
AST (U/L)	39–92	114 ± 14	108 ± 13	114 ± 22	100 ± 19	116 ± 32	95 ± 14
ALT (U/L)	17–50	30 ± 5	33 ± 4	40 ± 10	31 ± 6	31 ± 5	26 ± 3
ALP (U/L)	39–216	103 ± 19	93 ± 4	96 ± 10	87 ± 23	75 ± 18	66 ± 16
Total bil (mg/dL)	0.05–0.17	0.07 ± 0.01	0.07 ± 0.01	0.05 ± 0.02	0.06 ± 0.01	0.08 ± 0.01	0.06 ± 0.02
Direct bil (mg/dL)	0.03–0.07	0.03 ± 0.01	0.02 ± 0.01	0.02 ± 0.00	0.03 ± 0.01	0.04 ± 0.01	0.04 ± 0.01
Uric acid (mg/dL)	1.4–3.7	5.3 ± 1.9	5.1 ± 1.4	7.3 ± 1.7	5.7 ± 0.8	2.8 ± 0.7	1.9 ± 0.6
Sodium (mmol/L)	140–150	145 ± 2	143 ± 4	144 ± 3	144 ± 1	146 ± 1	144 ± 1 #
Potassium (mmol/L)	4–5.9	5.7 ± 0.5	6.1 ± 0.8	6.7 ± 0.7	7.1 ± 0.7 *	5.2 ± 0.7	4.9 ± 0.6
Chloride (mmol/L)	96–107	100 ± 1	100 ± 3	101± 3	101 ± 2	101 ± 1	101 ± 1
TCO_2 (mmol/L)	13–27.1	25.8 ± 2.0	27.1 ± 1.6	25.6 ± 2.1	25.6 ± 1.6	25.0 ± 1.1	27.4 ± 1.6 #

Values are expressed as Mean ± SD. * $p < 0.05$, significantly different when compared with the control. # $p < 0.05$, significantly different when compared with the satellite control. (BUN: blood urea nitrogen, AST: aspartate aminotransferase, ALT: alanine aminotransferase, ALP: alkaline phosphatase, bil: bilirubin, TCO_2: total carbon dioxide test).

3.2. Clastogenic Effect of Iron-Tannic Molecular Nanoparticles in Rats

The clastogenic effect of Fe-TA NPs was assessed through rat liver micronucleus assay, which served as the primary test in a battery of genotoxicity tests. Micronuclei can be formed during cell division from chromosome fragmentation or as a consequence of a failure to be incorporated into the daughter nuclei [25]. A single or repeated treatment of Fe-TA NPs did not affect water and food intake (data not shown) or result in body weight gains (Table 4). No evidence of a significant increase was observed in the number of micronucleated hepatocytes, binucleated hepatocytes, and mitotic index values after either a single dose or repeated doses of 5.5 mg/kg bw of Fe-TA NPs. In addition, no statistical differences were observed in these parameters from a single treatment of Fe-TA NPs in terms of its LD_{50} value at a dose of 55 mg/kg bw in rats of both genders. On the other hand, an injection of DEN, a positive control, produced a significant increase in the number of micronucleated hepatocytes and the mitotic index value (7.30 + 2.19 and 2.56 + 0.28 per 1000 hepatocytes) when compared to the negative control. These findings indicate that Fe-TA NPs did not result in incidences of clastogenicity in rats.

Table 4. Effect of Fe-TA NPs on micronucleus formation in rat livers.

Fe-TA NPs Treatment (mg/kg bw)	Body Weight Gain (%)	Total Number of (Per 1000 Hepatocytes)			Mitotic Index
		MN	MNH	BH	
5% Tween 80	30.8 ± 4.6	0.29 ± 0.39	0.29 ± 0.39	16.43 ± 2.82	0.72 ± 0.12
Fe-TA NPs 55 mg/kg bw (Single dose)	29.1 ± 4.8	0.68 ± 0.72	0.68 ± 0.72	18.60 ± 1.95	0.93 ± 0.12
Fe-TA NPs 5.5 mg/kg bw (Single dose)	32.4 ± 6.9	1.00 ± 0.93	1.00 ± 0.93	13.80 ± 0.84	0.79 ± 0.10
Fe-TA NPs 5.5 mg/kg bw (Repeated doses)	29.4 ± 4.2	0.30 ± 0.45	0.30 ± 0.45	16.00 ± 1.58	0.78 ± 0.13

Values are expressed as Mean ± SD. MN: Micronucleus, MNH: Micronucleated hepatocyte, BH: Binucleated hepatocytes.

3.3. Carcinogenicity of Iron (III)–Tannic Acid Nanoparticles in Rats

The medium-term carcinogenicity test is a reliable and practical tool to predict the carcinogenic potential of chemicals compared to the others. Immunohistochemically detected glutathione S-transferase placental form (GST-P) positive focus as the end-point marker can be used as early indicator of preneoplastic lesions in rat liver [26]. Although no abnormal clinical features were observed in the rats, one rat died after receiving 7 doses of 17.5 mg/kg bw of Fe-TA NPs. No significant changes in body weight, the weight of some vital organs, and serum ALT levels were observed (Table 5). With regard to the liver preneoplastic marker, diethylnitrosamine, a hepatocarcinogen, could induce the number and area of GST-P positive foci (14.22 ± 3.77 foci and 1.55 ± 0.56 mm^2 per liver area (cm^2), respectively), whereas Fe-TA NPs at various doses did not induce GST-P positive foci formation in rat livers (Table 5). Furthermore, the administration of Fe-TA NPs ranged from 0.55–17.5 mg/kg bw did not alter the number of PCNA-positive and apoptotic cells in the livers. Accordingly, it might be suggested that the administration of intraperitoneal Fe-TA NPs up to 17.5 mg/kg bw, once a week for 10 times, was not carcinogenic to the livers of rats.

3.4. Effect of Iron (III)–Tannic Acid Nanoparticles on Some Biochemical Parameters in Serum and Livers of Rats

Iron is an essential trace element, and its systemic and cellular levels are tightly regulated. Following Fe-TA NPs administration in vehicle control rat groups, serum iron, TIBC, and transferrin saturation values were not determined to be different among experimental groups in comparisons with the control group. The administration of Fe-TA NPs at dosages of 0.55 and 5.5 mg/kg bw significantly increased iron content in the liver. However, no alteration in the ferritin levels of rat livers was found (Figure 3). In addition, there was no significant alteration of total, oxidized, and reduced glutathione contents in rat livers at 0.55 and 5.5 mg/kg bw dose of Fe-TA NPs. Notably, a high dose, 17.5 mg/kg bw, increased both the oxidized and reduced glutathione contents; however, this dosage did not alter their ratios when compared with the control group (Figure 4). This would indicate that an antioxidant response occurred at the above-mentioned dose in rat livers. The antioxidant enzyme activities in rat livers were further studied and results are presented in Figure 5. It was observed that 0.55 mg/kg bw doses of Fe-TA NPs decreased catalase activity, whereas doses of 17.5 mg/kg bw increased these levels in rat livers. Heme oxygenase activity was also increased by a dose of 17.5 mg/kg bw of Fe-TA NPs. Remarkably, GR and GPx activities were not found to be altered in rat livers (Figure 5).

Table 5. Effect of iron (III)–tannic acid nanoparticles on general observation, serum ALT level, preneoplastic lesions formation, cellular apoptosis, and proliferation in rat livers.

Parameters	Fe-TA NPs Treatment (mg/kg bw)			
	0	0.55	5.5	17.5
Final Body Weight (g)	495 ± 33	463 ± 19	466 ± 30	478 ± 49
Absolute Organ weight (g)				
Liver	15.77 ± 1.96	14.00 ± 1.86	14.76 ± 2.67	17.27 ± 2.28
Spleen	0.88 ± 0.06	0.82 ± 0.13	0.77 ± 0.06	0.80 ± 0.17
Kidneys	3.40 ± 0.35	3.25 ± 0.23	3.39 ± 0.14	3.40 ± 0.74
Serum ALT levels (U/L)	41.7 ± 14.8	37.5 ± 8.6	32.4 ± 6.0	39.0 ± 14.5
Number of GST-P positive foci/Liver area (cm^2)	0.00 ± 0.00	0.00 ± 0.00	0.00 ± 0.00	0.00 ± 0.00
Area of GST-P positive foci (mm^2)/Liver area (cm^2)	0.00 ± 0.00	0.00 ± 0.00	0.00 ± 0.00	0.00 ± 0.00
Number of apoptotic cells/Liver area (mm^2)	5.27 ± 4.50	6.19 ± 1.48	6.17 ± 3.43	5.14 ± 1.01
Number of PCNA-positive cells/Liver area (mm^2)	8.31 ± 2.09	7.66 ± 1.87	8.95 ± 2.53	7.15 ± 2.16

Values are expressed as mean ± SD. ALT, alanine aminotransferase.

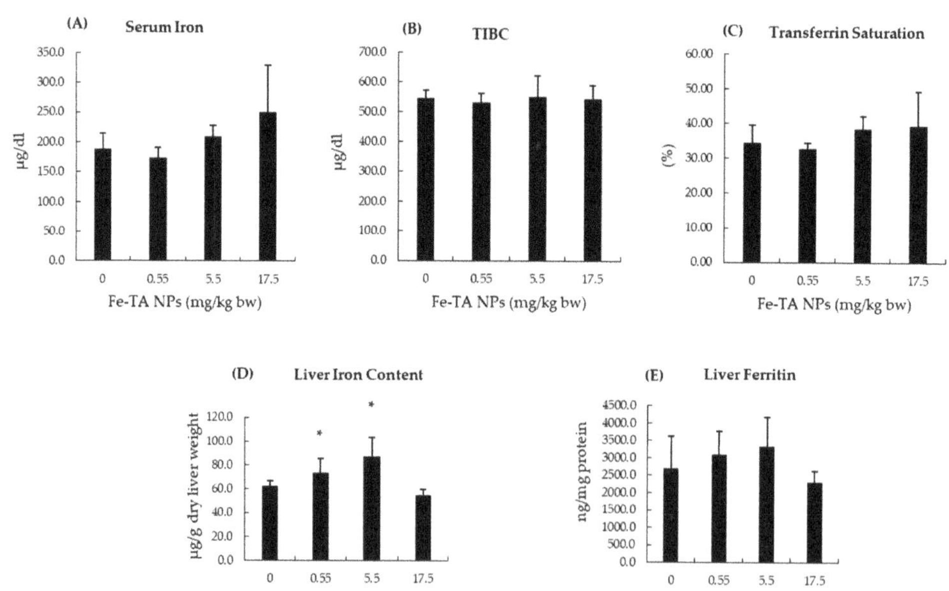

Figure 3. Effect of 10-week experiment of Fe-TA NPs on serum and liver iron parameters in rats. (**A**): Serum iron, (**B**): TIBC (Total iron binding capacity), (**C**): Transferrin saturation, (**D**): Liver iron content, and (**E**) Liver ferritin. Data are expressed as Mean ± SD. * $p < 0.05$, significantly different when compared with a control group.

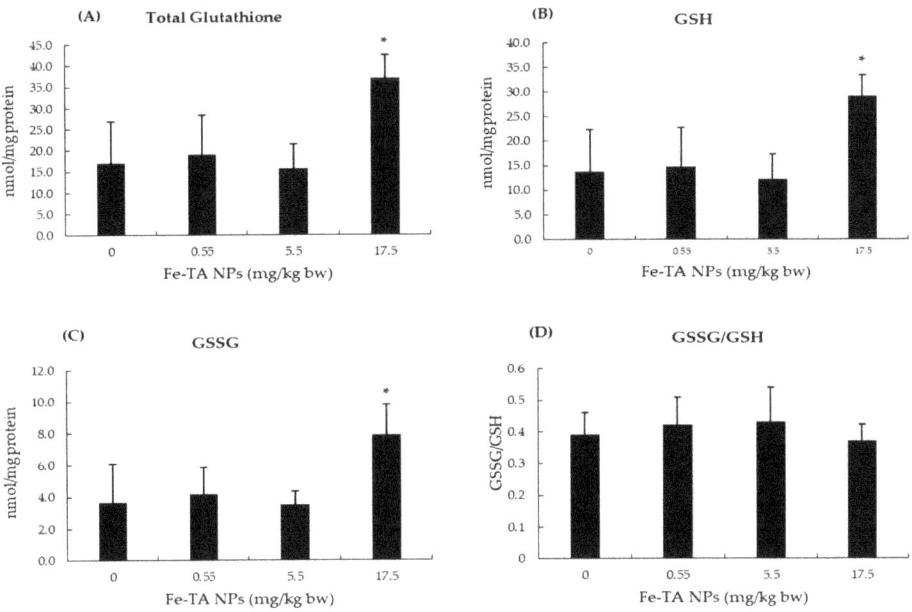

Figure 4. Effect of 10-week experiment of Fe-TA NPs on hepatic glutathione content in rats. (**A**): Total glutathione, (**B**): GSH (Reduced glutathione), (**C**): GSSG (Glutathione disulfide), and (**D**): GSSG/GSH (Glutathione disulfide to reduced glutathione ratio). Data are expressed as Mean ± SD. * $p < 0.05$, significantly different when compared with a control group.

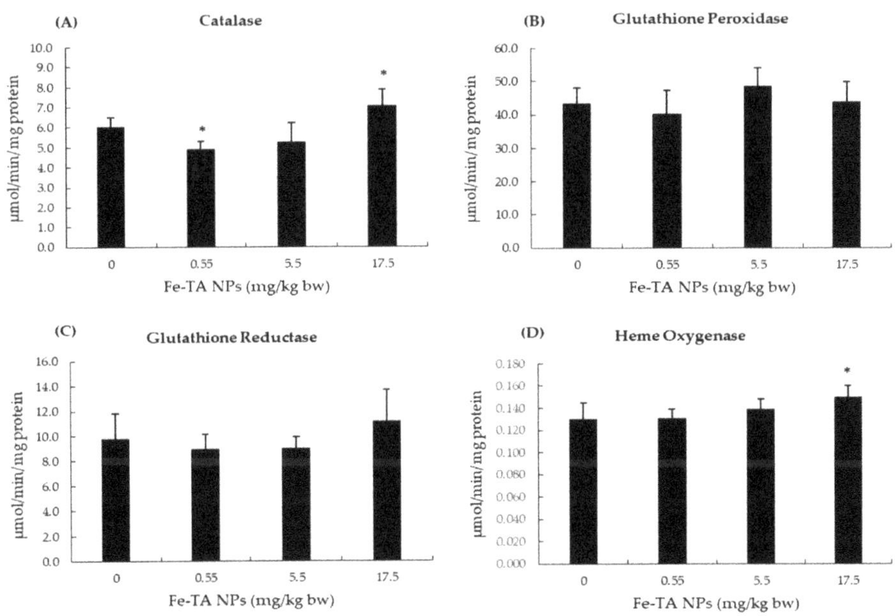

Figure 5. Effect of 10-week experiment of Fe-TA NPs on the activities of hepatic antioxidant enzymes in rats. (**A**): Catalase, (**B**): Glutathione peroxidase, (**C**): Glutathione reductase, and (**D**): Heme oxygenase. Data are expressed as Mean ± SD. * $p < 0.05$, significantly different when compared with a control group.

4. Discussion

Theranostic nanoparticles allow for the simultaneous application of diagnostic and therapeutic functionalities in a single platform and aim to improve better clinical efficiency for patients. These particles can offer great promise in the diagnostic imaging, monitoring, and treatment of cancer. The integration of iron and TA as a self-assembled nanoparticle can facilitate iron in providing an MRI signal, as well as to inhibit tumor cell growth and enable TA to exert anticancer effects. Fe–TA NPs enhance an MRI contrast signal in hepatocellular carcinoma cell lines and liver preneoplasia in rats, as well as to exhibit an anti-proliferative effect on hepatocellular carcinoma cells via enhanced autophagic cell death [10,11]. The outcomes of the present study have indicated that the LD_{50} value for intraperitoneal administration of Fe–TA NPs in rats was more than 55 mg/kg bw. Furthermore, repeated various doses of Fe–TA NPs did not result in incidences of subacute toxicity, genotoxicity, and hepatocarcinogenicity in rats.

Toxicological effects of NPs are dependent upon certain physicochemical properties such as size, shape, and surface charge; as well as composition, stability, and potential interactions with other biological systems [27].The in vivo toxicity assessments considered the uptake, biodistribution, metabolism, and clearance of NPs. While in circulation, NPs can be coated by serum proteins and then engulfed by both circulating and tissue macrophages, mainly in the liver, spleen, lymph nodes, and bone marrow of rats. We found no systemic toxicity following Fe–TA NPs administration at a dose of 0.55–5.5 mg/kg bw in rats. Importantly, body and organ weight changes are an indicator of substance toxicity [28]. Fe–TA NPs in the present study were found to increase the relative weights of the ovaries and seminal vesicles by 5.5 and 1.1 mg/kg bw dose, respectively; however, they did not induce any histopathological changes in these organs. Previous studies have reported that NPs can exert a toxic effect on reproductive organs [29]. The growth and activity of these organs are regulated by sex hormones that can alter organ weight and histology [30,31]. Therefore, it could be assumed that Fe–TA NPs treatments may be associated with weak endocrine activity.

The hematopoietic system is one of the sensitive parameters for toxicity assessment. Bone marrow and lymphoid organs are major hemopoietic sites in adults, while the liver and spleen can take part in this process under certain conditions. NPs can accumulate and exert toxic effects in these organs leading to impaired hematopoiesis [32]. A previous study involving intraperitoneal injections of Fe_3O_4 NPs particles at a dose of 500 mg/kg 3 times a week for 5 weeks in rats resulted in erythropoiesis stimulation, neutrophil leukocytosis, and increased numbers of Kupffer cells [33]. However, Fe–TA NPs treated rats of both sexes revealed no abnormal alterations in extramedullary hematopoietic sites and hematology parameters.

Iron is transported in the plasma as transferrin and approximately 30% of transferrin are saturated with iron under normal conditions. Cells take up iron by transferrin receptor (TfR)-1 on the plasma membrane. Depending upon cellular iron requirements, excess iron can be stored as ferritin and can hold about 4500 iron atoms per molecule. The expression of ferritin becomes increased when cellular iron concentrations rise [34]. NPs mainly accumulate in the liver particularly in Kupffer cells [35]. NPs generally stay within these cells and then are slowly released. It was found that an intraperitoneal repeated-dose administration ranging from 0.22–17.5 mg/kg bw of Fe–TA NPs did not alter the serum iron status parameters that were evaluated in the present study. Although elevated liver iron content was observed in 0.55 and 5.5 mg/kg bw of Fe–TA NPs treated rats, neither iron overload nor hepatotoxicity occurred as was indicated by normal serum ALT levels. However, at a dose of 17.5 mg/kg bw, liver iron content was not altered. This outcome might be related to the iron chelating activity of TA in Fe–TA NPs [36]. The hormesis of various biological endpoints has been reported to be associated with TA and iron levels, as well as the presence of nanoparticles [12,37,38]. It appears that Fe–TA NPs exhibited a hormetic dose response that provided iron at low doses and chelated iron at high doses in the liver.

Moreover, NPs are known to be able to damage the glomerular membranes causing nephrotoxicity [35]. Glomerular filtration and renal excretory function appeared not to be affected by Fe–TA NPs in this study. Different degrees of proteinuria in male rats in both the control and Fe–TA NPs treated groups could be due to excretion of sex-dependent proteins since it has been described that proteinuria was observed in male rats during sexual maturation at about 8 weeks of age [39,40]. In addition, a probable reason for increased serum potassium levels in male rats that were given a dose of 5.5 mg/kg bw of Fe–TA NPs may be associated with the interaction of NPs with ion channels such as K^+ and Na^+ channels, as well as Na^+/K^+-ATPase, thereby altering channel activities and ion homeostasis [41].

Carcinogenesis is a multistep process that includes initiation, promotion, and progression steps [42]. A chemical substance can introduce carcinogenesis by causing unrepairable DNA damage in normal cells resulting in mutated or initiated cells. Under certain circumstances, this could progress to malignant neoplasm. NPs can interact with macromolecules including DNA, proteins, and lipids. DNA alkylation and oxidative damage can result in either gene mutation or chromosomal abnormalities such as numerical alteration, aneuploidy, and chromosome fragmentation. An assessment of the genotoxic potential of Fe–TA NPs through rat liver micronucleus test has demonstrated that Fe–TA NPs are not clastogenic, indicating non-genotoxicity. Fe–TA NPs was further evaluated in terms of hepatocarcinogenicity through a medium-term carcinogenicity test, which is known to be a reliable method used to predict the carcinogenic potential of chemicals. Since the proliferation rate of an adult liver is low, a two-third partial hepatectomy was carried out to enhance proliferation. Moreover, the exposure of certain chemicals to the liver after a partial hepatectomy could increase the sensitivity of the assay in detecting carcinogenicity [26]. The present study found that Fe–TA NPs did not produce glutathione *S*-transferase placental form positive foci, a hepatic preneoplastic lesion in rats [26], while also indicating a lack of evidence for sustained cell proliferation and the ability to resist cell apoptosis in the liver. Hence, these outcomes indicate that Fe–TA NPs are not associated with hepatocarcinogenicity in rats.

Oxidative damage or modifications of macromolecules can lead to toxicity and carcinogenicity. NPs toxicity is mainly mediated through ROS generation [27]. TA has been reported to be associated with both antioxidant and pro-oxidant effects [43,44]. Moreover, TA can reduce ferric iron to ferrous iron which can then generate deleterious hydroxyl radicals through a Fenton reaction. Cellular antioxidants can respond to oxidative stress through enzymatic and non-enzymatic mechanisms [45]. Catalase and glutathione peroxidase catalyze the removal of hydrogen peroxide. They show different affinities to H_2O_2. Catalase has low affinity to H_2O_2, while glutathione peroxidase has high affinity to H_2O_2. Catalase could undergo irreversible or reversible inactivation depending upon the H_2O_2 concentration. However, glutathione peroxidase exhibited lesser sensitivity to H_2O_2 induced inactivation. Baud et al. reported that after being exposed to H_2O_2 (100 μM for 1 h) in oligodendrocytes, catalase activity was reduced but glutathione peroxidase activity was not changed [46]. In the present study, the decreased catalase activity at low dose of Fe-TA NPs would likely be due to its inactivation by H_2O_2 [21]. At higher doses, increased total glutathione content and heme oxygenase activity were observed, which would indicate alleviation of the oxidative environment while also indicating that catalase could have taken part in the antioxidant response. These results are in line with the findings of Canli et al. who reported alterations in antioxidant enzymes activities, including those of superoxide dismutase, CAT, GPx, GR, and glutathione *S*-transferase, in rat livers after oral administration of some metal-oxide NPs at doses of 0.5–50 mg/kg/day for 14 days. Importantly, this would not necessarily indicate a dose-response relationship and correlation among the enzymes at each dose [47].

5. Conclusions

According to a toxicological evaluation of Fe-TA NPs, the LD_{50} value was greater than 55 mg/kg bw. Repeated intraperitoneal administration of Fe-TA NPs at up to 17.5 mg/kg bw demonstrated no obvious signs of systemic toxicity, genotoxicity, or hepatocarcinogenicity. Consequently, Fe-TA NPs might be safe for use in diagnostic and therapeutic purposes. Since this study emphasized only the early phase of hepatocarcinogenicity, further studies are needed to explore the effects on other phases of hepatocarcinogenesis.

Author Contributions: Conceptualization, R.W.; methodology, C.B.H., A.C., C.P. (Charatda Punvittayagul) and S.S.; investigation, C.B.H. and A.C.; writing—original draft preparation, C.B.H. and A.C.; writing—review and editing, R.W.; supervision, R.W.; project administration, R.W.; funding acquisition, R.W. and C.P (Chalermchai Pilapong). All authors have read and agreed to the published version of the manuscript.

Funding: This work was supported by the Faculty of Medicine Research Fund and Chiang Mai University (grant number: 142-2563) and the National Research Council of Thailand (grant code: NRCT2561).

Institutional Review Board Statement: This study was conducted according to the guidelines of the Declaration of Helsinki and approved of by the Animal Ethics Committee of Faculty of Medicine, Chiang Mai University (Protocol No. 39/2560, approved on 14 February 2018, and No. 06/2563, approved on 17 April 2020).

Informed Consent Statement: Not applicable.

Data Availability Statement: All data presented in this study are available upon request from the corresponding author.

Acknowledgments: The authors are grateful to the Chiang Mai University Presidential Scholarship on the post-doctoral fellowship to Arpamas Chariyakornkul. We would also like to acknowledge the Animal Unit and Medical Science Research Equipment Center, Faculty of Medicine, Chiang Mai University for the generous provision of animal and equipment facilities.

Conflicts of Interest: The authors declare that they hold no conflicts of interest with regard to the publication of this manuscript.

References

1. Sung, H.; Ferlay, J.; Siegel, R.L.; Laversanne, M.; Soerjomataram, I.; Jemal, A.; Bray, F. Global cancer statistics 2020: GLOBOCAN estimates of incidence and mortality worldwide for 36 cancers in 185 countries. *CA Cancer J. Clin.* **2021**, *71*, 209–249. [CrossRef]
2. Daher, S.; Massarwa, M.; Benson, A.A.; Khoury, T. Current and future treatment of hepatocellular carcinoma: An updated comprehensive review. *J. Clin. Transl. Hepatol.* **2018**, *6*, 69. [CrossRef]
3. Ghanaati, H.; Alavian, S.M.; Jafarian, A.; Daryani, N.E.; Nassiri-Toosi, M.; Jalali, A.H.; Shakiba, M. Imaging and imaging-guided interventions in the diagnosis and management of hepatocellular carcinoma (HCC)-review of evidence. *Iran J. Radiol.* **2012**, *9*, 167. [CrossRef]
4. Marrero, J.A.; Kulik, L.M.; Sirlin, C.B.; Zhu, A.X.; Finn, R.S.; Abecassis, M.M.; Roberts, L.R.; Heimbach, J.K. Diagnosis, staging, and management of hepatocellular carcinoma: 2018 Practice Guidance by the American Association for the Study of Liver Diseases. *Hepatology* **2018**, *68*, 723–750. [CrossRef] [PubMed]
5. Silva, C.O.; Pinho, J.O.; Lopes, J.M.; Almeida, A.J.; Gaspar, M.M.; Reis, C. Current trends in cancer nanotheranostics: Metallic, polymeric, and lipid-based systems. *Pharmaceutics* **2019**, *11*, 22. [CrossRef]
6. Alshehri, S.; Imam, S.S.; Rizwanullah, M.; Akhter, S.; Mahdi, W.; Kazi, M.; Ahmad, J. Progress of cancer nanotechnology as diagnostics, therapeutics, and theranostics nanomedicine: Preclinical promise and translational challenges. *Pharmaceutics* **2021**, *13*, 24. [CrossRef]
7. Ruman, U.; Fakurazi, S.; Masarudin, M.J.; Hussein, M.Z. Nanocarrier-based therapeutics and theranostics drug delivery systems for next generation of liver cancer nanodrug modalities. *Int. J. Nanomed.* **2020**, *15*, 1437. [CrossRef] [PubMed]
8. Liu, P.; Shi, X.; Zhong, S.; Peng, Y.; Qi, Y.; Ding, J.; Zhou, W. Metal-phenolic networks for cancer theranostics. *Biomater. Sci.* **2021**, *9*, 2825–2849. [CrossRef] [PubMed]
9. Dai, Q.; Geng, H.; Yu, Q.; Hao, J.; Cui, J. Polyphenol-based particles for theranostics. *Theranostics* **2019**, *9*, 3170–3190. [CrossRef] [PubMed]
10. Saowalak, K.; Titipun, T.; Somchai, T.; Chalermchai, P. Iron (III)-tannic molecular nanoparticles enhance autophagy effect and T 1 MRI contrast in liver cell lines. *Sci. Rep.* **2018**, *8*, 6647. [CrossRef]

11. Phatruengdet, T.; Intakhad, J.; Tapunya, M.; Chariyakornkul, A.; Hlaing, C.B.; Wongpoomchai, R.; Pilapong, C. MRI contrast enhancement of liver pre-neoplasia using iron–tannic nanoparticles. *RSC Adv.* **2020**, *10*, 35419–35425. [CrossRef]
12. Jodynis-Liebert, J.; Kujawska, M. Biphasic dose-response induced by phytochemicals: Experimental evidence. *J. Clin. Med.* **2020**, *9*, 718. [CrossRef]
13. Fukushima, S.; Kinoshita, A.; Puatanachokchai, R.; Kushida, M.; Wanibuchi, H.; Morimura, K. Hormesis and dose–response-mediated mechanisms in carcinogenesis: Evidence for a threshold in carcinogenicity of non-genotoxic carcinogens. *Carcinogenesis* **2005**, *26*, 1835–1845. [CrossRef] [PubMed]
14. Vaiserman, A.; Cuttler, J.M.; Socol, Y. Low-dose ionizing radiation as a hormetin: Experimental observations and therapeutic perspective for age-related disorders. *Biogerontology* **2021**, *22*, 145–164. [CrossRef] [PubMed]
15. Phatruengdet, T.; Khuemjun, P.; Intakhad, J.; Krunchanuchat, S.; Chariyakornkul, A.; Wongpoomchai, R.; Pilapong, C. Pharmacokinetic/pharmacodynamic determinations of iron-tannic molecular nanoparticles with its implication in MR imaging and enhancement of liver clearance. *Nanotheranostics* **2022**, *6*, 195–204. [CrossRef] [PubMed]
16. Chariyakornkul, A.; Punvittayagul, C.; Taya, S.; Wongpoomchai, R. Inhibitory effect of purple rice husk extract on AFB 1-induced micronucleus formation in rat liver through modulation of xenobiotic metabolizing enzymes. *BMC Compl. Altern. Med.* **2019**, *19*, 1–11. [CrossRef] [PubMed]
17. Dokkaew, A.; Punvittayagul, C.; Insuan, O.; Limtrakul Dejkriengkraikul, P.; Wongpoomchai, R. Protective effects of defatted sticky rice bran extracts on the early stages of hepatocarcinogenesis in rats. *Molecules* **2019**, *24*, 2142. [CrossRef] [PubMed]
18. Koonyosying, P.; Kongkarnka, S.; Uthaipibull, C.; Svasti, S.; Fucharoen, S.; Srichairatanakool, S. Green tea extract modulates oxidative tissue injury in beta-thalassemic mice by chelation of redox iron and inhibition of lipid peroxidation. *Biomed. Pharmacother.* **2018**, *108*, 1694–1702. [CrossRef] [PubMed]
19. Lowry, O.H.; Rosebrough, N.J.; Farr, A.L.; Randall, R.J. Protein measurement with the Folin phenol reagent. *J. Biol. Chem.* **1951**, *193*, 265–275. [CrossRef]
20. Rahman, I.; Kode, A.; Biswas, S.K. Assay for quantitative determination of glutathione and glutathione disulfide levels using enzymatic recycling method. *Nat. Protoc.* **2006**, *1*, 3159–3165. [CrossRef] [PubMed]
21. Aebi, H. Catalase in vitro. *Methods Enzymol.* **1984**, *105*, 121–126. [CrossRef] [PubMed]
22. Carlberg, I.; Mannervik, B. Glutathione reductase. *Methods Enzymol.* **1985**, *113*, 484–490. [CrossRef]
23. Nagalakshmi, N.; Prasad, M.N. Responses of glutathione cycle enzymes and glutathione metabolism to copper stress in *Scenedesmus bijugatus*. *Plant Sci.* **2001**, *160*, 291–299. [CrossRef]
24. Punvittayagul, C.; Wongpoomchai, R.; Taya, S.; Pompimon, W. Effect of pinocembrin isolated from *Boesenbergia pandurata* on xenobiotic-metabolizing enzymes in rat liver. *Drug Metab. Lett.* **2011**, *5*, 1–5. [CrossRef] [PubMed]
25. Krishna, G.; Hayashi, M. In vivo rodent micronucleus assay: Protocol, conduct and data interpretation. *Mutat. Res.* **2000**, *455*, 155–166. [CrossRef]
26. Ito, N.; Tamano, S.; Shirai, T. A medium-term rat liver bioassay for rapid in vivo detection of carcinogenic potential of chemicals. *Cancer Sci.* **2003**, *94*, 3–8. [CrossRef] [PubMed]
27. Yang, W.; Wang, L.; Mettenbrink, E.M.; DeAngelis, P.L.; Wilhelm, S. Nanoparticle toxicology. *Annu. Rev. Pharmacol. Toxicol.* **2021**, *61*, 269–289. [CrossRef] [PubMed]
28. Andersen, H.; Larsen, S.; Spliid, H.; Christensen, N.D. Multivariate statistical analysis of organ weights in toxicity studies. *Toxicology* **1999**, *136*, 67–77. [CrossRef]
29. Wang, R.; Song, B.; Wu, J.; Zhang, Y.; Chen, A.; Shao, L. Potential adverse effects of nanoparticles on the reproductive system. *Int. J. Nanomed.* **2018**, *13*, 8487–8506. [CrossRef]
30. Rudmann, D.G.; Foley, G.L. Female Reproductive System. In *Haschek and Rousseaux's Handbook of Toxicologic Pathology*; Haschek, W.M., Rousseaux, C.G., Wallig, M.A., Eds.; Academic Press: Cambridge, MA, USA, 2013; Volume 3, pp. 2599–2633.
31. Creasy, D.M.; Chapin, R.E. Male reproductive system. In *Haschek and Rousseaux's Handbook of Toxicologic Pathology*, 3rd ed.; Haschek, W.M., Rousseaux, C.G., Wallig, M.A., Eds.; Academic Press: Cambridge, MA, USA, 2013; Volume 3, pp. 2493–2598.
32. Sharma, V.; Aneja, B.; Yata, V.K.; Malakar, D.; Mohanty, A.K. Systemic nanotoxicity and its assessment in animal models. In *Nanopharmaceuticals: Principles and Applications*; Yata, V.R.S., Dasgupta, N., Lichtfouse, E., Eds.; Springer: Cham, Switzerland; New York, NY, USA, 2020; Volume 3, pp. 201–243.
33. Katsnelson, B.A.; Degtyareva, T.D.; Minigalieva, I.I.; Privalova, L.I.; Kuzmin, S.V.; Yeremenko, O.S.; Kireyeva, E.P.; Sutunkova, M.P.; Valamina, I.I.; Khodos, M.Y.; et al. Subchronic systemic toxicity and bioaccumulation of Fe_3O_4 nano- and microparticles following repeated intraperitoneal administration to rats. *Int. J. Toxicol.* **2011**, *30*, 59–68. [CrossRef] [PubMed]
34. Anderson, G.J.; Frazer, D.M. Current understanding of iron homeostasis. *Am. J. Clin. Nutr.* **2017**, *106*, 1559S–1566S. [CrossRef] [PubMed]
35. Chrishtop, V.V.; Mironov, V.A.; Prilepskii, A.Y.; Nikonorova, V.G.; Vinogradov, V.V. Organ-specific toxicity of magnetic iron oxide-based nanoparticles. *Nanotoxicology* **2020**, *15*, 167–204. [CrossRef] [PubMed]
36. Phiwchai, I.; Yuensook, W.; Sawaengsiriphon, N.; Krungchanuchat, S.; Pilapong, C. Tannic acid (TA): A molecular tool for chelating and imaging labile iron. *Eur. J. Pharm. Sci.* **2018**, *114*, 64–73. [CrossRef] [PubMed]
37. Saul, N.; Pietsch, K.; Menzel, R.; Sturzenbaum, S.R.; Steinberg, C.E. The longevity effect of tannic acid in *Caenorhabditis elegans*: Disposable Soma meets hormesis. *J. Gerontol. A Biol. Sci. Med. Sci.* **2010**, *65*, 626–635. [CrossRef] [PubMed]

38. Iavicoli, I.; Leso, V.; Fontana, L.; Calabrese, E.J. Nanoparticle exposure and hormetic dose-responses: An update. *Int. J. Mol. Sci.* **2018**, *19*, 805. [CrossRef]
39. Perry, S. Proteinuria in the Wistar rat. *J. Pathol. Bacteriol.* **1965**, *89*, 729–733. [CrossRef] [PubMed]
40. Alt, J.M.; Hackbarth, H.; Deerberg, F.; Stolte, H. Proteinuria in rats in relation to age-dependent renal changes. *Lab. Anim.* **1980**, *14*, 95–101. [CrossRef]
41. Yin, S.; Liu, J.; Kang, Y.; Lin, Y.; Li, D.; Shao, L. Interactions of nanomaterials with ion channels and related mechanisms. *Br. J. Pharmacol.* **2019**, *176*, 3754–3774. [CrossRef]
42. Tanaka, T.; Shimizu, M.; Kochi, T.; Moriwaki, H. Chemical-induced carcinogenesis. *J. Exp. Clin. Med.* **2013**, *5*, 203–209. [CrossRef]
43. Khan, N.S.; Ahmad, A.; Hadi, S.M. Anti-oxidant, pro-oxidant properties of tannic acid and its binding to DNA. *Chem.-Biol. Interact* **2000**, *125*, 177–189. [CrossRef]
44. Cai, Y.; Zhang, J.; Chen, N.G.; Shi, Z.; Qiu, J.; He, C.; Chen, M. Recent advances in anticancer activities and drug delivery systems of tannins. *Med. Res. Rev.* **2017**, *37*, 665–701. [CrossRef] [PubMed]
45. Birben, E.; Sahiner, U.M.; Sackesen, C.; Erzurum, S.; Kalayci, O. Oxidative stress and antioxidant defense. *World Allergy Organ. J.* **2012**, *5*, 9–19. [CrossRef] [PubMed]
46. Baud, O.; Greene, A.E.; Li, J.; Wang, H.; Volpe, J.J.; Rosenberg, P.A. Glutathione peroxidase-catalase cooperativity is required for resistance to hydrogen peroxide by mature rat oligodendrocytes. *J. Neurosci.* **2004**, *24*, 1531–1540. [CrossRef] [PubMed]
47. Canli, E.G.; Ila, H.B.; Canli, M. Response of the antioxidant enzymes of rats following oral administration of metal-oxide nanoparticles (Al_2O_3, CuO, TiO_2). *Environ. Sci. Pollut. Res. Int.* **2019**, *26*, 938–945. [CrossRef] [PubMed]

nanomaterials

MDPI

Article

An Assessment of the Oral and Inhalation Acute Toxicity of Nickel Oxide Nanoparticles in Rats

Tara Lyons-Darden [1,*], Jason L. Blum [2], Mark W. Schooley [2], Melissa Ellis [2], Jennifer Durando [2], Daniel Merrill [2] and Adriana R. Oller [3]

1 NiPERA Inc., 2525 Meridian Pkwy Ste 240, Durham, NC 27713, USA
2 Product Safety Labs, 2394 US Highway 130, Dayton, NJ 08810, USA
3 Oller Consulting, 722 Gaston Manor Dr, Durham, NC 27703, USA
* Correspondence: tlyons-darden@nipera.org

Abstract: Nickel oxide nanoparticles (NiO NPs) have been the focus of many toxicity studies. However, acute toxicity studies that identify toxicological dose descriptors, such as an LC_{50} or LD_{50}, are lacking. In this paper, the acute toxicity of NiO NPs was evaluated in albino-derived Sprague-Dawley rats through OECD guideline studies conducted by both the oral and inhalation routes of exposure. The animals were assessed for mortality, body weight, behavioral observations, and gross necropsy. Results from previously conducted (unpublished) acute inhalation studies with larger NiO microparticles (MPs) are also included for comparison. Mortality, the primary endpoint in acute toxicity studies, was not observed for rats exposed to NiO NPs via either the oral or inhalation exposure routes, with a determined LD_{50} of >5000 mg/kg and an LC_{50} > 5.42 mg/L, respectively. Our results suggest that these NiO NPs do not exhibit serious acute toxicity in rats or warrant an acute toxicity classification under the current GHS classification criteria. This aligns with similar results for NiO MPs from this and previously published studies.

Keywords: nickel; nickel oxide; nanoparticle; acute toxicity; nanotoxicity

1. Introduction

Nickel (Ni) is an abundant natural element that is essential for plants and thus is naturally found in many of the foods that we safely consume [1–4]. Additionally, nickel's physical and chemical properties contribute to the widespread utilization of Ni, Ni compounds, and Ni alloys in many industrial and commercial applications. With this level of exposure, it is no surprise that general Ni toxicity has been widely studied, with the release of Ni^{2+} ions and their ability to interact with specific biological targets (cells, DNA, etc.) identified as the primary source of toxicity. Many studies have shown nickel's primary association with allergic contact dermatitis after skin contact as well as respiratory effects such as toxicity and/or carcinogenicity after inhalation exposure [5,6]. While established toxicological profiles are available for the traditional Ni-containing particles of various water insoluble Ni substances, this is not the case for much smaller Ni-containing nanomaterials (NMs).

Nanomaterials are defined as having at least one dimension between 1 and 100 nm [7,8]. Their unique size-dependent properties have contributed to a considerable increase in the production and use of engineered nanoparticles (NPs), including Ni NPs, for various applications (e.g., electronics, textiles, aviation, catalysis, etc.) in recent decades [8–11]. At the nanoscale, many of the physicochemical properties differ from larger bulk-sized or microscale for the same materials [7,8]. Greater surface area, higher surface energy, and magnetism are just some of the properties that make NMs advantageous in a wide variety of products such as electronic circuits, surface coatings, and other technologies [8,12–14]. Within these specialized consumer products, NMs are usually tightly embedded, with minimal exposure risk expected for the general public [12,15,16]. However, the increased utilization of these NMs has raised some concern about their potential effects on health and safety with regard to occupational exposures arising from the handling, processing, and analyzing of NMs [13,15,17].

The health effects data are overwhelmingly lacking for many NMs, specifically OECD guideline studies used to establish toxicological dose descriptors. This is true for nickel-based NPs, including nickel oxide (NiO) NPs. Many in vitro and in vivo studies evaluating a multitude of health effects have been conducted with NiO NPs. Acute and chronic studies have shown that NiO NPs are associated with increased lung injury, inflammation, oxidative stress, apoptosis, and altered biochemical indices [18–25]. A detailed analysis of these particular endpoints with considerations of various exposure routes, exposure durations, animal species, etc., is beyond the scope of this paper. However, several recently published literature reviews have evaluated many of these studies and the potential health effects of oxidic Ni NPs [26–28]. The acute in vivo studies evaluated in these reviews primarily focused on inflammation, histopathology, cytotoxicity, genotoxicity, oxidative stress, apoptosis, hematopoietic effects, and functional/biochemical indices. Yet, most of these reports do not provide enough information to determine toxicological dose descriptors for key acute toxicity endpoints that can be used for the eventual hazard classification of NPs.

This paper presents data assessing the acute toxicity of engineered NiO NPs and identifies dose descriptors for acute toxicity via both the oral and inhalation routes of exposure. The exposure concentrations selected for these in vivo animal studies are admittedly much higher than any expected acute human exposure [29–31]. However, the selected exposure concentrations are aligned with the current criteria for the Globally Harmonized System of Classification and Labelling of Chemicals (GHS), which would likely be used to assign a hazard classification to these NPs. Additionally, the results from the acute oral and inhalation animal studies with NiO NPs are compared to previously conducted studies that evaluated larger non-nanoscale (micron) NiO microparticles (MPs), which are more widely used and have a more established toxicological profile. Specifically, NiO MP data from a previous acute oral study [32] as well as data from acute inhalation studies (performed in 2010 and 2013 but not previously published) are presented in this paper for comparison with the NiO NP results. The study designs and protocols implemented for the NPs are very similar to the previous studies conducted with the MPs.

2. Materials and Methods

2.1. Test Substances

A commercially available 20 nm engineered NiO NP was chosen and purchased from Nanoshel, LLC (County Cavan, Ireland; Product # NS6130-03-337). A non-nanoscale (micron) NiO black powder (provided by a member of the Nickel REACH Consortia) was also evaluated. The NiO NPs and MPs were stored at room temperature in amber glass bottles topped with nitrogen. Table 1 lists the physicochemical properties of these NiO test substances.

Table 1. Physico-chemical characteristics of the NiO test samples.

Test Substance	Purity (%)	BET Surface Area (m^2/g)	Hydrodynamic Particle Size [a] (μm)	Shape	Reactive Oxygen Species [b] (nM)	Zeta Potential [b] (mV)	Dispersity Index [b]
NiO NPs	99.9	74	0.210	Rectangular prism	23.8	−14.6	0.735
NiO MPs	96.4	96	15.6	NA	NA	NA	NA

[a] Particle diameter (d0.5) was measured with Malvern Zetasizer (Malvern, Westborough, MA, USA; dynamic light scattering); NPs were measured in water; MPs had been previously measured in Isopar G, an organic solvent.
[b] Analyses were evaluated in water. NA: not available.

2.2. Animal Care and Observations

All animal studies were conducted at Product Safety Labs (Dayton, New Jersey, USA) under animal use protocols approved by Product Safety Labs' Institutional Animal Care and Use Committee. Albino-derived Sprague-Dawley rats between 8 and 12 weeks of age were obtained from several different sources (see Table 2), as these studies were conducted in different time periods. The rats were housed according to the *Guide for the Care and Use of Laboratory Animals* [33,34], with a 12 h light/dark photoperiod cycle and acclimated to their new environment for 7–9 days or 26–39 days for the MP and NP studies, respectively. Laboratory conditions were within the OECD guideline recommendations during exposures, with reported temperatures between 20 and 25 °C, and 50–70% relative humidity. During the 14-day observation period following exposures, in most cases the temperature and humidity remained within OECD guideline recommendations. However, the humidity was elevated up to 86% for 6 days of the 14-day observation period following NP inhalation exposure. A portable dehumidifier was utilized to lower the high humidity levels resulting from facility deviations during the summer months. The elevated humidity conditions were considered minor, with no impact on the NP study. The rats were provided ad libitum access to food and filtered tap water, except no access to food was available when the rats were fasted overnight prior to oral exposure or during inhalation exposure.

Table 2. Details about the animals used in the acute studies.

Test Substance	Animal Species Strain	Source	Age (Weeks)	Initial Body Weight (Grams)	Acclimation Period (Days)
NiO NPs (oral and inhalation studies)	Rat Sprague-Dawley, derived albino	Charles River Laboratories	10–12	320–463, males 202–271, females	26–39
NiO MPs (low-dose inhalation study)	Rat Sprague-Dawley, derived albino	Ace Animals, Inc.	8–9	257–277, males 180–211, females	9
NiO MPs (high-dose inhalation study)	Rat Sprague-Dawley, derived albino	Harlan Laboratories, Inc.	8–9	222–246, males 158–172, females	7

During the 14-day observation period, animals were examined daily for mortality, signs of gross toxicity, and behavioral changes. Individual animal body weight was recorded prior to administration of NiO NPs or MPs on observation days 7 and 14 for the oral study and on observation days 1, 3, 7, and 14 for the inhalation studies. At the end of the 14-day observation period, the rats were euthanized via CO_2 inhalation and their internal organs were visually examined for gross abnormalities.

2.3. Acute Oral Study

The acute oral toxicity study was conducted with six female rats (nulliparous and non-pregnant) according to the OECD 425 Up-And-Down Procedure. Single treatment dose progressions via oral gavage were conducted one at a time with careful evaluation of the animal(s) before moving on to a lower or higher dose. An initial administration of 175 mg/kg (administered as 50% *w*/*w* in distilled water) was given to a female rat, with the upper limit for higher doses capped at 5000 mg/kg. LD_{50}s above this dose do not result in acute oral toxicity classification according to GHS hazard classification criteria. All animals were observed for 14 days following oral administration of NiO NPs. A median lethal oral dose (LD_{50}), indicating 50% mortality, was determined.

2.4. Acute Inhalation Studies

The nose-only exposure studies were conducted using male and female (nulliparous and non-pregnant) rats according to the OECD 403 Acute Inhalation Toxicity guidelines. A total of 40 rats were exposed to the NiO NPs and 20 rats were exposed to the NiO MPs. The NiO NPs and high dose MP powders (8 mg/L) were aerosolized with a Fluid Energy Jet Mill (Schenck Process, Whitewater, WI, USA; Serial J2724E), while a modified Wright Dust

Generator (Product Safety Labs, Dayton, NJ, USA; Model # 4Z538A) was used to generate the lower dose (5 mg/L) of MPs. Chamber concentration measurements and particle size distributions were obtained from weighing glass fiber filters (and filter paper in the case of particle size distribution for the MPs) before and after collection of breathing zone samples from the animals at various intervals per exposure. The inhalation study with NiO NPs included exposure to aerosolized NiO NPs for 4 h at a 5 mg/L target concentration. LC_{50}s above this concentration do not result in acute inhalation toxicity classification according to GHS hazard classification criteria. The study was carried out in two separate rounds that included a total of 40 animals. The inhalation studies with NiO MPs evaluated single exposures for 4 h at target concentrations of 5 and 8 mg/L. The studies were carried out in two rounds conducted 3 years apart (initial round with 5 mg/L and second round with 8 mg/L). All animals were observed for 14 days following inhalation. A median lethal inhalation concentration (LC_{50}), indicating 50% mortality, was determined.

3. Results

3.1. Acute Oral Toxicity

Following a single oral exposure to NiO NPs at increasing doses from 175 to 5000 mg/kg, no mortality was observed following the 14-day observation period, resulting in an LD_{50} of >5000 mg/kg (Table 3). Additionally, no significant differences in body weights were reported. Individual body weight observations following acute oral exposure to NiO NPs are presented in Supplemental Information, Table S1. All animals appeared to be active and healthy during in-life behavioral observations, with no gross abnormalities observed during necropsy (Supplemental Information, Table S3).

Table 3. Acute oral exposure results.

Test Substance	Dose Level (mg/kg)	Mortality (# Dead/# Treated)	LD_{50} (mg/kg)
NiO NPs	175	0/1	>5000
	550	0/1	
	1750	0/1	
	5000	0/3	
Data below previously published in Henderson et al., 2012 [32]			
NiO MPs	5000	0/1	9990
	6300	0/1	
	7930	0/3	
	9990	2/4	
	11,000	2/2	

Previously published data with NiO MPs [32] can be compared to the NP data. Following oral exposure to NiO MPs at increasing doses from 5000 to 11,000 mg/kg, mortality was observed during the 14-day observation period at doses of 9990 mg/kg and higher, resulting in an LD_{50} of 9990 mg/kg (Table 3). Individual body weight observations following acute oral exposure to NiO MPs are presented in Supplemental Information, Table S2. At doses below 9990 mg/kg, all animals appeared to be active and healthy during in-life behavioral observations with no gross abnormalities observed during necropsy. However, at doses of 9990 mg/kg and above, several behavioral and gross abnormalities were observed including: reduced fecal volume, hypoactivity, facial staining, and discolored intestines (Supplemental Information, Table S4).

3.2. Acute Inhalation Toxicity

Following a single inhalation exposure (4 hrs) to NiO NPs at 5 mg/L in two subsequent rounds, no mortality was observed during the 14-day observation period, with the aggregated data resulting in an LC_{50} of >5.42 mg/L (Table 4). Similarly, single inhalation exposures (4 hrs) to NiO MPs at target levels of 5 and 8 mg/L did not result in mortality, leading to an LC_{50} of >8.30 mg/L. Despite the small nanoscale size of the primary NPs, the mass median aerodynamic diameter (MMAD) of the aerosolized NPs ranged from 3 to 4 μm, similar in size to the MMAD of the MPs. Reduced individual body weights were observed during the first 7 days following inhalation for most animals, with resumed weight gain by the end of the observation period (Supplemental Information, Tables S5 and S6 for NPs and MPs, respectively). For exposures to NiO NPs and MPs, the most common in-life behavioral observations were irregular respiration and hypoactivity during the 14-day observation period and slight to moderate lung discoloration was often observed during necropsy (Supplemental Information, Tables S7 and S8 for NPs and MPs, respectively).

Table 4. Acute inhalation exposure results.

Test Substance	Nominal Conc. (mg/L)	Measured Conc. (mg/L)	MMAD (μm)	Mortality (# Dead/# Treated)		LC_{50} (mg/L)
				Male	Female	
NiO NPs	5.0	5.41–5.42	3.01–4.32	0/20	0/20	>5.42
NiO MPs	5.0	5.15	2.4–2.5	0/5	0/5	>8.30
	8.0	8.30	2.95–3.01	0/5	0/5	

4. Discussion

Acute oral and inhalation studies with NiO NPs were conducted to determine their classification for acute toxicity within the current regulatory framework. These studies evaluated doses that align with those needed for GHS hazard classifications for oral and inhalation toxicity. Both the oral and inhalation studies with NPs resulted in no mortality at any of the exposure levels tested. These results are similar to the previously reported oral [32] and currently reported inhalation results for non-nanoscale NiO MPs. Exposure to NiO MPs also did not result in mortality at similar concentrations, though NiO MPs did induce mortality at much higher oral concentrations. Our results indicate that NiO NPs (as with their micron size counterparts) would not warrant an acute toxicity classification under the current GHS classification criteria, an indication of no acute toxicity or hazard concern for single oral or 4 h inhalation exposures.

The results of our oral and inhalation studies, tested at much higher exposure concentrations than expected for human exposure [29–31], did not indicate acute toxicity in the form of mortality or other serious effects for NPs, although irregular respiration and discolored lungs were noted in the inhalation study. These results align with our observations for MPs at similar doses. However, at even higher doses than evaluated for the NPs, additional adverse effects were observed for the MPs, including mortality. An important point to note is that the behavioral observations and gross abnormalities noted for the NPs did not differ from those observed for the MPs, indicating no new nano-specific effects of concern. This aligns with other research that suggested similar toxicity profiles for NPs and their non-nanoscale counterparts [26,35]. Additionally, many other studies reporting adverse effects (e.g., inflammation, cytotoxicity, reactive oxygen species or ROS, altered biochemical indices, etc.) for nickel NPs did not always put the results into perspective to indicate if or how they differed from their non-nanoscale counterparts. Further, many in vivo studies that reported adverse effects for nickel NPs used exposure routes that are not relevant for potential human exposure (e.g., intratracheal instillation, oropharyngeal aspiration, and intraperitoneal injection). Our results show that relevant exposure routes with very high exposure concentrations (based on GHS hazard criteria) of NiO NPs do not result in severe acute toxicity or new nano-specific effects. Similar results are to be expected for the much lower exposure concentrations relevant for potential human exposure.

One possible explanation for the similarity of the in vivo results observed for the MPs and NPs may be the tendency of NPs to agglomerate or aggregate to sizes similar to those of MPs [17,36]. Nanoparticle agglomeration is also supported by the measured zeta potential (−14.6 mV) of our NiO NPs, which is within the range to suggest colloidal instability and the increased likelihood of agglomeration or aggregation [37]. Additionally, typical nickel toxicity for non-nanoscale MPs is related to the release of nickel ions and the general low solubility of NiO renders it less likely to release nickel ions compared to more soluble nickel compounds [38,39]. This could be the same for NiO NPs and explain the comparable in vivo results with NiO MPs, even though there is often speculation that NPs have the potential to be more toxic than their non-nanoscale counterparts due to unique size-specific characteristics. There is a general assumption of higher surface reactivity (as measured by ROS) and larger surface area for NPs [40–43], yet the latter did not hold true for the NiO NPs evaluated in this study. The characterization of the NiO particles (as received) based on a nitrogen adsorption technique indicated that the MPs and NPs had similar surface areas; however, this information did not match the quite different median hydrodynamic diameters measured in water (NPs) and isopar G (MPs). It is uncertain if the surface areas remain similar in vivo once ingested or inhaled, but the similar MMADs indicate that this may be the case at least for inhalation. Interestingly, the high surface reactivity of our NiO NPs (23.8 nM), as measured by ROS, would seem to suggest that the nanoparticles are quite reactive with possibly increased dissolution that would in turn result in a greater likelihood for toxicity. However, similar to the Kunc et al. study, which reported that ROS did not appear to impact cytotoxicity [44], ROS did not appear to impact the acute toxicity results in our studies. Thus, as mentioned above, it appears that a factor that could be influential in determining acute toxicity is agglomeration/aggregation, which may have been responsible for the similar MMAD values of the MPs and NPs in the inhalation studies.

Although our results suggest that NiO NPs, similarly to NiO MPs, do not exhibit acute or severe toxicity in rats via oral or inhalation exposure routes, there are still questions to be answered. For instance, what is the fate of these NPs (e.g., the uptake in local and secondary organs, macrophage phagocytosis, rate of excretion, etc.) once inhaled or ingested at the doses employed in our studies? What size are the NPs once inhaled or ingested? If they become agglomerated once inhaled or ingested, do they remain the same size or does further agglomeration or dissolution occur over time by physiological processes in the body? Future studies evaluating these questions could provide additional information on NiO NPs.

As a preliminary assessment, our studies were conducted according to the standard OECD guidelines for acute toxicity tests that were developed with non-nanoscale particles in mind. The OECD previously deemed "the approaches for the testing and assessment of traditional chemicals are in general appropriate for assessing the safety of nanomaterials but may have to be adapted to the specificities of nanomaterials" [45]. However, it is important to consider that nano-specific modifications (such as criteria/cutoff values for GHS classification, target MMAD for inhalation studies, adjustments to study protocols, etc.) for the acute toxicity testing of NMs could prompt the need for future follow-up studies.

Supplementary Materials: The following supporting information can be downloaded at: https://www.mdpi.com/article/10.3390/nano13020261/s1: Table S1 contains the individual animal body weight observations following acute oral exposure to NiO NPs. Table S2 contains the individual animal body weight observations following acute oral exposure to NiO MPs (Henderson et al., 2012). Table S3 contains the individual in-life behavioral and necropsy observations following acute oral exposure to NiO NPs. Table S4 contains the individual in-life behavioral and necropsy observations following acute oral exposure to NiO MPs (Henderson et al., 2012). Table S5 contains the individual body weight observations following acute inhalation exposure to NiO NPs. Table S6 contains the individual body weight observations following acute inhalation exposure to NiO MPs. Table S7 contains the individual in-life behavioral and necropsy observations following acute inhalation exposure to NiO NPs. Table S8 contains the individual in-life behavioral and necropsy observations following acute inhalation exposure to NiO MPs.

Author Contributions: Conceptualization, Formal analysis, Writing, Review: T.L.-D. and A.R.O.; Methodology (acute toxicity), Investigation, Data Analyses, Review: J.L.B., M.W.S., M.E., J.D. and D.M. All authors have read and agreed to the published version of the manuscript.

Funding: This research was funded by NiPERA, Inc. (Durham, NC, USA; https://nickelinstitute.org/en/science/), the science division of the Nickel Institute, an association of global nickel-producing companies.

Data Availability Statement: The data presented in this study are available upon request from the corresponding author.

Acknowledgments: The studies were performed at Product Safety Labs in Dayton, NJ, USA. The Authors wish to thank all the staff at Product Safety Labs who helped and contributed to these studies.

Conflicts of Interest: Authors Lyons-Darden and Oller is and was, respectively, employed by the funding organization, NiPERA Inc. Authors Blum, Schooley, Ellis, Durando, and Merrill are employees of Product Safety Labs and were involved in conducting the studies. The authors declare no conflict of interest.

References

1. Denkhaus, E.; Salnikow, K. Nickel essentiality, toxicity, and carcinogenicity. *Crit. Rev. Oncol./Hematol.* **2002**, *42*, 35–56. [CrossRef]
2. Shahzad, B.; Tanveer, M.; Rehman, A.; Cheema, S.A.; Fahad, S.; Rehman, S.; Sharma, A. Nickel; whether toxic or essential for plants and environment—A review. *Plant Physiol. Biochem.* **2018**, *132*, 641–651. [CrossRef] [PubMed]
3. Li, Y.; Zamble, D.B. Nickel Homeostasis and Nickel Regulation: An Overview. *Chem. Rev.* **2009**, *109*, 4617–4643. [CrossRef] [PubMed]
4. Harasim, P.; Filipek, T. Nickel in the environment. *J. Elem.* **2015**, *20*, 525–534. [CrossRef]
5. Buxton, S.; Garman, E.; Heim, K.E.; Lyons-Darden, T.; Schlekat, C.E.; Taylor, M.D.; Oller, A.R. Concise Review of Nickel Human Health Toxicology and Ecotoxicology. *Inorganics* **2019**, *7*, 89. [CrossRef]
6. Genchi, G.; Carocci, A.; Lauria, G.; Sinicropi, M.S.; Catalano, A. Nickel: Human Health and Environmental Toxicology. *Int. J. Environ. Res. Public Health* **2020**, *17*, 679. [CrossRef]
7. Auffan, M.; Rose, J.; Bottero, J.-Y.; Lowry, G.V.; Jolivet, J.-P.; Wiesner, M.R. Towards a definition of inorganic nanoparticles from an environmental, health and safety perspective. *Nat. Nanotechnol.* **2009**, *4*, 634–641. [CrossRef]
8. Barhoum, A.; García-Betancourt, M.L.; Jeevanandam, J.; Hussien, E.A.; Mekkawy, S.A.; Mostafa, M.; Omran, M.M.; Abdalla, M.S.; Bechelany, M. Review on Natural, Incidental, Bioinspired, and Engineered Nanomaterials: History, Definitions, Classifications, Synthesis, Properties, Market, Toxicities, Risks, and Regulations. *Nanomaterials* **2022**, *12*, 177. [CrossRef]
9. Jaji, N.-D.; Lee, H.L.; Hussin, M.H.; Akil, H.; Zakaria, M.R.; Othman, M.B.H. Advanced nickel nanoparticles technology: From synthesis to applications. *Nanotechnol. Rev.* **2020**, *9*, 1456–1480. [CrossRef]
10. Sana, S.S.; Singh, R.P.; Sharma, M.; Srivastava, A.K.; Manchanda, G.; Rai, A.R.; Zhang, Z.-J. Biogenesis and Application of Nickel Nanoparticles: A Review. *Curr. Pharm. Biotechnol.* **2021**, *22*, 808–822. [CrossRef]
11. Joudeh, N.; Linke, D. Nanoparticle classification, physicochemical properties, characterization, and applications: A comprehensive review for biologists. *J. Nanobiotechnol.* **2022**, *20*, 262. [CrossRef] [PubMed]
12. Oberdörster, G.; Maynard, A.; Donaldson, K.; Castranova, V.; Fitzpatrick, J.; Ausman, K.; Carter, J.; Karn, B.; Kreyling, W.; Lai, D.; et al. Principles for characterizing the potential human health effects from exposure to nanomaterials: Elements of a screening strategy. *Part. Fibre Toxicol.* **2005**, *2*, 8. [CrossRef] [PubMed]
13. Zhao, J.; Shi, X.; Castranova, V.; Ding, M. Occupational Toxicology of Nickel and Nickel Compounds. *J. Environ. Pathol. Toxicol. Oncol.* **2009**, *28*, 177–208. [CrossRef]
14. Baig, N.; Kammakakam, I.; Falath, W. Nanomaterials: A review of synthesis methods, properties, recent progress, and challenges. *Mater. Adv.* **2021**, *2*, 1821–1871. [CrossRef]
15. Biskos, G.; Schmidt-Ott, A. Airborne Engineered Nanoparticles: Potential Risks and Monitoring Challenges for Assessing their Impacts on Children. *Paediatr. Respir. Rev.* **2012**, *13*, 79–83. [CrossRef]
16. Schimpel, C.; Resch, S.; Flament, G.; Carlander, D.; Vaquero, C.; Bustero, I.; Falk, A. A methodology on how to create a real-life relevant risk profile for a given nanomaterial. *ACS Chem. Health Saf.* **2018**, *25*, 12–23. [CrossRef]
17. Bierkandt, F.S.; Leibrock, L.; Wagener, S.; Laux, P.; Luch, A. The impact of nanomaterial characteristics on inhalation toxicity. *Toxicol. Res.* **2018**, *7*, 321–346. [CrossRef] [PubMed]
18. Roach, K.A.; Anderson, S.E.; Stefaniak, A.B.; Shane, H.L.; Kodali, V.; Kashon, M.; Roberts, J.R. Surface area- and mass-based comparison of fine and ultrafine nickel oxide lung toxicity and augmentation of allergic response in an ovalbumin asthma model. *Inhal. Toxicol.* **2019**, *31*, 299–324. [CrossRef]
19. Sutunkova, M.P.; Privalova, L.I.; Minigalieva, I.A.; Gurvich, V.B.; Panov, V.G.; Katsnelson, B.A. The most important inferences from the Ekaterinburg nanotoxicology team's animal experiments assessing adverse health effects of metallic and metal oxide nanoparticles. *Toxicol. Rep.* **2018**, *5*, 363–376. [CrossRef]
20. Jeong, M.-J.; Jeon, S.; Yu, H.-S.; Cho, W.-S.; Lee, S.; Kang, D.; Kim, Y.; Kim, Y.-J.; Kim, S.-Y. Exposure to Nickel Oxide Nanoparticles Induces Acute and Chronic Inflammatory Responses in Rat Lungs and Perturbs the Lung Microbiome. *Int. J. Environ. Res. Public Health* **2022**, *19*, 522. [CrossRef]

21. Abdel-Ghany, H.S.M.; Abdel-Shafy, S.; Abuowarda, M.M.; El-Khateeb, R.M.; Hoballah, E.; Hammam, A.M.M.; Fahmy, M.M. In vitro acaricidal activity of green synthesized nickel oxide nanoparticles against the camel tick, Hyalomma dromedarii (Ixodidae), and its toxicity on Swiss albino mice. *Exp. Appl. Acarol.* **2021**, *83*, 611–633. [CrossRef] [PubMed]
22. Hussain, M.F.; Ashiq, M.N.; Gulsher, M.; Akbar, A.; Iqbal, F. Exposure to variable doses of nickel oxide nanoparticles disturbs serum biochemical parameters and oxidative stress biomarkers from vital organs of albino mice in a sex-specific manner. *Biomarkers* **2020**, *25*, 719–724. [CrossRef] [PubMed]
23. Ali, A.A.-M. Evaluation of some biological, biochemical, and hematological aspects in male albino rats after acute exposure to the nano-structured oxides of nickel and cobalt. *Environ. Sci. Pollut. Res.* **2019**, *26*, 17407–17417. [CrossRef] [PubMed]
24. Dumala, N.; Mangalampalli, B.; Kamal, S.S.K.; Grover, P. Repeated oral dose toxicity study of nickel oxide nanoparticles in Wistar rats: A histological and biochemical perspective. *J. Appl. Toxicol.* **2019**, *39*, 1012–1029. [CrossRef] [PubMed]
25. Dumala, N.; Mangalampalli, B.; Kamal, S.S.K.; Grover, P. Biochemical alterations induced by nickel oxide nanoparticles in female Wistar albino rats after acute oral exposure. *Biomarkers* **2017**, *23*, 33–43. [CrossRef]
26. More, S.; Kovochich, M.; Lyons-Darden, T.; Taylor, M.; Schulte, A.; Madl, A. Review and Evaluation of the Potential Health Effects of Oxidic Nickel Nanoparticles. *Nanomaterials* **2021**, *11*, 642. [CrossRef]
27. Chang, X.; Zhao, H.; Gao, J.; Chen, L.; Zhu, A.; Wang, C.; Yu, S.; Ren, X.; Ge, P.; Sun, Y. Pulmonary toxicity of exposure to nano nickel oxide. *Micro Nano Lett.* **2018**, *13*, 733–738. [CrossRef]
28. Gmoshinski, I.V.; Khotimchenko, S.A. Assessing risks caused by nickel-based nanomaterials: Hazardous factor identification. *Health Risk Anal.* **2021**, *2*, 177–191. [CrossRef]
29. Schulte, P.A.; Murashov, V.; Zumwalde, R.; Kuempel, E.D.; Geraci, C.L. Occupational exposure limits for nanomaterials: State of the art. *J. Nanoparticle Res.* **2010**, *12*, 1971–1987. [CrossRef]
30. Pietroiusti, A.; Magrini, A. Engineered nanoparticles at the workplace: Current knowledge about workers' risk. *Occup. Med.* **2014**, *64*, 319–330. [CrossRef]
31. Dumala, N.; Mangalampalli, B.; Chinde, S.; Kumari, S.I.; Mahoob, M.; Rahman, M.F.; Grover, P. Genotoxicity study of nickel oxide nanoparticles in female Wistar rats after acute oral exposure. *Mutagenesis* **2017**, *32*, 417–427. [CrossRef] [PubMed]
32. Henderson, R.G.; Durando, J.; Oller, A.R.; Merkel, D.J.; Marone, P.A.; Bates, H.K. Acute oral toxicity of nickel compounds. *Regul. Toxicol. Pharmacol.* **2012**, *62*, 425–432. [CrossRef] [PubMed]
33. *Guide for the Care and Use of Laboratory Animals*, 7th ed.; National Research Council. National Academies Press (US): Washington, DC, USA, 1996.
34. *Guide for the Care and Use of Laboratory Animals*, 8th ed.; National Research Council. National Academies Press (US): Washington, DC, USA, 2011.
35. Donaldson, K.; Poland, C.A. Nanotoxicity: Challenging the myth of nano-specific toxicity. *Curr. Opin. Biotechnol.* **2013**, *24*, 724–734. [CrossRef] [PubMed]
36. Choi, S.-J.; Lee, J.K.; Jeong, J.; Choy, J.-H. Toxicity evaluation of inorganic nanoparticles: Considerations and challenges. *Mol. Cell. Toxicol.* **2013**, *9*, 205–210. [CrossRef]
37. Kumar, A.; Dixit, C.K. Methods for characterization of nanoparticles. In *Advances in Nanomedicine for the Delivery of Therapeutic Nucleic Acids*; Woodhead Publishing: Sawston, UK, 2017; pp. 43–58.
38. Henderson, R.G.; Cappellini, D.; Seilkop, S.K.; Bates, H.K.; Oller, A.R. Oral bioaccessibility testing and read-across hazard assessment of nickel compounds. *Regul. Toxicol. Pharmacol.* **2012**, *63*, 20–28. [CrossRef]
39. Oller, A.R.; Cappellini, D.; Henderson, R.G.; Bates, H.K. Comparison of nickel release in solutions used for the identification of water-soluble nickel exposures and in synthetic lung fluids. *J. Environ. Monit.* **2009**, *11*, 823–829. [CrossRef]
40. Bakand, S.; Hayes, A. Toxicological Considerations, Toxicity Assessment, and Risk Management of Inhaled Nanoparticles. *Int. J. Mol. Sci.* **2016**, *17*, 929. [CrossRef]
41. Fröhlich, E. Cellular Targets and Mechanisms in the Cytotoxic Action of Non-biodegradable Engineered Nanoparticles. *Curr. Drug Metab.* **2013**, *14*, 976–988. [CrossRef]
42. Nel, A.; Xia, T.; Mädler, L.; Li, N. Toxic Potential of Materials at the Nanolevel. *Science* **2006**, *311*, 622–627. [CrossRef]
43. Xia, T.; Li, N.; Nel, A.E. Potential Health Impact of Nanoparticles. *Annu. Rev. Public Health* **2009**, *30*, 137–150. [CrossRef]
44. Kunc, F.; Bushell, M.; Du, X.; Zborowski, A.; Johnston, L.J.; Kennedy, D.C. Physical Characterization and Cellular Toxicity Studies of Commercial NiO Nanoparticles. *Nanomaterials* **2022**, *12*, 1822. [CrossRef]
45. *Recommendation of the Council on the Safety Testing and Assessment of Manufactured Nanomaterials*; OECD: Paris, France, 2013.

nanomaterials

MDPI

Article

Form-Dependent Toxicity of Silver Nanomaterials in Rainbow Trout Gills

Joëlle Auclair [1], Patrice Turcotte [1], Christian Gagnon [1], Caroline Peyrot [2], Kevin James Wilkinson [2] and François Gagné [1,*]

[1] Aquatic Contaminants Research Division, Environment and Climate Change Canada, Montreal, QC H2Y 2E7, Canada; joelle.auclair@ec.gc.ca (J.A.)
[2] Chemistry Department, Montréal University, Montreal, QC H3C 3J7, Canada
* Correspondence: francois.gagne@ec.gc.ca

Abstract: The toxicity of the form of nanoparticles is presently not well understood. The purpose of this study consists in comparing the toxicity of various forms of silver nanoparticles (nAg) in juvenile rainbow trout *Oncorhynchus mykiss*. Juveniles were exposed to various forms of polyvinyl-coated nAg of similar size for 96 h at 15 °C. After the exposure period, the gills were isolated and analyzed for Ag uptake/distribution, oxidative stress, glucose metabolism, and genotoxicity. Higher levels of Ag were detected in gills in fish exposed to dissolved Ag followed by spherical, cubic, and prismatic nAg. Size-exclusion chromatography of gill fractions revealed that the dissolution of nAg was observed for all forms of nAg where prismatic nAg released more important levels of Ag in the protein pool as in fish exposed to dissolved Ag as well. The aggregation of nAg was more important for cubic nAg in respect of the other forms of nAg. The data revealed that lipid peroxidation was closely associated with protein aggregation and viscosity. Biomarkers revealed changes in lipid/oxidative stress and genotoxicity, which were related to the loss of protein aggregation and inflammation (NO_2 levels), respectively. In general, the observed effects were found for all forms of nAg where the effects from prismatic nAg were generally higher than for spherical and cubic nAg. The strong relationship between genotoxicity and inflammation response suggests the participation of the immune system in the observed responses of juvenile fish gills.

Keywords: silver nanoparticle; form; availability; dissolution; oxidative stress; lipids; genotoxicity

1. Introduction

Silver nanoparticles (nAg) represent an important part of the nanotechnology industry because of their antimicrobial properties [1]. They are found in many consumer and medical products such as clothes (shirts, pants, socks etc.), tubing, and various devices as aseptic agents and odor prevention [2]. These applications resulted in increased contamination of various aquatic ecosystems from the inadvertent release through solid waste leaching and municipal/industrial wastewaters [3]. Municipal wastewater is an important source of nAg in the aquatic environment where they are found mostly in suspended matter and sludges [4]. While most of the nAg are removed from the effluents (95%), some nAg remain in the effluent at the low ng/L range [5]. Silver nanoparticles come in many sizes (1–100 nm), coatings (citrate, polyvinylpyrrolidone, silicate), and forms (sphere, cube, plate/prism, and tube), which can influence the water–solid partition, half-lives, and reactivity [6]. Transmission electron microscopic observation of aged uncoated spherical and cubic nAg in marine and freshwater revealed that cubic nAg were eroded around the edges and that spherical nAg formed large aggregates. In another study, the catalytic activity of nAg wires, prisms, cubes, and spheres toward the plasmon-resonance-induced oxidation of p-aminothiophenol differed in the following: triangular prisms ~ spheres > wires cubes [7]. The geometry of nanoparticles could influence the thickness and composition of the protein corona at the surface, a process usually occurring in both the extra- and intracellular environments, and could change the resulting toxicity of these nanoparticles. In *Staphylococcus aureus*, prismatic nAg was more toxic than spherical, tubular, and cubic nAg were [8]. The thickness of fetal bovine serum proteins on the nanoparticles was lower for prismatic and tubular nAg compared to spherical and cubic nAg, with the spherical nAg containing six times more albumin at the surface. Although tubular nAg had low albumin content, the lower toxicity was explained by very long dimensions (5 μm) of the wires and a reduced availability for bacteria. Protein corona thickness was lower in the most toxic forms of nAg, indicating that a tenuous protein shell thickness at the surface of nAg could favor dissolution more quickly. In addition, the amount of low-molecular-weight proteins tended to be higher in steep-edged nanostructures such as prismatic and cubic nAg. This suggests that cubes and prisms have more surface tensions/constraints as shown by reduced protein thickness (for large proteins) and an increased amount of low-molecular-weight proteins.

The toxicity of various shapes of polyvinylpyrrolidone-coated nAg at the same size range was recently examined in freshwater organisms [9]. The study revealed that the form of nAg produced different effects, not always attributable to ionic Ag. The formation of liquid crystals in cells was increased as follows: prismatic > cubic > spherical nAg, suggesting again that steep-edged nAg produce more steric tensions and could disrupt the physical properties in cells (i.e., liquid crystal formation and viscosity). It is well known that the oxidative properties of ionic Ag released from nAg could lead to oxidative stress and damage such as lipid peroxidation (LPO) and DNA damage at relatively low exposure concentrations between 10 and 100 μg/L in Nile tilapia [10]. In respect of genotoxicity, the exposure of Nile tilapia to nAg increased DNA strand breaks in liver cells within the first 24 h and decreased afterward, suggesting effectiveness in DNA repair mechanisms [11]. Nevertheless, cytogenetic damage (irreversible) was observed in peripheral blood leucocytes as shown by nuclear abnormalities and micronuclei formation. Moreover, these studies were conducted with ovoid-shaped nAg, but other forms were not investigated in these studies. In this respect, the toxicity of various forms of nAg could result not only from the release of toxic Ag ions but from uptake differences, stability in the cell's environment, and crowding effects from the various geometries of nAg. In this study, a novel size-exclusion chromatography methodology [12] was used to better understand the stability (aggregation vs. dissolution of Ag) of various forms of nAg in the intracellular environment.

The purpose of this study was therefore to compare the toxic effects of various forms of PVP-coated nAg of similar dimensions in juvenile rainbow trout in an attempt to address the null hypothesis of this study: the geometry of nAg has no influence toward toxicity. The nAg were all coated with polyvinylpyrrolidone and were of a similar size range (50–70 nm) to optimize form-related effects. As gills are the first target tissues exposed to suspended matter and dissolved components in surface water, the accumulation of Ag in gill tissues and the intracellular distribution of Ag were examined by low-pressure size-exclusion chromatography. The toxic effects were investigated by following the changes in viscosity, protein aggregation, oxidative stress, DNA damage, and lipid and glucose metabolisms in juvenile fish gills. An attempt was made to identify critical events leading to toxicity in fish gill tissues.

2. Materials and Methods

2.1. Exposure of Rainbow Trout Juveniles

Rainbow trout juveniles (length 4 ± 0.2 cm; weight: 3 ± 0.3 g) were maintained for one month at 15 °C using a 16–8 h light and dark cycle with constant aeration according to a standardized method [13]. The aquarium consisted of UV-treated and charcoal-filtered tap water of the city of Montréal (conductivity: 280 μScm^{-1}, total organic carbon: 1–3 mg/L, total suspended solids < 1 mg/L). Juveniles were exposed to 5 and 50 μg/L of total Ag as polyvinylpyrrolidone (PVP)-coated silver nanoparticles (nAg) of a similar size range (70–80 nm) but of different form: sphere, cube, and prism/platelet (NanoXact,

NanoComposix Inc., San Diego, CA, USA). Prismatic nAg was a triangular prism with rounded edges with a thickness between 5 and 10 nm. The characterization of the 3 forms of nAg was provided in a previous study [9]. Dynamic light scattering and zeta potential analyses revealed that the mean sizes of the nanoparticles were 73, 70, and 86 nm for the spherical, cubic, and prismatic nAg, respectively. The zeta potential of spherical and cubic PVP-coated nAg was −38 and −40 mV, respectively. The surface area was 31,416, 60,000, and 6000 nm^2 for nAg spheres, cubes, and prisms, respectively. Fish were also exposed to 5 µg/L of dissolved Ag ($AgNO_3$) for comparison purposes. These concentrations were selected based on the upper range for total Ag found in municipal effluents [14,15]. For each nAg and dissolved Ag, stock solutions (1 mg/L) were prepared on the day of exposure in MilliQ water to prevent aggregation based on previous studies [16]. Silver concentrations were prepared at 5 and 50 µg/L as the total Ag for each nanoform of Ag in 60 L containers lined with polyethylene bags at 15 °C with aeration. The negative control consisted of tap water only. Water pH, conductivity, and dissolved oxygen were measured daily and monitored for any signs of distress, anormal behavior, and mortality. At the end of the exposure time (96 h), the fish were anesthetized in 0.1% MS222 buffered to pH 7.4 with 1 mM $NaHCO_3$ (Sigma-Aldrich, Oakville, ON, Canada) following the ethical committee guidelines for aquatic vertebrates. The length and weight of fish were taken, and gills were quickly removed and homogenized in ice-cold 250 mM sucrose containing 20 mM Tris-acetate, pH 7.5, 1 µg/mL of apoprotinin B (protease inhibitor), and 0.1 mM EDTA. Homogenization was achieved using a Polytron tissue grinder at a 30 s burst (low speed setting) on ice to prevent heating of the sample. The homogenates were centrifuged at 3000× *g* for 10 min and the supernatant was stored at −85 °C until biomarker analysis.

2.2. Stability of Ag in the Exposure Media and Detection Gill Tissues

Preliminary experiments were performed to examine the stability of the nAg form and dissolved Ag in the exposure media. Water samples were taken at 1 h and 96 h in the exposure media (without fish) and the dissolved levels of Ag (following filtration at 0.45 µm on a nanocellulose membrane) were determined using ion-coupled plasma quadrupole mass spectrometry as previously described [17]. The water samples (10 mL) were acidified in 1% HNO_3 prior to analysis. The relative levels of Ag in gills and in the chromatographic fractions (see below) were determined using a chemiluminescent (CL) methodology [18]. Briefly, the supernatant (50 µL) was mixed with 200 µL of alkaline 0.1 M Na_2CO_3 containing 0.5 mM luminol (freshly prepared) and luminescence readings were integrated for 1 min. Positive controls were prepared with dissolved Ag (0.5 µM) and nAg (2 µM as Ag) for calibration. Dissolved Ag emitted 8–10 times more light than an equimolar concentration of Ag such as nAg. The chemiluminescence (CL) data were expressed as µmole Ag/mg proteins.

2.3. Size-Exclusion Chromatography

The distributions of Ag in the gill homogenate fractions were analyzed using a recent methodology for nanoparticles' separation by low-pressure size-exclusion chromatography [12]. Briefly, a 40 × 1 cm column was prepared with Sephacryl S500 (capable of resolving nanoparticles from the protein pools based on size) using 1 mM KH_2PO_4, pH 7.4, containing 0.2% Tween-20 (non-ionic detergent) as the elution buffer. The elution buffer was found to prevent nanoparticle aggregation, the adsorption of proteins, and solubilization of any lipid structures (vesicles) present in the sample. The gill homogenates were centrifuged at 3000× *g* for 10 min, and a 100 µL sample of the supernatant was added to 100 µL of 0.2% Tween-20 and was injected directly to the column. The column was eluted at a flow rate of 0.75 mL/min and 1 mL fractions were collected after passing through 280 nm UV and conductivity detectors. The collected fractions were analyzed for Ag using the alkaline luminol CL-assay as described above [18] and at 410 nm for nAg detection using plasmon resonance spectroscopy [19]. The total volume of the column (Vt) was determined with 1 µg/mL of NaCl (conductivity), 25 nm citrate-coated nAg spherical and polystyrene fluorescently labeled nanoparticles (50 and 100 nm diameter), serum albumin (67 kdA) and rat metallothionein (6–7 kDa) for the protein pool range, and NaCl for low-molecular-weight salts. The void volume was estimated at 7 mL (Ve/Vt = 0.21) representing the upper range limit of the column estimated at 140 nm diameter particles. The spherical nAg eluted at a volume of 20 mL (Ve/Vt = 0.6) with some presence of aggregates at Ve/Vt = 0.21–0.25. The elution data were normalized to 1 to facilitate visual changes between chromatograms.

2.4. Biochemical Assessments

Rainbow trout (n = 8) were analyzed for fork length, body, and gill mass. The following endpoints were performed using compact, hand-held instruments for fluorescence (TBS-380, Turner Biosystem, Sunnyvale, CA, USA) and colorimetric analyzers on dried pads embedded with the appropriate reagents (EMP, URS-11 Solid phase pads, Wenzhou, China). The pads were purchased commercially as test strips for chemical analysis in biological samples (EMP, test strips, Wenzhou, China). The gill homogenate fraction (S3) was diluted with 2 volumes of MilliQ water and were analyzed for soluble proteins, bilirubin (Bil; heme degradation), glucose, acetoacetate (ketone body), nitrite (NO_2), peroxidase, esterase activities, and antioxidant capacity. Esterase activity was determined by blotting 100 µL of the S3 fraction to a solid phase pad containing 4 mg of N-toluene sulfonyl alanine ester/g pad as described by the manufacturer (EMP, URS-11 reader). Peroxidase (Perox) activity was determined using cumene hydroperoxide (6 mg/g) and tetramethylbenzidine reagent (40 mg/g pad) as described above. Total soluble proteins were determined using a pad embedded with 3 mg of tetrabromophenol blue/g pad and read by the hand-held colorimetric analyzer. The levels of acetoacetate (a ketone body) were determined using 77 mg of nitroprusside/g pad and analyzed by the reader. The levels of glucose were determined using a pad embedded with 0.16 mg of glucose oxidase/g pad containing 70 mg of KI/g pad and 6 mg of peroxidase and analyzed by the reader as above. The levels of bilirubin (Bil) were determined using 0.5% 2,4-dichloroanilide diazonium/g pad in 1 mg of KH_2PO_4 buffer at pH 7.4/g pad. The antioxidant potential was determined as ascorbic acid equivalents using a pad embedded with 3 mg of 2,5-dichlorophenolindophenol buffered at pH 7.4 with 1 mg of KH_2PO_4 at pH 7.4/g pad and analyzed by the colorimeter as above. The levels of nitrites (NO_2) were determined using a pad embedded with 45 mg of p-arsanilic acid and 0.1 mg of 3-hydroxy-1,2,3,4 tetrahydrobenzo-(h)-quinolin/g pad. All solid phase readings were validated with standard additions of each chemical or with varying amounts of S3 fraction for enzyme assessments (esterase and peroxidase), and solid phase readings were normalized to protein contents in the homogenate fraction as determined above. Lipid peroxidation (LPO) and DNA damage were determined by the malonaldehyde (thiobarbituric acid reactants) and DNA strand assays using fluorometry as previously described [20,21]. The levels of LPO were expressed as thiobarbituric acid reactants (TBARS) fluorescence units determined at 485 nm excitation and 540 nm emission (TBS-380, Turner Biosystems, Sunnyvale, CA, USA) with constant instrument calibration with the internal standard (fluorescein). The levels of DNA strand breaks were determined on the supernatant containing small, protein-free DNA strands from the precipitated genomic protein-rich DNA fraction using fluorometry at 350 nm excitation and 460 nm emission (TBS-380, Turner Biosystems, Sunnyvale, CA, USA) using the Hoescht dye. The dye was prepared in Tris-base (0.1 M) in high salt (0.4 M NaCl) and 4 mM deoxycholate detergent to prevent interferences from SDS and high pH during fluorescence readings [22,23]. The instrument was calibrated with quinine sulfate internal standard and 1 µg/mL of salmon sperm DNA. The data were expressed as DNA equivalents/mg proteins. Total lipids were determined using the Nile Red methodology [24]. Briefly, a 50 µL volume of the S3 fraction was mixed with 200 µL of 10 µM Nile Red in MilliQ water and fluorescence readings were taken at 485 nm excitation and >540 nm emission (TBS-380, Turner Biosystems, Sunnyvale, CA, USA). Standard solutions of triton X-100 (10 mg/mL) were used as positive controls and the instrument was calibrated with the fluorescein internal standard. The data were expressed as fluorescence units/mg proteins.

3. Data Analysis

Trout (n = 8) were exposed in 60 L containers for each of the 8 treatment groups (controls, 5, and 50 μg/L for spherical, cubic, and prismatic PVP-nAg, and 5 μg/L for dissolved Ag) and the exposure experiments were repeated twice. The data were analyzed using the non-parametric analysis of variance with the Conover–Inman test as the multiple comparison *post hoc* test to highlight differences between treatment groups. Correlation analysis was performed using the Pearson-moment procedure followed by principal components to identify the most important biomarkers explaining the data variance and interrelationships between them. Significance was set at $p \leq 0.05$ and all tests were performed using the SYSTAT software package (version 13).

4. Results

Rainbow trout juveniles were exposed to two concentrations (5 and 50 μg/L) of different forms of nAg (sphere, cube, and prism/plate) and to 5 μg/L for dissolved Ag for 96 h at 15 °C. The stability of these suspensions was determined in the aquarium water after 1 and 96 h of dissolution time (Table 1). Dissolved Ag remained stable over the exposure time. The forms of nAg were fairly stable (60–70% of Ag remaining after 96 h) in exposure water with relatively low changes in a decreasing manner: prismatic > cubic and spherical nAg. The relative levels of Ag were determined using a chemiluminescence assay that can detect both nAg and dissolved Ag in gill tissues (Figure 1). As expected, the levels of Ag were highest in fish exposed to the dissolved Ag form and were 3 times higher than the reported Ag concentrations from the nAg forms at the same concentration (5 μg/L). Spherical nAg was significantly increased from controls for the 50 μg/L exposure group. The relative levels of Ag were significantly increased from controls for each exposure concentration (5 and 50 μg/L) group for cubic and prismatic forms of nAg. Morphometric changes in fish were determined by following the condition factor (fish weight/fork length), hepatosomatic index (HSI), and the gill index (Table 2). Only the HSI was significantly affected by some forms of nAg: at 5 and 50 μg/L for spherical nAg and at 5 μg/L for prismatic nAg and dissolved Ag. The gill homogenate supernatant was further resolved by size-exclusion chromatography to gain some insights into the stability of nAg in tissues and the distribution of Ag in the cellular protein pool (Figure 2). In the following chromatograms, the proteins were analyzed by the absorbance at 280 nm, the nanoparticles were analyzed at 410 (plasmon resonance) nm, and Ag was analyzed by the chemiluminescent methodology. The total volume (Vt) was estimated with NaCl (conductivity), giving Ve/Vt = 1, and the void volume was evaluated with Ve/Vt = 0.25. For spherical nAg, the presence of nAg eluted at Ve/Vt = 0.4 corresponded to a molecular size of 40–60 nm based on the column calibration with 100, 50, and 20 nm polystyrene nanobeads (Figure 2A). Two other Ag peaks were found, corresponding to Ve/Vt = 0.62 (corresponding to 10–20 nm size) and Ve/Vt = 0.8–0.9, corresponding to low-molecular-weight proteins and ligands such as glutathione and amino acids. Indeed, the protein pools eluted at Ve/Vt = 0.6–0.95 were dissolved components (KCl) eluted at the total column volume of Ve/Vt = 1. Metallothionein, a silver binding protein with 7000 d mass, eluted at Ve/Vt = 0.75, suggesting that Ag was not only sequestered by metallothioneins but also with other low-molecular-weight proteins based on the UV absorption at 280 nm. For cubic nAg, a similar pattern was found to spherical nAg, but a peak (410 nm and Ag) was found at the void volume at Ve/Vt = 0.25–0.35, suggesting large aggregates > 200 nm (Figure 2B). For prismatic nAg, a higher proportion of Ag was found in the low-molecular-weight range (Ve/Vt = 0.7–1) and was observed with no evidence of large aggregates. Based on the absorbance at 280 nm, a significant number of proteins was associated with prismatic nAg, while little to no proteins were associated with spherical and cubic nAg. For dissolved Ag, most of the Ag was in the low-molecular-size range overlapping the proteins (A280 nm), suggesting that ionic Ag was mostly bound to protein pools.

Table 1. Stability of Ag in the exposure media.

Ag Form (5 μg/L)	Dissolved Ag 1 h (μg/L)	Dissolved Ag 96 h (μg/L)	Decreasing Rate (h^{-1}) /Half Life (h)
Sphere	4.9 ± 0.4	3.1 ± 0.3	0.018/126
Cube	5.2 ± 0.2	2.9 ± 0.2	0.024/114
Prismatic	6.5 ± 0.5	3.4 ± 0.3	0.033/150
Dissolved	4.7 ± 0.4	4.9 ± 0.4	~0

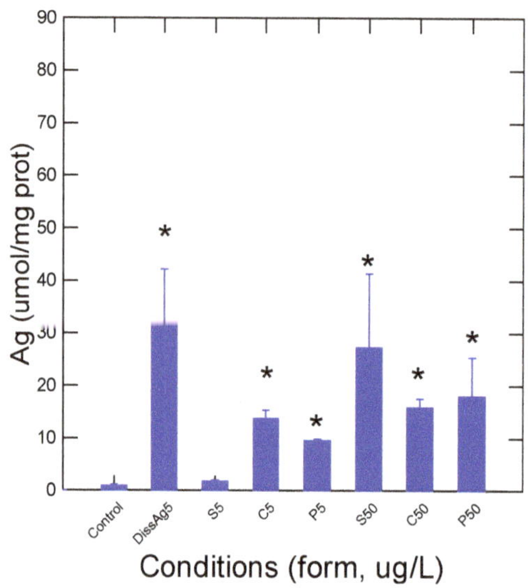

Figure 1. Relative Ag levels in fish gills. The relative levels of Ag were determined by an enhanced luminol/Na_2CO_3 chemiluminescence in gill homogenates. The data represent the mean with standard error. Abbreviations for x-axis (and for the following figures): DissAg5 (5 μg/L of dissolved Ag), S5 (spherical nAg; 5 μg/L), C5 (cubic nAg; 5 μg/L), P5 (prismatic nAg; 5 μg/L), S50 (spherical; nAg, 50 μg/L). The star * symbol indicates significance at $p \leq 0.05$ from controls.

Table 2. Change in fish morphology following exposure to the various forms of Ag.

Ag	CF	HIS	GI
Controls	1.02 ± 0.01	0.98 ± 0.07	0.05 ± 0.02
Dissolved	1 ± 0.01	1.19 ± 0.07 *	0045 ± 0.017
Sphere			
5	0.98 ± 0.02	0.9 ± 0.07 *	0.048 ± 0.017
50	0.97 ± 0.01	1.3 ± 0.08 *	0.052 ± 0.018
Cube			
5	0.93 ± 0.009	0.93 ± 0.05	0.05 ± 0.018
50	1.09 ± 0.03	1.2 ± 0.1	0.048 ± 0.017
Prism			
5	1 ± 0.02	0.76 ± 0.02 *	0.051 ± 0.02
50	1 ± 0.02	1.1 ± 0.09	0.054 ± 0.019

The star symbol * indicates significance from controls. CF: condition factor (wet/fork length); HIS: hepatic-somatic index (liver weight/fish weight), GI: gill index (gill weight/fish weight.

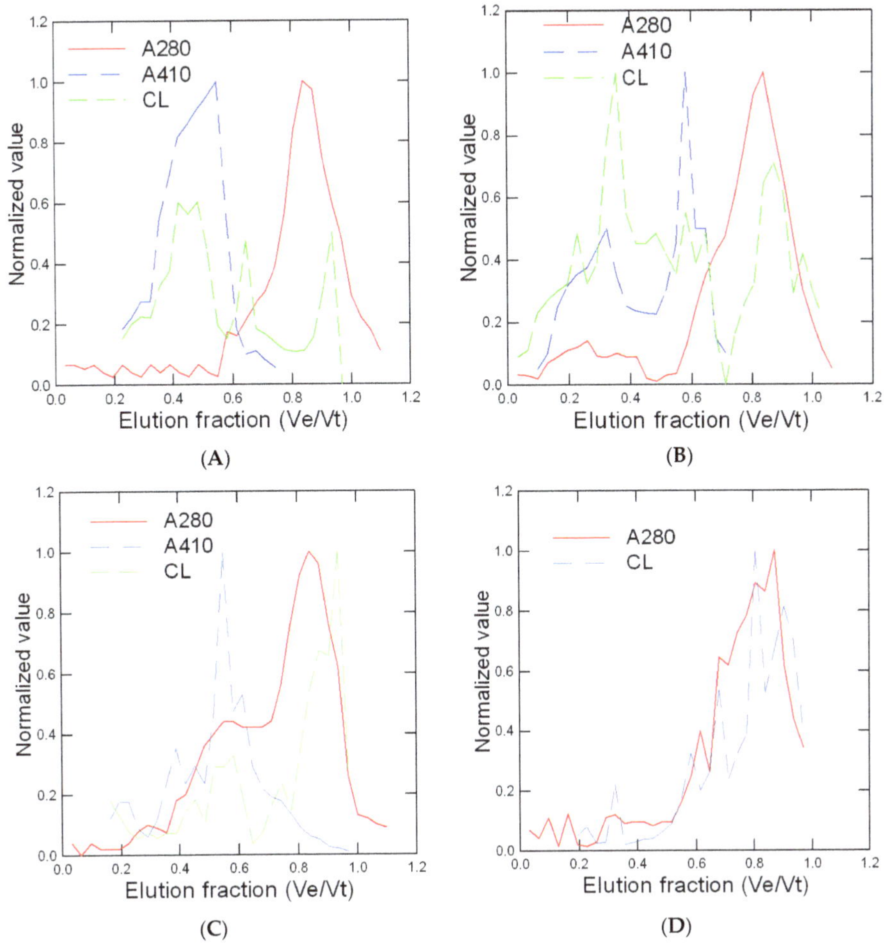

Figure 2. Size-exclusion chromatography analysis of gills in fish exposed to various forms of Ag. The gill homogenate fractions were analyzed by size-exclusion chromatography on a 40 × 1 cm Sephacryl S500 column using 1 mM NaCl and 0.2% Tween-20 as the elution media for spherical (**A**), cubic (**B**), and prismatic (**C**) nAg and dissolved Ag (**D**). The samples were analyzed by a size-exclusion chromatography column pre-calibrated with polystyrene NP (100, 50, and 20 nm diameter), albumin, and NaCl (total volume). The abscissa axis is defined by Ve/Vt, where Ve is the elution volume and Vt is the total column volume (as determined by NaCl). The data were normalized to 1 (fraction signal − min signal)/(max signal − min signal) for ease of measurement for comparisons of each chromatogram.

The energy status of juvenile rainbow trout was determined by following the glucose, acetoacetate, and esterase (Est) activity and total lipids in gill tissues (Figure 3). Glucose and acetoacetate levels were not affected by the forms of Ag but Est activity was increased by spherical nAg at 50 μg/L. Lipids were increased by all the forms of nAg but not with dissolved Ag. Correlation analysis revealed that Est activity was significantly correlated with the relative Ag levels in gills (r = 0.32). Total lipids were negatively correlated with glucose levels (r = −0.30). Oxidative stress (inflammation) and damage were determined by following the changes in peroxidase, lipid peroxidation, NO_2 levels, and the antioxidant potential in gills from juvenile rainbow trout exposed to the various forms of Ag (Figure 4).

Peroxidase activity was significantly increased by dissolved Ag, and only a marginal increase was observed with prismatic nAg. Correlation analysis revealed that peroxidase activity was significantly correlated with acetoacetate levels (r = 0.49). The levels of LPO tended to be higher in the exposure group but was only significantly higher for spherical nAg at 50 µg/L. The correlation analysis of LPO was significantly related to total lipids (r = 0.39). Globally, the NO_2 levels were also higher with all forms of nAg including dissolved Ag but were significant with dissolved and spherical nAg. Correlation analysis revealed that NO_2 levels were significantly correlated with peroxidase activity (r = 0.61), acetoacetate (r = 0.54), and glucose (r = 0.43). The antioxidant potential was also significantly decreased by all forms of nAg and Ag. The correlation revealed that the antioxidant potential was significantly correlated with NO_2 levels (r = −0.33), peroxidase activity (r = −0.31), and glucose levels (r = 0.31).

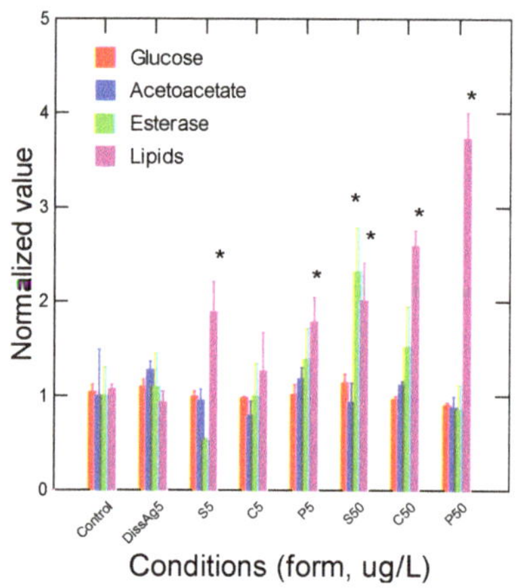

Figure 3. Energy status in trout exposed to nAg forms. Energy status was determined by following the changes in glucose, acetoacetate (a ketone body), esterases, and lipids in trout gills exposed to various forms of Ag. The data represent the mean with standard error. The star * symbol indicates significance at $p \leq 0.05$ from controls.

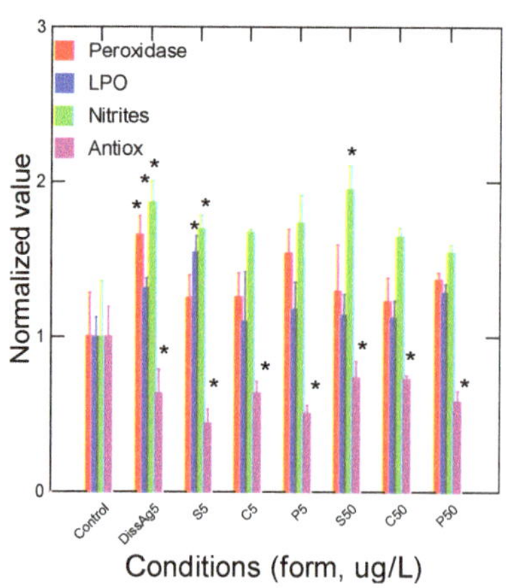

Figure 4. Oxidative stress and damage in trout exposed to nAg forms. Oxidative stress was determined by following the changes in peroxidase, lipid peroxidation, nitrites (NO_2), and antioxidant (Antiox) potential in gills of rainbow trout exposed to Ag forms. The data represent the mean with standard error. The star * symbol indicates significance at $p \leq 0.05$ relative to controls.

Tissue damage in trout exposed to various forms of Ag was determined by following the levels of DNA breaks, protein aggregation, and heme degradation product bilirubin (Figure 5). Viscosity was also included although not strictly a marker of damage. DNA damage was present in all Ag forms where the genotoxic potential of dissolved Ag is similar to all forms of nAg. Correlation analysis of DNA damage revealed a significant trend with NO_2 levels (r = 0.68), peroxidase activity (r = 0.45), glucose (r = 0.87), and acetoacetate (r = 0.39). Protein aggregation was significantly reduced by cubic and spherical nAg for the low and high concentrations. A reduction in protein aggregation also occurred for spherical nAg but only at high concentrations. Correlation analysis revealed significant relationships with glucose (r = 0.36) and LPO (r = 0.46). The levels of Bil were determined as an indicator of heme damage. Bil levels were decreased in fish exposed to the lowest concentration of cubic nAg and at the highest concentration of prismatic nAg. Correlation analysis revealed that Bil levels were correlated with glucose (r = 0.62), antioxidant potential (r = 0.67), and protein aggregation (r = 0.3). Viscosity changes tended to be lower by the forms of nAg but not significantly. However, a significant increase in viscosity was observed for the low concentration of spherical nAg. The correlation analysis of viscosity revealed a significant correlation with esterase activity (r = −0.36), LPO (r = 0.72), protein aggregation (r = 0.63), and total lipids (r = 0.37). In an attempt to gain a global view of the various responses, a principal component analysis was performed (Figure 6). The total variance was 50% and the following biomarkers had the highest component weight (>0.7): viscosity, glucose, DNA proteins, NO_2 levels, and gill protein levels. Changes in viscosity were closely associated with protein aggregation and LPO. DNA damage and NO_2 levels were closely related with each other, peroxidase activity, and bilirubin, suggesting the involvement of oxidative stress and increased hemoprotein turnover. Gill Ag levels were closely related with acetoacetate and esterase activity, suggesting impacts on gluconeogenesis and lipid mobilization.

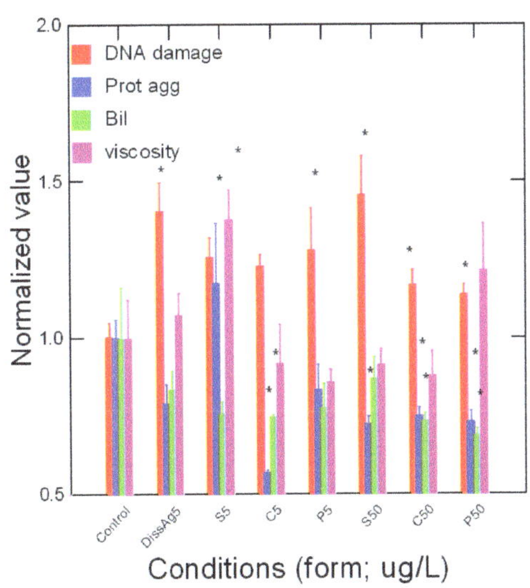

Figure 5. Tissue damage in trout exposed to nAg forms. Tissue damage was evaluated by DNA strand breaks, protein aggregation, and heme degradation (bilirubin) in trout gills exposed to forms of Ag. The data represent the mean with standard error. The star * symbol indicates significance at $p \leq 0.05$ compared to controls.

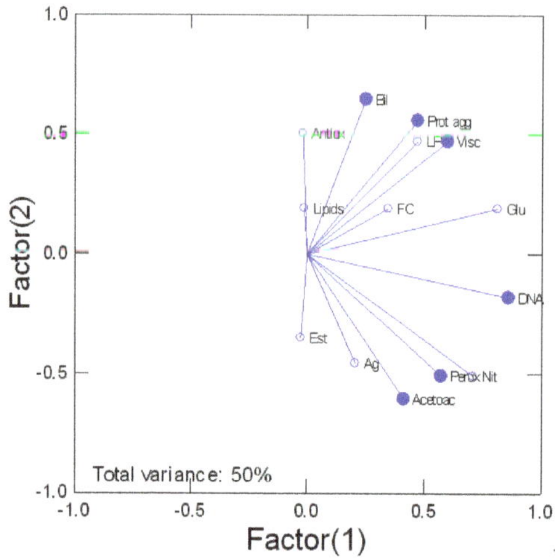

Figure 6. Principal component analysis of biomarker data. The principal component analysis of biomarker data is shown. The total variance explained was at 50% and the most important biomarkers (weight > 0.7) are highlighted with the filled circles. The relative level of Ag is shown as the filled square.

5. Discussion

The fish accumulated 3–5 times more Ag in fish exposed to ionic/dissolved Ag compared to the nAg forms. At low concentrations, cubic and prismatic nAg accumulated more in gills compared to spherical nAg. A possible explanation for this could be that a more reactive form of Ag is released from cubic and prismatic nAg than for spherical nAg in tissues. Indeed, the CL assay used in the present study could detect both nanoparticulate and dissolved Ag but was reported eight times more sensitive with the latter [18]. The same accumulation rate (threefold accumulation in gills) of *Oreochromis mossambicus* fish was exposed to ionic compared to spherical nAg [10]. The half-lives of the selected forms of nAg were within the order of $\geq$114 h with an estimated time for 126, 114, and 150 h for spherical, cubic, and prismatic nAg, respectively. This suggests that fish were readily exposed to suspensions of nAg during the exposure period of 96 h and could pose a risk to predators feeding on fish (including humans) in the vicinity of urban pollution. This is in keeping with a previous study that examined the stability of citrate-coated nAg in various waters and wastewaters [25]. The half-life of spherical nAg was of the order of 100 days in untreated wastewaters and the proportion of ionic Ag+ did not increase over time. This study also showed that the size distribution of nAg was fairly stable during that time and the free Ag+ that occurred represented 40–50% of the total Ag levels at time 0 where this proportion remained fairly constant in time in untreated water samples. This is consistent with the previous observation where the levels of suspended Ag were significantly correlated with nAg in treated municipal effluents [5]. In some conditions, the nanoparticles were more quickly degraded in treated municipal effluents, synthetic effluent, and reclaimed waters with nAg half-lives < 10 h. This suggests that nAg could be relatively stable in complex media (untreated wastewaters) but could quickly dissolve in treated effluents and reclaimed waters. The coating of the nAg forms was PVP, which confers more stability than citrate, which is electrostatic in nature (i.e., exchangeable with other anions). Indeed, it was found that PVP-coated nAg did not aggregate in water, even at an ionic strength equivalent to 10 mM $CaCl_2$, compared to citrate-coated nAg [26]. In this study, we examined the stability of nAg in gill tissue extracts by size-exclusion chromatography where Ag was found in the low-molecular-weight range (Figure 2). For spherical and cubic nAg, about 50% of Ag was found in the low-molecular-weight range, and this proportion rose to about 70% for prismatic nAg, suggesting increased degradation in cells. Detectable amounts of proteins (A280 nm) were present in the size range of prismatic nAg, suggesting the adsorption of proteins to the nanoparticles. The in vitro exposure of human mesenchymal stem cells and human keratinocytes to PVP-coated nAg spheres and prisms lost more than 90% of their volume after 24 h at 37 °C [27]. It was found that the cell accumulation of prismatic nAg was facilitated by membrane fluidity in mesenchymal stem cells compared to keratinocytes. In this study, the edges of prismatic (platelet) nAg became quickly eroded, showing more surface tension favoring the liberation of silver to the low-molecular-weight pool in keeping with the increased dissolution of prismatic nAg compared to spherical and cubic PVP-coated nAg in the present study. Protein corona formation at the surface of metallic nanoparticles in cells is a common occurrence [3,28]. The liberation of Ag from interactions with metallothioneins, a thiol-rich metal binding proteins with high affinity for Ag, led to nAg dissolution [28]. The formation of the protein corona with a non-metal binding protein ceruloplasmin led to a more stable corona–nanoparticle complex. Based on the chromatographic elution profiles of the various forms, Ag was detected in fractions consistent with metallothioneins (Ve/Vt = 0.75–0.8) with the prismatic nAg and dissolved Ag in fish gills. It is noteworthy that metallothioneins are devoid of aromatic amino acids and therefore do not absorb at 280 nm. The toxicity of prismatic (plate) nAg was higher than those of spherical and tubular nAg in the fish cell line and zebrafish embryos [29]. It was suggested that the toxicity of nAg plates was attributed to surface roughness/heterogenicity (point defects), and treatment of nAg plates with cysteine reduced the toxicity and oxidative stress induced by the nanoparticle and dissolved Ag.

The toxicity of the forms of PVP-coated nAg was coated in freshwater unionids and revealed the following order of toxicity: prism < cube < sphere based on protein-ubiquitin (damage) levels [9]. In another study, nAg prisms reduced the air survival time in dreissenids compared to PVP-coated spherical and cubic nAg [16]. Reduced acetylcholinesterase activity was somewhat reduced more strongly with nAg plates (prisms). Prismatic and spherical nAg reduced growth and reproduction in *Caenorhaditis elegans* than tubular/wire nAg [30]. In keratinocytes, truncated prismatic nAg was more toxic that the spheric nanoparticles [31]. This suggests that asymmetric and elongated structures (cylinders, rectangles, etc.) contribute to more surface tension, favoring toxicity in part at least from the liberation of dissolved components. At some point, the toxicity could be reduced when dimensions tend to wire-like geometries, limiting absorbance and cell internalization. The toxicity of prismatic nAg was also more toxic than spherical nAg in zebra fish embryos [32]. Toxicity was not necessarily related to the release of ionic Ag and the particle form contributed to toxicity as well.

It is noteworthy that genotoxicity was one of the principal components associated with each form of Ag. The extent of DNA damage was associated, in part at least, to Ag levels in gills (r = 0.46), suggesting that internalization in cells is a contributing factor to effects. The levels of NO_2, an indicator of inflammation from the formation of peroxynitrite in phagosomes and the activation of the NO_2 pathway [33], were highly correlated with DNA strand breaks (r = 0.96), suggesting the contribution of the immune function in fish exposed to the forms of Ag. Increased NO_2 levels could form 3-nitrotyrosine in proteins and contribute to protein aggregation and viscosity. This was supported by the significant correlations between NO_2 levels and viscosity (r = 0.88) and protein aggregation (r = 0.65). The addition of NO_2 groups on tyrosine increases the polarity of the tyrosine, leading to protein aggregation and viscosity. The genotoxicity of nAg was reported in fish in the liver and muscle at the level of DNA breaks and the formation of micronuclei occurring after 48 h [11]. The genotoxicity was higher in citrate-coated spherical nAg compared to PVP-coated nAg in rainbow trout [34], suggesting the involvement of surface interactions (citrate coating involves electrostatic interactions while PVP coating involves encapsulation by neutral PVP) and perhaps dissolution of nAg from citrate coated nAg. Encapsulated PVP-coated nAg could interact at the surface of the cell membrane and enter by pinocytosis and fuse with lysosomes/phagosomes, leading to peroxynitrite and

NO_2 production, hence the strong correlation with NO_2 levels. This was supported by transmission electron microscopy of gilthead seabream head-kidney leucocytes showing phagocytosis and pinocytosis of silver nanospheres [35]. Significant reductions in leucocyte viability were observed at concentrations between 1 and 10 µg/mL in addition to reduced phagocytosis activity and respiratory burst. The presence of nAg in cells was confirmed by transmission electron microscopy following pinocytosis and phagocytosis.

In conclusion, exposure to selected forms of nAg led to an accumulation of Ag in gills but to a lower degree than ionic Ag. Size-exclusion chromatography data revealed some degree of dissolution of Ag to the low-molecular-weight range and was highest with prismatic nAg compared to spherical and cubic nAg. Cubic nAg showed a higher aggregation as revealed by increased nAg at the void volume of the column. Effect biomarkers revealed that forms of nAg induced important changes in glucose levels, LPO, and genotoxicity. Oxidative stress (LPO) was closely associated with decreased protein aggregation and viscosity, while DNA damage was strongly related to NO_2 levels, suggesting that oxidative stress was mediated by inflammation. Although toxicity involved the dissolution of Ag from the various forms of nAg into the low-molecular-weight-range protein/peptide pool, the toxicity was higher for prismatic nAg, leading to stronger effects toward lipid levels, oxidative stress, and DNA damage. This suggests that the release of more complex forms of nAg in respect of the usual spheres could produce more impacts in fish in respect of the release of dissolved Ag in tissues.

Author Contributions: J.A.: formal analysis, methodology, data curation. P.T.: formal analysis, methodology, data curation. C.G.: supervision, writing—editing. C.P.: data curation, methodology, data curation. K.J.W.: investigation, methodology (chemical aspects). F.G.: conceptualization, funding acquisition, formal analysis, writing—original draft preparation, writing—review and editing. All authors have read and agreed to the published version of the manuscript.

Funding: This research received no external funding only through internal funding from the Chemical management plans of Environment and Climate Change Canada.

Institutional Review Board Statement: Ethical review and approval were provided internally for this study due to program requirements in using fish in toxicity studies.

Informed Consent Statement: Not applicable.

Data Availability Statement: Data are available upon request to the corresponding author.

Acknowledgments: This work was funded by the Chemical Management Plan of Environment and Climate Change Canada.

Conflicts of Interest: The authors declare no conflict of interest.

References

1. Gottardo, S.; Mech, A.; Drbohlavová, J.; Małyska, A.; Bøwadt, S.; Riego Sintes, J.; Rauscher, H. Towards safe and sustainable innovation in nanotechnology: State-of-play for smart nanomaterials. *NanoImpact* **2021**, *21*, 100297. [CrossRef]
2. McGillicuddy, E.; Murray, I.; Kavanagh, S.; Morrison, L.; Fogarty, A.; Cormican, M.; Dockery, P.; Prendergast, M.; Rowan, N.; Morris, D. Silver nanoparticles in the environment: Sources, detection and ecotoxicology. *Sci. Total Environ.* **2017**, *575*, 231–246. [CrossRef] [PubMed]
3. Zhang, C.; Hu, Z.; Deng, Z. Silver nanoparticles in aquatic environments: Physicochemical behavior and antimicrobial mechanisms. *Water Res.* **2016**, *88*, 403–427. [CrossRef] [PubMed]
4. Bolaños-Benítez, V.; McDermott, F.; Gill, L.; Knappe, J. Engineered silver nanoparticle (Ag-NP) behaviour in domestic on-site wastewater treatment plants and in sewage sludge amended-soils. *Sci. Total Environ.* **2020**, *722*, 137794. [CrossRef]
5. Gagnon, C.; Turcotte, P.; Gagné, F.; Smyth, S.A. Occurrence and size distribution of silver nanoparticles in wastewater effluents from various treatment processes in Canada. *Environ. Sci. Pollut. Res. Int.* **2021**, *28*, 65952–65959. [CrossRef]
6. Formo, E.V.; Potterf, C.B.; Yang, M.; Unocic, R.R.; Leonard, D.N.; Pawel, M. How a Nanostructure's Shape Affects its Lifetime in the Environment: Comparing a Silver Nanocube to a Nanoparticle When Dispersed in Aqueous Media. *Environ. Sci. Technol.* **2016**, *50*, 7082–7089. [CrossRef]
7. Da Silva, A.G.; Rodrigues, T.S.; Wang, J.; Yamada, L.K.; Alves, T.V.; Ornellas, F.R.; Ando, R.A.; Camargo, P.H. The Fault in Their Shapes: Investigating the Surface-Plasmon-Resonance-Mediated Catalytic Activities of Silver Quasi-Spheres, Cubes, Triangular Prisms, and Wires. *Langmuir* **2015**, *31*, 10272–10278. [CrossRef]

8. Ashkarran, A.A.; Ghavami, M.; Aghaverdi, H.; Stroeve, P.; Mahmoudi, M. Bacterial Effects and Protein Corona Evaluations: Crucial Ignored Factors in the Prediction of Bio-Efficacy of Various Forms of Silver Nanoparticles. *Chem. Res. Toxicol.* **2012**, *25*, 1231–1242. [CrossRef]
9. Auclair, J.; Peyrot, C.; Wilkinson, K.J.; Gagné, F. The geometry of the toxicity of silver nanoparticles to freshwater mussels. *Comp. Biochem. Physiol. C Toxicol. Pharm.* **2021**, *239*, 108841. [CrossRef]
10. SIbiya, A.; Gopi, N.; Jeyavani, J.; Mahboob, S.; Al-Ghanim, K.A.; Sultana, S.; Mustafa, A.; Govindarajan, M.; Vaseeharan, B. Comparative toxicity of silver nanoparticles and silver nitrate in freshwater fish *Oreochromis mossambicus*: A multi-biomarker approach. *Comp. Biochem. Physiol.* **2022**, *259*, 109391. [CrossRef]
11. Abdel-Khalek, A.A.; Al-Quraishy, S.; Abdel-Gaber, R. Silver Nanoparticles Induce Time- and Tissue-Specific Genotoxicity in Oreochromis niloticus: Utilizing the Adsorptive Capacities of Fruit Peels to Minimize Genotoxicity. *Bull. Environ. Contam. Toxicol.* **2022**, *108*, 300–308. [CrossRef] [PubMed]
12. Gagné, F. Isolation and quantification of polystyrene nanoplastics in tissues by low pressure size exclusion chromatography. *J. Xenobiot.* **2022**, *12*, 109–121. [CrossRef] [PubMed]
13. Biological test method: Acute lethality test using rainbow trout. In *Environmental Protection Series*; Report EPS 1/RM/9; Method Development and Applications Section, Environmental Technology Centre, Environment Canada: Ottawa, ON, Canada, 1990.
14. Polesel, F.; Farkas, J.; Kjos, M.; Carvalho, P.A.; Flores-Alsina, X.; Gernaey, K.V.; Hansen, S.F.; Plósz, B.G.; Booth, A.M. Occurrence, characterisation and fate of (nano)particulate Ti and Ag in two Norwegian wastewater treatment plants. *Water Res.* **2018**, *141*, 19–31. [CrossRef] [PubMed]
15. Azodi, M.; Sultan, Y.; Ghoshal, S. Dissolution Behavior of Silver Nanoparticles Formation of Secondary Silver Nanoparticles in Municipal Wastewater by Single-Particle, I.C.P.-M.S. *Environ. Sci. Technol.* **2016**, *50*, 13318–13327. [CrossRef]
16. Auclair, J.; Peyrot, C.; Wilkinson, K.J.; Gagné, F. The influence of silver nanoparticle form on the toxicity in freshwater mussels. *Appl. Sci.* **2022**, *12*, 1429. [CrossRef]
17. Auclair, A.; Turcotte, P.; Gagnon, C.; Peyrot, C.; Wilkinson, K.J.; Gagné, F. Toxicological effects of inorganic nanoparticle mixtures in freshwater mussels. *Environments* **2021**, *7*, 109. [CrossRef]
18. Li, S.; Sun, H.; Wang, D.; Hong, J.; Tao, S.; Yu, H.; Wang, X.; Wei, X. Enhanced chemiluminescence of the luminol–$AgNO_3$ system by Ag nanoparticles. *Luminescence* **2012**, *27*, 211–216. [CrossRef]
19. Zielinska, A.; Skwarek, E.; Zaleska, A.; Gazdac, M.; Hupka, J. Preparation of silver nanoparticles with controlled particle size. *Procedia. Chem.* **2009**, *1*, 1560–1566. [CrossRef]
20. Wills, E.D. Evaluation oflipid peroxidation in lipids and biological membranes. In *Biochemical Toxicology. A Practical Approach*; Snell, K., Mullock, B., Eds.; IRL Press: Oxford, UK, 1987; pp. 127–152.
21. Olive, P.L. DNA precipitation assay: A rapid and simple method for detecting DNA damage in mammalian cells. *Environ. Mol. Mutagen.* **1988**, *11*, 487–495. [CrossRef]
22. West, D.C.; Sattar, A.; Kumar, S. A simplified in situ solubilization procedure for the determination of DNA and cell number in tissue cultured mammalian cells. *Anal. Biochem.* **1985**, *147*, 289–295. [CrossRef]
23. Bester, M.J.; Potgieter, H.C.; Vermaak, W.J.H. Cholate and pH reduce interference by SDS in the determination of DNA with Hoescht. *Anal. Biochem.* **1994**, *223*, 299–305. [CrossRef]
24. Greenspan, P.; Mayer, E.P.; Fowler, S.D. Nile Red: A selective fluorescent stain for intracellular lipid droplets. *J. Cell. Biol.* **1985**, *100*, 965–973. [CrossRef]
25. Cervantes-Avilés, P.; Huang, Y.; Keller, A.A. Incidence and persistence of silver nanoparticles throughout the wastewater treatment process. *Water Res.* **2019**, *156*, 188–198. [CrossRef]
26. Jang, M.-H.; Bae, S.-J.; Lee, S.-K.; Lee, Y.-J.; Hwang, Y.S. Effect of material properties on stability of silver nanoparticles in water. *J. Nanosci. Nanotechnol.* **2014**, *14*, 9665–9669. [CrossRef]
27. Graf, C.; Nordmeyer, D.; Sengstock, C.; Ahlberg, S.; Diendorf, J.; Raabe, J.; Epple, M.; Köller, M.; Lademann, J.; Vogt, A.; et al. Shape-dependent dissolution and cellular uptake of silver nanoparticles. *Langmuir* **2018**, *34*, 1506–1519. [CrossRef]
28. Liu, W.; Worms, I.A.M.; Herlin-Boime, N.; Truffier-Boutry, D.; Michaud-Soret, I.; Mintz, E.; Vidaud, C.; Rollin-Genetet, F. Interaction of silver nanoparticles with metallothionein and ceruloplasmin: Impact on metal substitution by Ag(I), corona formation and enzymatic activity. *Nanoscale* **2017**, *9*, 6581–6594. [CrossRef]
29. George, S.; Lin, S.; Ji, Z.; Thomas, C.R.; Li, L.J.; Mecklenburg, M.; Meng, H.; Wang, X.; Zhang, H.; Xia, T.; et al. Surface Defects on Plate-Shaped Silver Nanoparticles Contribute to Its Hazard Potential in Fish Gill Cell Line and Zebrafish Embryos. *ACS Nano.* **2012**, *6*, 3745–3759. [CrossRef]
30. Moon, J.; Kwak, J.I.I.; An, Y.-J. The effects of silver nanomaterial shape and size on toxicity to *Caenorhabditis elegans* in soil media. *Chemosphere* **2019**, *215*, 50–56. [CrossRef]
31. Holmes, A.M.; Lim, J.; Studier, H.; Roberts, M.S. Varying the morphology of silver nanoparticles results in differential toxicity against micro-organisms, HaCaT keratinocytes and affects skin deposition. *Nanotoxicology* **2016**, *10*, 1503–1514. [CrossRef]
32. Abramenko, N.B.; Demidova, T.B.; Abkhalimov, E.V.; Ershov, B.G.; Krysanov, E.Y.; Kustov, L.M. Ecotoxicity of different-shaped silver nanoparticles: Case of zebrafish embryos. *J. Hazard. Mater.* **2018**, *347*, 89–94. [CrossRef]
33. Campolo, N.; Issoglio, F.M.; Estrin, D.A.; Bartesaghi, S.; Radi, R. 3-Nitrotyrosine and related derivatives in proteins: Precursors, radical intermediates and impact in function. *Essays Biochem.* **2020**, *64*, 111–133. [CrossRef] [PubMed]

34. Auclair, J.; Turcotte, P.; Gagnon, C.; Peyrot, C.; Wilkinson, K.J.; Gagné, F. The influence of surface coatings on the toxicity of silver nanoparticle in rainbow trout. *Comp. Biochem. Physiol. C Toxicol. Pharmacol.* **2019**, *226*, 108623. [CrossRef] [PubMed]
35. Guardiola, F.A.; Logothetis, P.; Meseguer, J.; Esteban, M.A. Evaluation of silver nanospheres on viability and innate cellular parameters of gilthead seabream (*Sparus aurata L.*) head-kidney leucocytes. *Fish Shellfish. Immunol.* **2017**, *69*, 99–107. [CrossRef] [PubMed]

nanomaterials

Article

Titanium Dioxide Nanoparticles: Effects on Development and Male Reproductive System

Elena Maria Scalisi [1,*], Roberta Pecoraro [1], Antonio Salvaggio [2], Fabiano Capparucci [3], Cosimo Gianluca Fortuna [4], Massimo Zimbone [5], Giuliana Impellizzeri [5] and Maria Violetta Brundo [1]

1 Department of Biological, Geological and Environmental Science, University of Catania, 95124 Catania, Italy; roberta.pecoraro@unict.it (R.P.); mariavioletta.brundo@unict.it (M.V.B.)
2 Experimental Zooprophylactic Institute of Sicily "A. Mirri", 90129 Palermo, Italy; antonio.salvaggio@izssicilia.it
3 Department of Chemical, Biological, Pharmacological and Environmental Science, University of Messina, 98166 Messina, Italy; fcapparucci@unime.it
4 Department of Chemical Sciences, University of Catania, 95124 Catania, Italy; cg.fortuna@unict.it
5 CNR-IMM, 95123 Catania, Italy; massimo.zimbone@ct.infn.it (M.Z.); giuliana.impellizzeri@ct.infn.it (G.I.)
* Correspondence: elenamaria.scalisi@unict.it; Tel.: +39-095-7306040

Abstract: Titanium dioxide nanoparticles (TiO_2-NPs) are used intensively. Thanks to their extremely small size (1–100 nm), TiO_2-NPs are more absorbable by living organisms; consequently, they can cross the circulatory system and then be distributed in various organs including the reproductive organs. We have evaluated the possible toxic effect of TiO_2-NPs on embryonic development and the male reproductive system using *Danio rerio* as an organism model. TiO_2-NPs (P25, Degussa) were tested at concentrations of 1 mg/L, 2 mg/L, and 4 mg/L. TiO_2-NPs did not interfere with the embryonic development of *Danio rerio*, however, in the male gonads the TiO_2-NPs caused an alteration of the morphological/structural organization. The immunofluorescence investigation showed positivity for biomarkers of oxidative stress and sex hormone binding globulin (SHBG), both confirmed by the results of qRT-PCR. In addition, an increased expression of the gene responsible for the conversion of testosterone to dihydrotestosterone was found. Since Leydig cells are mainly involved in this activity, an increase in gene activity can be explained by the ability of TiO_2-NPs to act as endocrine disruptors, and, therefore, with androgenic activity.

Keywords: TiO_2-NPs; *Danio rerio*; male infertility; embryonic development; endocrine system; testis

1. Introduction

The technology-based industry realized that nanoparticles presented potential opportunities such as increased energy efficiency and cleaning up industrial contaminants [1]. The use of nanotechnology helps decrease production costs by reducing energy consumption, attenuating environmental pollution, and increasing production efficiencies in developed countries.

The manufacturing of nanotechnology are the engineered nanoparticles (ENPs) in which the rearrangement of the atoms gives them new properties [2,3]. Titanium dioxide nanoparticles (TiO_2-NPs), especially, are among the most engineered metal oxide nanoparticles in the world [4,5] and are used to remove pollutants from wastewater [6]. Presently, their photocatalysis property is considered an efficient methodology in the area of wastewater treatment [7,8] because TiO_2-NPs are minimally selective; therefore, many types of contaminants such as polycyclic aromatic hydrocarbons [9], chlorinated organic compounds [10], dyes [11], pesticides, cyanide [12], phenols [13], arsenic [14], and heavy metals [15] can be degraded by TiO_2 nanoparticles. Moreover, TiO_2-NPs are used in a wide variety of products, such as food colorants (under E code number E171), nutritional supplements, personal care products (cosmetics, sunscreens), toothpaste [16,17], and paint [18].

The increasing production and use of manufactured nanoparticles, such as titanium dioxide nanoparticles (TiO_2-NPs), has inevitably led to their release into the aquatic environment; thereby posing a threat to aquatic organisms and humans alike [19,20]. Their nano size facilitates the penetration of different live tissues, into the body, through the blood circulatory system [21]; the NPs can be distributed in various organs thus accumulating as foreign bodies [22,23]. It has been shown that the TiO_2-NPs can pass the testicular blood barrier inducing an effect on the testis and then on male reproductive health, given their nano-size [24–26]. Several studies in rodents and mice have shown the toxicity effect of TiO_2-NPs on the male reproductive system; in addition, there is an increase in data about in vitro and in vivo studies on NPs that support the notion that different types of nanoparticles are capable of altering the normal and physiological activity of the endocrine system [27]. According to European Food Safety Authority EFSA (2010), any substance that has the ability to interact with one or more elements of the endocrine system (i.e., exhibit endocrine activity) falls into the category of "endocrine active substances" also called "endocrine disrupting chemicals" (EDCs). Endocrine disruptors have the potential to dysregulate hormone activity in exposed aquatic organisms [28] and their disruption is linked to reproduction and development dysregulation [29]. Although several chemical products are recognized as potential endocrine disruptors, the research has been limited to a few groups of chemical substances (disinfection byproducts, perfluoroalkyl and polyfluoroalkyl substances, bisphenol A, phthalates, pesticides, pharmaceutical agents, and heavy metals); consequently, the data on a number of other xenobiotics that may act as EDCs are still scant and incomplete [30].

In this regard, ENPs must be included among these chemical compounds since the intensity of ENP exposure is significantly increased due to their applications; moreover, they are persistent in the environment. The release of ENPs into the aquatic environment along with other environmental contaminants is common [31].

Thus, in this study, we investigated the possible toxic effect of commercial titanium dioxide nanoparticles (TiO_2-NPs) on embryonic development and the male reproductive system, evaluating markers that act in their possible role as endocrine disruptors.

2. Materials and Methods

2.1. Preparation of Work Solutions

The titanium dioxide nanoparticle powders supplied by CNR-IMM (Microelectronics and Microsystems of Catania-National Research Council, Italy), were purchased from Sigma Aldrich. (St. Louis, MO, USA)

According to the data in the literature and the suggestions by researchers of the National Research Council, we chose the following concentrations to test: 1 mg/L, 2 mg/L, and 4 mg/L of TiO_2-NPs [32]. Briefly, each mass (1 mg, 2 mg, 4 mg) was dispersed in 1 L of osmosis water which was reconstituted with the addition of inorganic salts (sea salt and red sea salt); this is optimal for the housing of zebrafish, according to Nüsselin-Volhard and Dahm [33] and prevents fungi growth. Four cycles of sonication were performed for each solution; each cycle duration was 20 min with a 10 min break using an ultrasonic bath (FALC Labsonic LBS2, Treviglio, BG, Italy) (with a frequency of 40 kHz under an extractor fan to disrupt any possible aggregates [34].

2.2. Breeding of Zebrafish and Experiment Design

Danio rerio fishes were raised in a fish room at the Fish Pathology and Experimental Centre of Sicily (CISS) of the Department of Veterinary Science (University of Messina). The embryos were used to perform the zebrafish embryo toxicity test (ZFET); whereas the male adults had been used for a 30-day chronic toxicity test (authorization n°1244/2015-PR approved by the Italian Health Ministry).

Male and female adults were useful for the supply of embryos and were kept in a breeding room in an optimal condition regarding photoperiod (light/dark cycle:14 h/10 h), water quality (27 ± 1 °C, pH 7.2 ± 0.3, 6.00 ppm dissolved oxygen content (DO)), and feeding. Subsequently, male and female fishes (ratio 2:1) were placed in a hatching tank to breed. The tank was equipped with steel grids for the eggs to fall through to the bottom of the tank and avoid predation by the adults. The eggs were collected by Pasteur pipettes, rinsed in aquarium water at 28 °C, and analyzed under a stereomicroscope (≥30-fold magnification). The infertile eggs were discarded, and the fertilized eggs at the blastula stage (about 3–3.5 h post-fertilization) were selected to perform the zebrafish embryo toxicity test (ZFET).

One- to two-year-old zebrafish wildtypes, not consanguineous, with a body weight of about 0.5 g and an average length of 3 cm, were used for a 30-day chronic toxicity test. Prior to TiO_2-NP treatment, fishes were acclimated to experimental conditions (26–28 °C; light: dark/14 h:10 h; daily water change), including daily manipulation and nutrition. After, the zebrafish were randomly divided into four groups of 10 fishes: three experimental groups (1 mg/L, 2 mg/L, and 4 mg/L TiO_2-NPs) and a control group (with osmosis water).

2.3. Acute Toxicity Experiment of Zebrafish Embryo

The assay was conducted in accordance with the guidelines of the fish embryo acute toxicity (FET) test with the zebrafish (*Danio rerio*) developed by the Organization for Economic Cooperation and Development (OECD TG 236) [35–37]. As suggested by protocol procedure, 24 eggs at the blastula stage were transferred into a 24-well multi-plate with one embryo per well, containing 2 mL of work solutions at 28 °C. Multi-well plates were set up for the 1 mg/L, 2 mg/L, and 4 mg/L TiO_2-NPs solutions, in addition, multi-well plates of positive controls (3,4-dichloroaniline at the concentration of 4 mg/L in water) and negative controls (water dilution) were made. Three replicates were performed for each experimental group. The maintenance of 26 ± 1 °C in wells was ensured by a control of room temperature; moreover, every 24 h each work solution was renewed in all wells (semi-static renewal) [35]. The exposure time was selected to be 96 h pos-fertilization (hpf) because even if most organs in the embryos are well developed at 96 hpf, the larvae are formed after 120 hpf [38,39]. We selected the end of the test to be at 144 h after fertilization to investigate the expression of protein markers associated with the endocrine disruption by immunohistochemical analysis.

2.4. Evaluation of Toxicological Endpoints and DanioScope™ Analysis

According to the OECD, every 24 h the acute toxicological endpoints (coagulated embryos, lack of somite formation, non-detachment of the tail, and lack of heartbeat) were assessed and quantified as observed or not observed; any positive outcome in one of these observations would indicate that the zebrafish embryo was dead. All endpoints can occur after 24hrs of exposure, except for heartbeat which in normal zebrafish development is visible after 48 h. In addition, using DanioScope™ software (Noldus Information Technology bv, Wageningen, the Netherlands), the following were evaluated: heartbeat, body length of larvae, and malformations (Supplementary Information Text S1).

2.5. Immunohistochemical Analysis on Zebrafish Larvae

An immunohistochemical analysis was performed to localize (in whole larvae) a marker of oxidative stress heat shock protein-70 (Hsp70), poly (ADP-ribose) polymerase-1 (PARP-1), and a marker that suggested the action of TiO_2-NPs such as endocrine disruptor sex hormone-binding globulin (SHBG) and prothymosin-α (PTMA). Moreover, metallothionein, a specific marker of exposure to TiO_2-NPs, was detected. The procedure was based on the immunofluorescence protocols of Pecoraro et Colleagues [37] (Supplementary Information Text S2).

2.6. *Adult Zebrafish Exposure Experiment*

As previously mentioned, adult zebrafish were randomly divided into four groups of 10 fishes: three experimental groups (1 mg/L, 2 mg/L, and 4 mg/L TiO_2-NPs) and a control group (with TiO_2-NPs osmosis water). The tanks were equipped with aerators and fishes were subjected to a semi-static exposure regime for 30 days (water was changed every 24 h with a new solution of TiO_2-NPs). The water parameters were monitored daily before and after the replacement of the solutions by the multiparameter probe (HI9829 Aquaprobe, Hanna instruments, Padua, Italy). During experimentation, the fishes were maintained as follows: photoperiod 10 h dark/14 h light with an intensity equal to 250 lux, 6.9–7.5 pH, 26–28 °C temperature, and dissolved oxygen ≥6.00 ppm; in addition, they were fed with commercial feed "GEMMA micro 300 Skretting Zebrafish" twice per day prior to the change of water TiO_2-NPs solution to avoid adsorption of nanoparticles by food particles. After adding the food, fishes were monitored for 10 min to verify that the food had been consumed. After the change of TiO_2-NPs solution, the fishes' behavior was evaluated for one-hour to highlight changes in swimming speed, respiration, loss of equilibrium, bottom stationing, and any other possible abnormal behavior. Until the end of the experimentation (30 days), the zebrafish were kept under the conditions mentioned and without sources of noise and/or vibrations. Upon completion of the experimentation, the fishes were euthanized by anesthesia with a dose of 0.7 g/L tricaine methane sulfonate (MS-222), buffered, then testis and gill tissues were dissected.

2.7. *TiO_2-NPs Accumulation*

Through the single particle inductively coupled plasma-mass spectrometer (spICP-MS), the concentration of TiO_2-NPs in organs of zebrafish was evaluated. TiO_2-NPs were analyzed using ICP-MS NexION® 350D (Perkin Elmer, Waltham, MA, USA) with the Syngistix Nano Application software (Perkin Elmer, Waltham, MA, USA). Supplementary Information Text S3.

2.8. *Histological Examination*

Histological examination was performed following our standard protocol (Supplementary Information Text S4). The sections were observed using an optical microscope (Set E200 Nikon, Amsterdam, Netherlands)) and the images were captured by a digital camera (CMOS Nikon, Amsterdam, Netherlands) connected to the microscope. Potential morphological alterations on the structure of the testis, and gill tissues were identified.

2.9. *Immunohistochemical Analysis*

An immunohistochemical analysis was performed on testis sections to detect SHBG and P540 expression. This followed our standard protocol of immunohistochemical used for several experiments on zebrafish [32]; furthermore, the same primary antibodies used for the immunohistochemical investigation on zebrafish larvae were used.

2.10. *Preparation of Semithin Sections and Electron Microscopy*

Tests were fixed with 2.5% glutaraldehyde (brand) for 90 min at +4 °C. Subsequently, the protocol for electron microscopy was followed (Supplementary Information Text S5) to obtain semi-thin (thickness 0.85 μm) and ultra-thin (thickness 0.085 μm) sections useful for transmission electron microscopy (TEM,). (Palackého třída, Brno, Czech Republic),

2.11. *RNA Extraction and qRT-PCR*

The qRT-PCR was used to evaluate the mRNA levels of marker genes in the testis. SHBG, SRD5A2, SOD2, and GPX4B were selected for analysis, and β-actin was used as a housekeeping gene (Supplementary Information Text S6).

2.12. Crystal Structure of Sex Hormone-Binding Globulin (SHBG)

The SHBG protein in zebrafish, is ortholog to human. Thanks to Fingerprint for Ligand and Protein FLAP, it was analyzed that SHBG bound to dihydrotestosterone (DHT) (Supplementary Information Text S7).

2.13. Statistical Analysis

Data on zebrafish embryo experimentation, namely the coagulate, survival, and hatching rate of the TiO_2-NP exposed groups and the unexposed group, were represented as the average percentage of the coagulate, survival, and hatching rate from three replicates. Statistical analysis was performed by one-way analysis of variance (ANOVA) test to compare differences between groups. A $p < 0.05$ was considered to be a statistically significant difference.

3. Results

3.1. Nanoparticles Characterization

According to the manufacturer's information, the purity of the TiO_2-NPs was 99.5% with metal traces. The nanopowder crystalline phase was mixed: 86% anatase, and 14% rutile. SEM analyses were performed in plan-view to characterize the morphology of the nanoparticles (Figure 1). As is apparent from the SEM analyses, powders are constituted by 'cobblestones' with a size in the order of 1–2 microns (Figure 1, at left); nevertheless, at higher magnification, SEM images show that these structures are composed by smaller particles aggregated to each other. Isolated nanoparticles had an average diameter of about 50 nm.

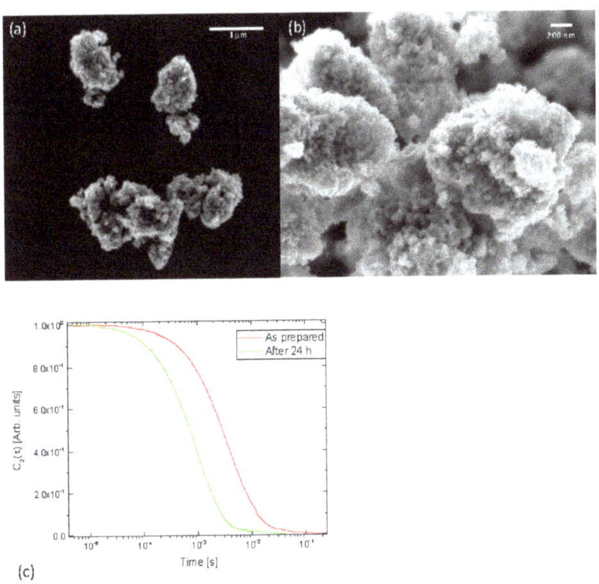

Figure 1. SEM images of TiO_2-NP. Marker is 1 µm (**a**), and 200 nm (**b**). (**c**) Autocorrelation function of the colloidal solution as prepared solution and after 24 h of sedimentation.

To characterize the nanoparticle solution, hydrodynamic diameter was measured by dynamic light scattering. Apparatus and methods are described elsewhere [40,41].

Figure 1c shows the dynamic light scattering DLS autocorrelation function for the "as prepared" solution and for the solution after 24 h of sedimentation. The hydrodynamic diameter was estimated to be 1100 nm for the "as prepared" solution (red curve). This is in perfect agreement with the SEM image of Figure 1a. Furthermore, solutions were allowed to sediment for 24 h and correlation functions were measured again. The hydrodynamic diameter decreased to about 283 nm (green curve). This indicates that the larger particles were deposited and only smaller aggregates remained in suspension during the experiment.

3.2. Embryonic Development of Zebrafish

The toxicity of TiO_2-NPs on zebrafish embryos was defined by observing specific toxicological endpoints, as mentioned. At 24 hpf, the rate of coagulated eggs was 11.3% for the 1 mg/L group, 28.3% for the 2 mg/L group, and 23.3% for the 4 mg/L group, whereas, in the unexposed embryo group was 5%. At 48 hpf, the coagulated egg rate increased for the 4 mg/L groups (25%); otherwise, it remained unchanged in the other experimental groups (see Supplementary Informations Figure S1).

At 24 hpf, and all the exposure period, TiO_2-NPs were evidently deposited on the bottom of the wells. However, TiO_2-NPs also adhered on the surface of embryonic chorion with increasing concentrations of TiO_2-NPs, as shown in Figure 2

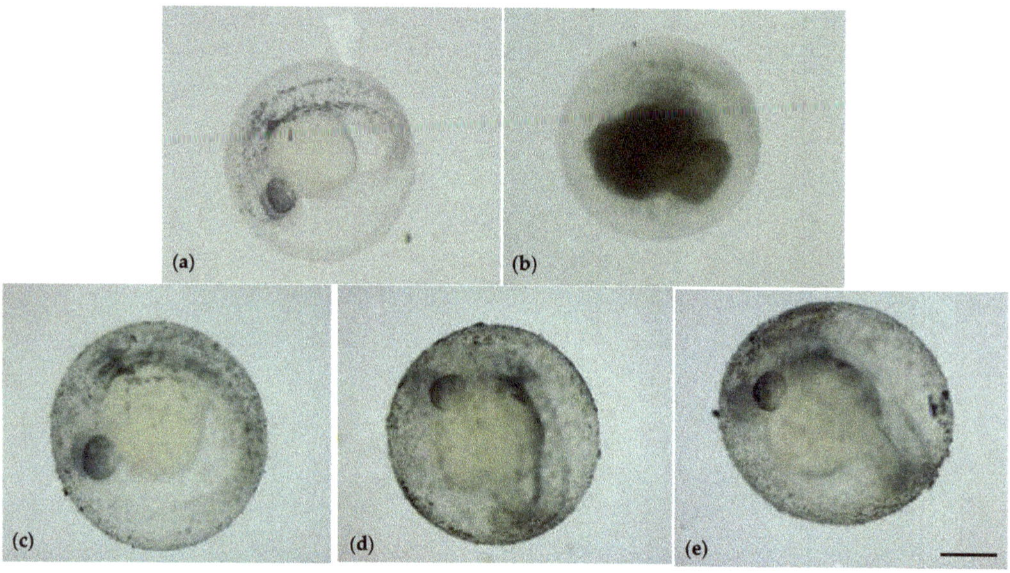

Figure 2. (**a**) Zebrafish embryo unexposed to TiO_2-NPs (negative controls); (**b**) zebrafish embryo exposed to 3,4-dichloroaniline (positive controls). Zebrafish embryo exposed to 1 mg/L (**c**); 2mg/L (**d**); and 4 mg/L (**e**) TiO_2- NPs. Scale bar 410 μm.

The TiO_2-NPs formed an external white layer on the chorion, that did not affect the hatching of embryos. At 48 hpf, the hatching rate was 36.7% (1 mg/L), 15% (2 mg/L), and 18.3% (4 mg/L); whereas the unexposed group showed a rate of 12%. The hatching rate was statistically significant between all exposed groups and the unexposed group ($p < 0.05$); no statistical significance had been observed at 72 hpf except for the 1 mg/L group (see Supplementary Information Figure S2). After 96 h of exposure to TiO_2-NPs, the survival of the hatched embryos for control was above 90%, as described by Kimmel et al. [38], while the survival of the exposed groups was below 90% but remained unchanged until the end of experimentation. No statistically significant increase in mortality rates for groups exposed to 2 mg/L and 4 mg/L TiO_2-NPs relative to the control treatments were observed ($p > 0.05$). At the end of the test (144 hpf), we observed a rate of mortality of 8.30% (unexposed group), 13.30% (1 mg/L), 31.7% (2 mg/L), and 25% (4 mg/L). All embryos showed a complete development of head, notochord, fin, pigmentation, and the organ's heart and eyes. There was no morphological malformation compared to the unexposed group (Figure 3).

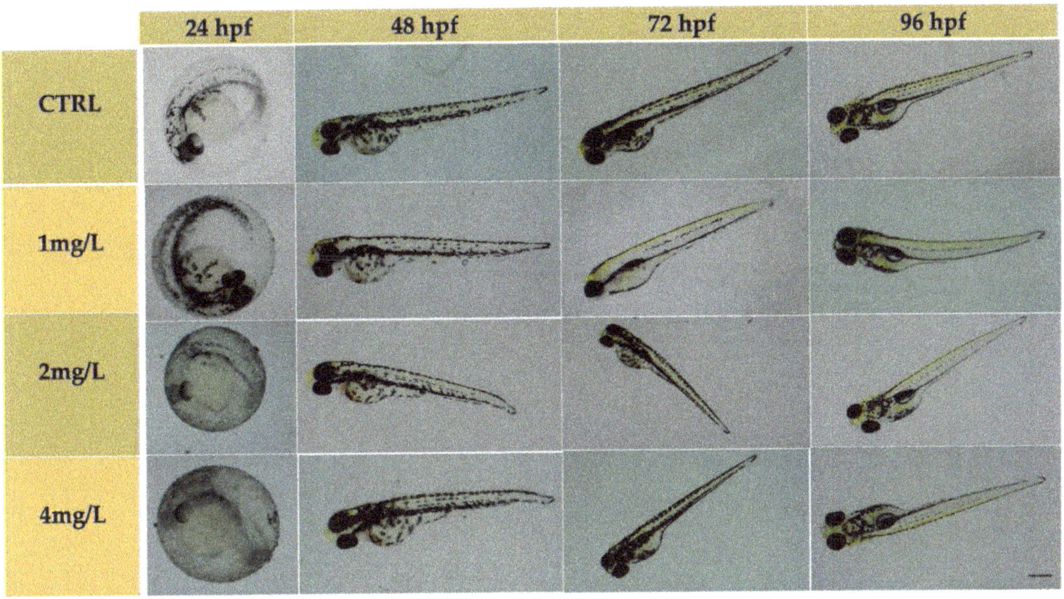

Figure 3. Phenotypes of larvae exposed to TiO_2-NPs and the unexposed group from 24 to 96 hpf. Scale bar 420 µm.

3.3. DanioScope Analysis

Through the DanioScope software (1.2 Wageningen, the Netherlands)), it was shown that TiO_2-NPs affected the body length of larvae because they exhibited reduced body length compared to the unexposed group ($p < 0.05$) at 96 hpf. The mean body length in the 4 mg/L group was 172 µm, while in the unexposed group was 215 µm. The heart rate was measured through the registration of beats per minute (BPM) by the DanioScope software which highlighted an increase in embryos exposed. An increase in BPM was observed in the 1 mg/L group (217.20 BPM) and a higher BPM (236.3) appeared in the 4 mg/L group. BPM in zebrafish is physiologically around 120–180 bpm. The exposure to TiO2-NPs resulted in a statistically significant increase in heart rates in exposed embryos ($p < 0.05$) (see Supplementary Figure S3).

3.4. Immunohistochemical Markers on Zebrafish Larvae

For markers of oxidative stress, the immunohistochemical investigation showed a positivity for the poly (ADP-ribose) polymerase-1 (PARP1). Positivity was observed at the lower concentration (1 mg/L) and increased at the higher concentration (4 mg/L) compared to the control. Moreover, a positivity for heat shock proteins-70 (HSP70) in the exposed group, especially at the higher concentration, suggests the ability of TiO_2-NPs to induce oxidative stress. Additionally, positivity for biomarker metallothioneins (MTs), which is linked to detoxification pathways in the presence of toxic substances, was found in the whole body of the embryo except for the end of the tail. Using Image J software, the intensity of fluorescence was quantified for each biomarker evaluated. Figure 4 shows the images of larvae and their average fluorescence intensity (AU).

Regarding the SHBG and PTMA, positivity was observed for both. Particularly for SHBG, positivity was observed on the head of the embryo, with a higher expression for the 4 mg/L TiO_2-NPs concentration compared to the control. Whereas, the positivity for prothymosin α (PTMA) occurred in the body of larvae, especially at the concentration of 4 mg/L. Figure 5 shows the images of larvae and their average fluorescence intensity (AU).

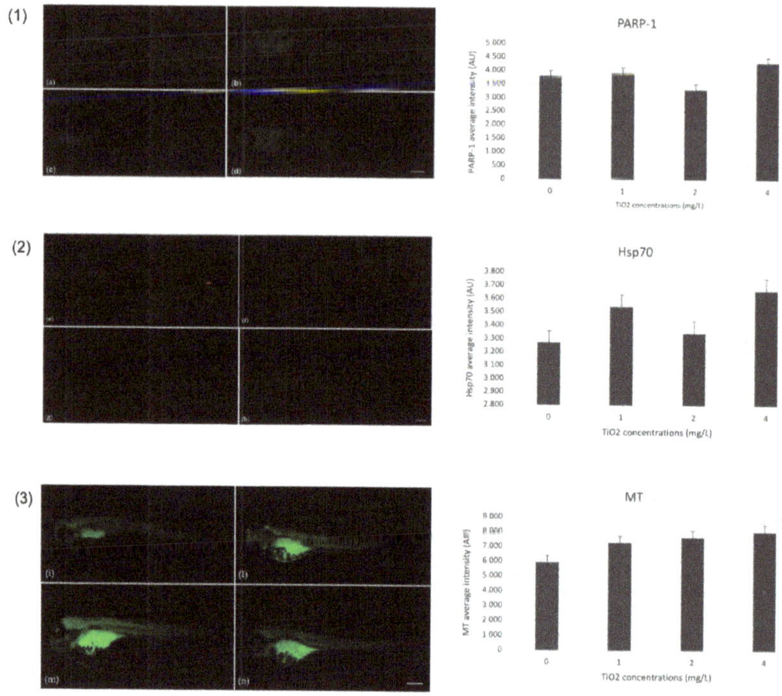

Figure 4. In order (1) PARP-1 antibody-staining, (2) Hsp70 antibody-staining, and (3) MT antibody-staining. (**a**,**e**,**i**) larva unexposed to TiO_2-NPs; (**b**,**f**,**l**) larva exposed to 1 mg/L TiO_2-NPs; (**c**,**g**,**m**) larva exposed to 2 mg/L TiO_2-NPs; and (**d**,**h**,**n**) larva exposed to 4 mg/L TiO_2-NPs. The histogram next to each photo represents the average fluorescence intensity (AU) of the corresponding biomarker. Scale bar 420 μm.

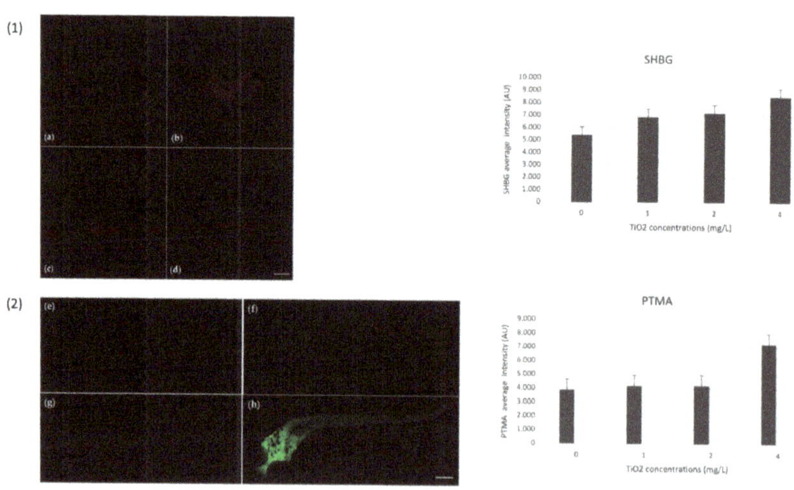

Figure 5. In order (1) SHBG antibody-staining and (2) PTMA antibody-staining. (**a**,**e**) larva unexposed to TiO_2-NPs; (**b**,**f**) larva exposed to 1 mg/L TiO_2-NPs; (**c**,**g**) larva exposed to 2 mg/L TiO_2-NPs; and (**d**,**h**) larva exposed to 4 mg/L TiO_2-NPs. The histogram next to each photo represents the average fluorescence intensity (AU) of the corresponding biomarker. Scale bar 420 μm.

3.5. Adult Exposure

Daily monitoring until the end of experimentation, revealed no fish died, no abnormal behaviors such as loss of equilibrium, refusal to feed, and no apparent abnormalities in the body of fish. An accumulation of TiO_2-NPs was found in zebrafish testicles of our experiment groups at concentrations of 1.18×10^{-3} mg/Kg and 8.14×10^{-3} mg/Kg under 2 mg/L and 4 mg/L doses, respectively. Using the Syngistix Nano Application software, it appeared that TiO_2-NPs had a size < 100 nm. In our study, 30-day exposure of zebrafish to TiO_2-NPs suggests that fish uptake TiO_2-NPs by breathing and feeding. The nanoparticles were constantly resuspended in the aqueous medium, so fish are highly likely to internalize them via their gills and mouth. We found an accumulation of TiO_2NPs in the gill tissue for the concentrations of 2 mg/L (1.50×10^{-3} mg/Kg) and 4 mg/L (2.39×10^{-3} mg/Kg); whereas, no accumulate TiO_2-NPs were found in the control as expected, but also at 1 mg/L concentration.

3.6. Histological Observations

Gill tissue from the control, 2 mg/L, and 4 mg/L groups showed alterations in their morphology. We observed increasing cellularity in the interlamellar space and hyperplasia brings a higher width of the secondary lamella with respect to control (indicated by blue arrows in Figure 6.

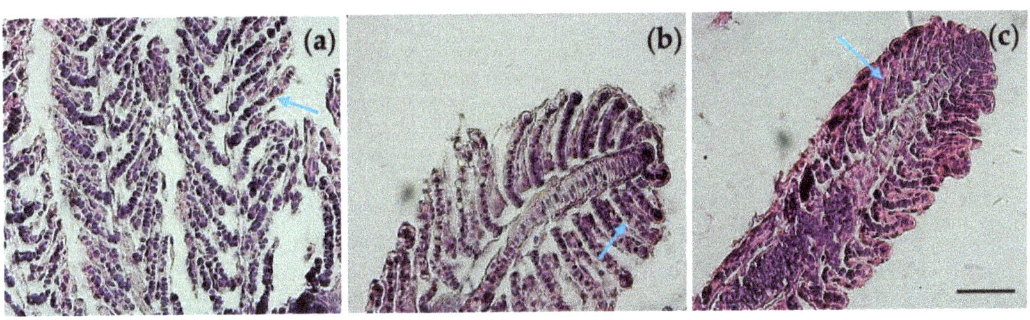

Figure 6. Histological section of gills. (**a**) CTRL gills. Exposed groups to (**b**) 2mg/L; and (**c**) 4mg/L TiO_2-NPs. Arrows indicate the secondary lamellae, that showed a hyperplasia in the exposed groups, while it is not appeared in the control groups. Scale bar 1500 μm.

Examination of the gonads tissue showed an alteration in the spermatogenic epithelium in the groups exposed to TiO_2-NPs. The exposure of TiO_2-NPs caused a detachment of the spermatogenic epithelium from the connective tissue, whereas, it had not been observed in the control group, respectively, blue arrow and green arrow in Figure 7. Moreover, the tubules (the area occupied by spermatogonia) were increased, compared to the area occupied by spermatozoa that were decreased (in Figure 8, red * indicates the area of spermatozoa, which have blue nuclei; whereas, the remaining are spermatogonia, which have light blue nuclei).

A disordered arrangement of spermatogonia was observed at the concentration of 4 mg/L (red arrow), and the connective tissue presented irregularities making it difficult to distinguish between Leydig cells and connective cells. Otherwise, the unexposed group was intact (green arrow) (Figure 9). Whereas, the testis tissues in the 1mg/L dosage group showed no changes compared with that of the control.

In the ultrathin sections of testis, we observed by transmission electron microscopy (TEM) the presence of vesiculation in Sertoli cells. It was evident at the concentration of 4 mg/L with the detachment of the cell membrane compared to the control, as shown by red * in Figure 10.

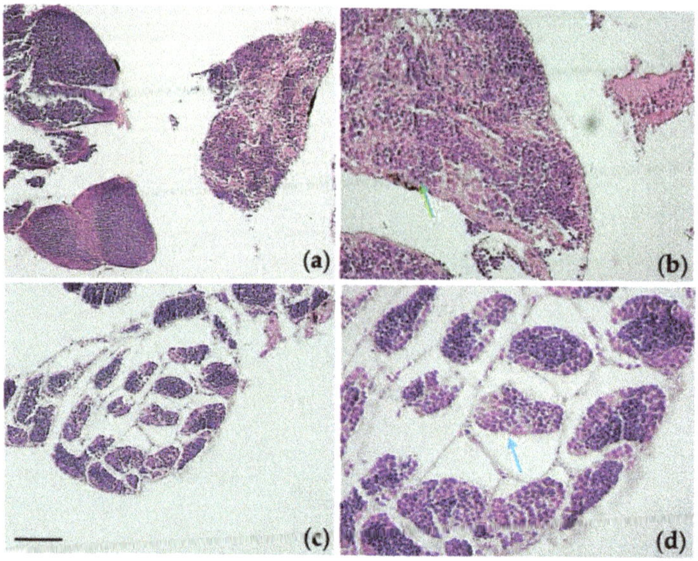

Figure 7. Histological sections of the testis. Unexposed group: 10× (**a**), and 40× (**b**). Group of 2mg/L TiO_2-NPs: (**c**) 10×, and (**d**) 40×. E-E staining, sections 4 μm. Blue arrow indicate the detachment of the spermatogenic epithelium from the connective tissue in the exposed group (2mg/L), while in the control group the spermatogenic epithelium maintains its contact with connective tissue (green arrow). Scale bar 260 μm.

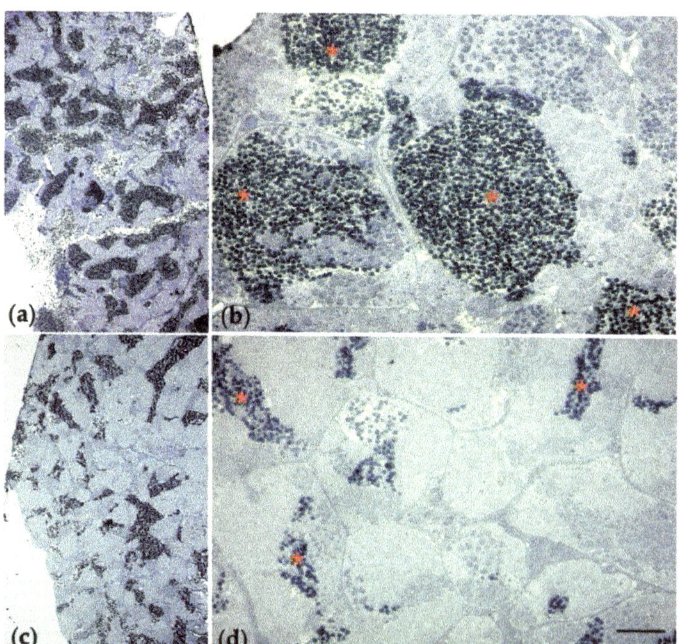

Figure 8. Histological sections of testis staining with toluidine blue (sections 0.85 μm). Unexposed group: 10× (**a**), and 40× (**b**). Group of 2mg/L TiO_2-NPs: (**c**) 10×, and (**d**) 40×. Red * indicates the area occupied by spermatozoa into the tubules. Scale bar 260 μm.

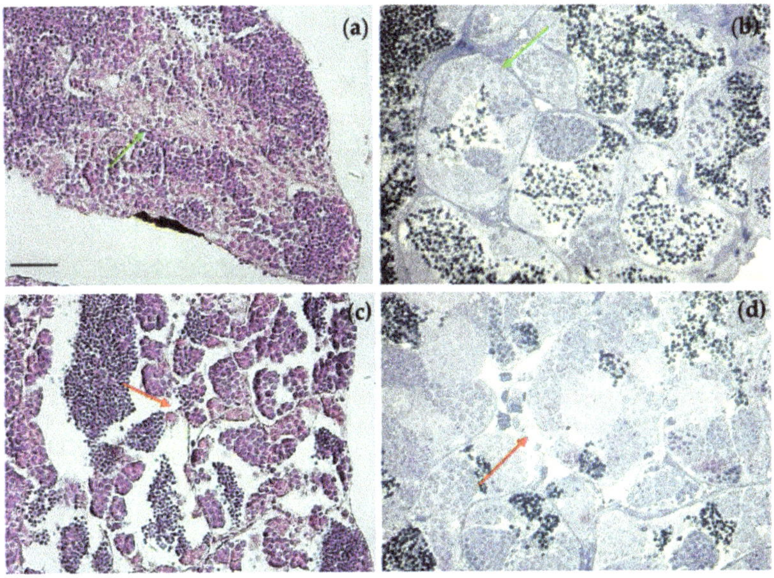

Figure 9. (**a**) (10×) and (**b**) (40×) unexposed group with good morphology and organization of tubules testis; (**c**) (10×) and (**d**) (40×) group exposed to 4mg/L TiO_2-NPs with disorganization of tubules testis. Green arrows indicate the intact and well organization of tubules testis, while the red arrows indicate their disordered organization. Scale bar 260 μm.

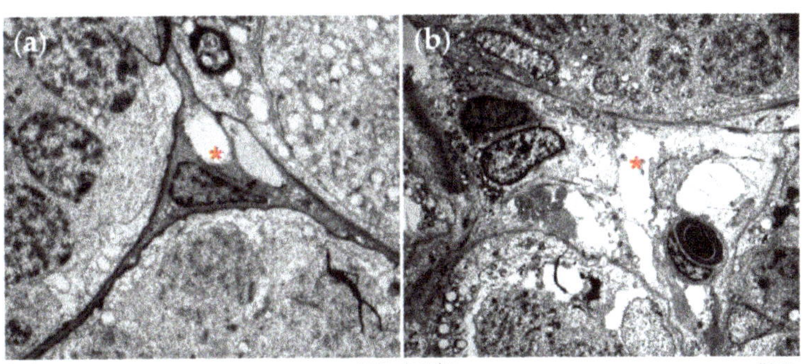

Figure 10. Ultrathin sections of zebrafish testis. (**a**) Section of testis unexposed; (**b**) section of testis exposed to 4 mg/L with evident vesiculation. Red * indicate the Sertoli cells that showed a evident vesiculation in the exposed group compared to control. Scale bar 2 μm.

3.7. Immunohistochemical Analysis and Gene Expression

We explored markers to investigate the action of TiO_2-NPs such as endocrine disruptors. In particular, the results of immunohistochemical analysis on SHBG reveal that it was expressed in control groups in a regular way because the cysts maintained their morphological organization; however, positive expression remained in the group exposed to 2 mg/L, than in 4 mg/L. As shown in Figure 11, positivity was related to the cells that were around the seminiferous tubule cysts.

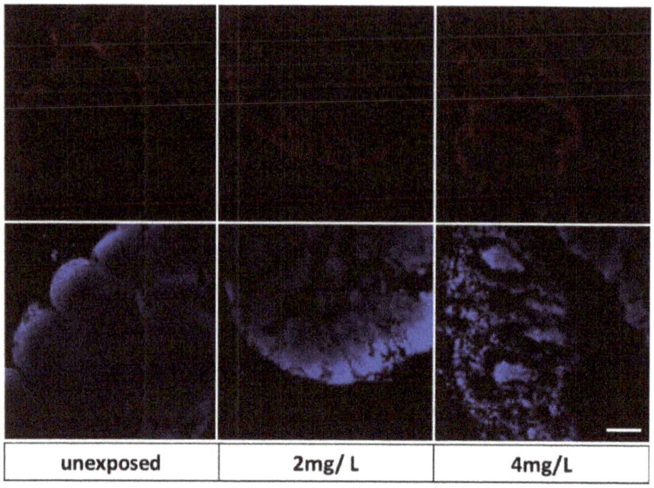

Figure 11. SHBG expression in testis tissue. Nuclei blue (DAPI) and red fluorescent of SHBG protein. Scale bar 265 μm.

The result of gene expression of SHB in the testis indicates an increased expression of the SHBG gene for the 2 mg/L and 4 mg/L concentrations compared to the control group. Increased transcription of antioxidant enzymes such as glutathione peroxidase (GPX), and especially superoxide dismutase (SOD), was observed. Figure 12 shows the results of qRT-PCR of all genes investigated.

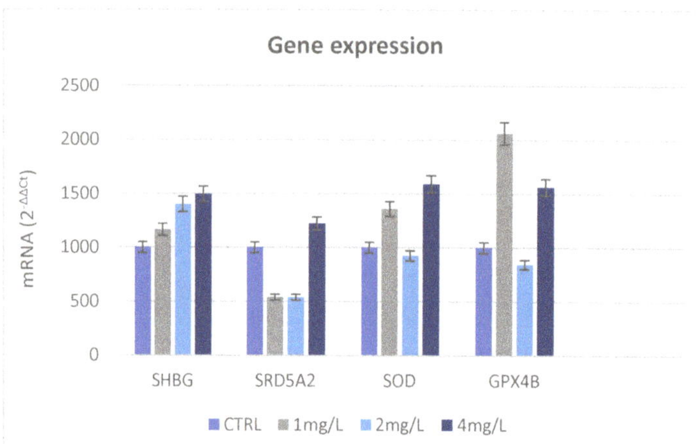

Figure 12. Results of qRT-PCR of all genes investigated.

In this way, oxidative stress is a common pathway of toxicity and disease that may be caused by many pollutants, such as TiO_2-NPs.

The immunohistochemical investigation of cytochrome P540 (Cyp19b) confirmed the ability of nanoparticles to induce oxidative stress. As shown in Figure 13, positivity was found in the expression of P540 in the groups exposed to TiO_2-NPs (2 mg/L and 4 mg/L concentrations) compared to the control which did not show positivity for the biomarker.

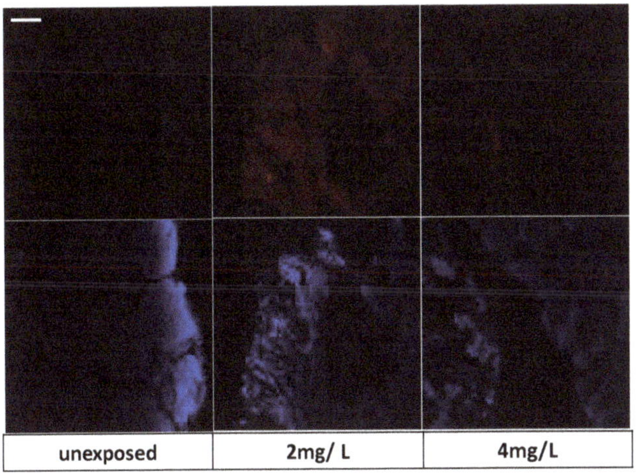

Figure 13. P540 expression in testis tissue. Nuclei blue (DAPI) and red fluorescent of P540 protein. Scale bar 265 µm.

3.8. Crystal Structure of Sex Hormone-Binding Globulin (SHBG)

The result of the crystal structure of the SHBG shows that the pocket is very large and many amino acid residues are involved. Figure 14 shows the SHBG interactions involved in binding, which can characterize the pocket. The interaction most involved is that of the hydrophobic character (green area), and there are also two areas of hydrogen bond acceptor characters (red). Thus, the hydrophobic interaction could explain the binding to NPs.

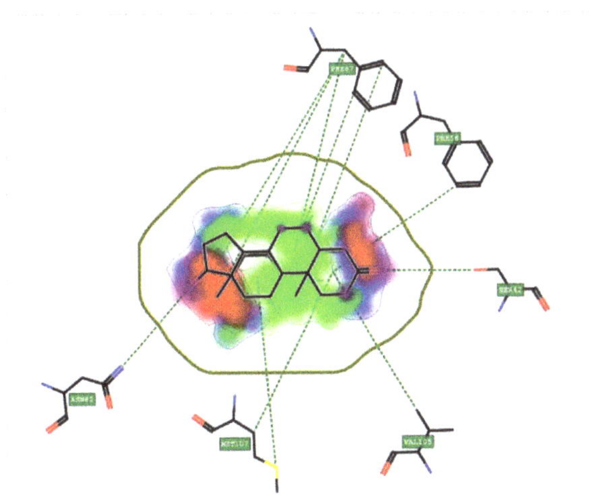

Figure 14. 2D-depiction shows the interactions that characterize the pocket.

4. Discussion

In this study, the toxicity of TiO_2 NPs on zebrafish embryos and adult fish has been investigated. With the zebrafish embryonic test (Z-FET), it was made evident that TiO_2-NPs did not interfere with embryonic development because zebrafish chorion acted as a special biological structure that covered the embryo until hatching. Then, it acted as a barrier that blocked the entry of various pollutants [42]; simultaneously due to its pores, the chorion ensured the transport of necessary oxygen, salt ions, and nutrients from the aquatic environment to the embryo and excretion of waste in the opposite direction [43]. The small diameter of pores (between 300 nm and 1 micron) can allow the entry of NPs adhering to the chorion [43,44]. NPs diffusion may be toxic to embryo development during the period of organogenesis [45]; however, the literature is scant regarding the interaction of NPs with the chorion, and how this structure interacts and affects the absorption, accumulation, and distribution of nanoparticles in the embryos [46].

Although it was evident that there was sedimentation of TiO_2-NPs during the experimentation, the embryos and larvae were constantly exposed to the TiO_2-NP aggregates because they were mostly located on the bottom of the wells, also after the hatching when they could freely swim. TiO_2-NPs have low acute toxicity to fish survival [47] and TiO_2-NPs concentrations higher than our experimental groups did not affect the survival rate in zebrafish [48]. In addition, the group control showed normal development [38] as well as the exposed groups with a low dose (1 mg/L) of TiO_2-NPs [49]. However, using the DanioScope software, it was shown that TiO_2-NPs caused alterations in the body length of larvae and the heart rate. These data are in accord with other studies on nanoparticle toxicity [50–54]. Regarding immunohistochemical markers on zebrafish larvae, the positivity of the biomarkers whose expression is regulated by environmental stressors highlights that the TiO_2-NPs are stressful stimuli for zebrafish embryos. Despite this, the zebrafish larvae are able to resist the presence of toxic substances and they can tolerate the presence of metal concentrations. In particular, the PARP-1 is involved in single-strand break (SSB) repair [55] induced by chemicals [56–58], the heat shock protein-70 (Hsp70) expression increases in response to environmental and physiological stressors [59] and protects the cells against induction of cell death by a variety of stresses; data in the literature have shown an increased expression of the Hsp70 due to exposure of chitosan nanoparticles, ZnO [60], and transition metals oxide (CuO, ZnO, NiO, and Co3O4) [61]. Finally, metallothioneins (MTs) are involved in homeostasis, protection against heavy metals and oxidant damages, and metabolic regulation, sequestration, and/or redox control [62]. Previous studies have shown the positivity of MTs in zebrafish embryos exposed to AuNPs [63].

Positivity for the SHBG and PTMA biomarkers can suggest that TiO_2-NPs act similar to endocrine disruption. The SHBG is a protein capable of binding steroids in the blood of fish and other vertebrate species. It is well characterized in humans [64,65] but only one ortholog (SHBG) has thus far been identified in zebrafish; being expressed in the digestive tract, liver, gills, pancreas, and testis [66,67]. Furthermore, on sex steroid transportation, regulation, and action [68], the SHBG has shown affinity to synthetic steroids such as ethinylestradiolor gestagens [66,68] and, in addition, it has been reported that it binds with phthalates and other environmental contaminants [69–71]. In this regard, SHBG is a potential target for environmental compounds found in the body. Consequently, increasing the risk of potential disruption in steroid homeostasis, experimental exposure to phthalate in rodents has shown adverse effects due to the deregulation of metabolic pathways by phthalate compounds [72]. Regarding fish, there is scant knowledge about SHBG structure and the site(s) of expression, although studies have highlighted that during the development of zebrafish, SHBG mRNA first appears within the liver and gut [68]. In addition, studies have shown SHBG affinity to xenobiotics able to act as endogenous sex steroids (testosterone and estradiol). Chen et al. [73] have found an increase in SHBG in zebrafish larvae exposed to bisphenol AF (BPAF), which is recognized as an endocrine disruptor. Prothymosin α (PTMA) is a small nuclear protein (109–113 amino acids depending on the species) with a potent nuclear localization signal (NLS), although data show cytoplasmic and extracellular presence as well, under specific physiological or pathological conditions [74,75]. It was observed that PTMA expression levels vary following stimulation by estrogen; thus, as well as SHBG expression, the positivity increment of PTMA in embryo zebrafish in our experimentation suggests that TiO_2-NPs act similarly to endocrine disruption.

In adult exposure, the accumulation of TiO_2-NPs in the testis can lead to impairment of the male reproductive system, as suggested by data in the literature. The accumulation of TiO_2-NPs in the testis induces cytotoxicity and gene expression changes [76–82]. The exposure to the TiO_2-NPs can occur via various routes; generally, the NPs enter and distribute in the exposed site, but due to the blood they are translocated to secondary organs such as the liver, spleen, kidneys, brain, ovaries, and testes [83–86]. As previously mentioned, it is highly likely that the fishes internalize the TiO_2-NPs via their gills and mouth. Gills are usually targeted organs for toxicity because they are in continuous contact with the water column; therefore, they are the main entrance route for all contaminants and for nanomaterials [83]. Into the intestine, bioaccumulation of TiO_2-NPs was lower than in gills. Since TiO_2-NPs after ingestion through the mouth were distributed in the digestive tract and then excreted through feces [85], there was no oral supply of nanoparticles because no oral administration was performed.

The gills' histological alterations that were observed are in accordance with the literature; the exposure to copper nanoparticles had reported a 3.5-fold increase in the gill filament of zebrafish, at already 24 h of exposure [86]; whereas, silver nanoparticles caused a slight change. In addition, the NPs are foreign substances taken by mononuclear phagocytic cells which become the entry route of NPs into the tissues and cells. The tissue macrophages phagocytose and sequester nanoparticles; for example, in the mouse model, it was demonstrated that there was a durability of Au, Ag, and SiO_2 NPs thanks to their action. This probably occurred in our experiment because, as previously mentioned, bioaccumulation of TiO_2-NPs was found in the testes of our experimental groups. Our results have shown an alteration in the spermatogenic epithelium in exposed groups to TiO_2-NPs. Previous studies on mice treated intragastrically with dosages of 10, 50, and 100 mg kg^{-1} of body weight (PC) anatase TiO_2-NPs for 28 days showed morphological changes in testes with a reduction in germ cell number, spherospermia, interstitial glands vacuole, malalignment, and vacuolization of spermatogenic cells [87]. In zebrafish, negative effects on the epithelium of testicular tubules have been observed with parenchyma degeneration, a decline in germinal epithelium cells and spermatozoa, or reduction in spermatogonial differentiation, particularly in exposure to chemical substances [88–90]. Few previous studies have evaluated the negative effect of TiO_2-NPs on zebrafish testis; however, Kotil et al. [32] evaluated the ultrastructure of zebrafish testis exposed to 1 mg/L, 2 mg/L, and 4 mg/L concentrations of TiO_2-NPs. Their results showed that TiO_2-NPs induced autophagy and necrosis at higher doses in Sertoli cells and consequently negatively affected spermatogenic cells and testicular morphology. Our results are in accordance with the literature.

Regarding the positivity of the SHBG biomarker in the seminiferous tubule cysts, it is known that SHBG is in the blood of all vertebrate species, apart from birds, and acts as a carrier of androgens and estrogens regulating their bioavailability [91]. In mammalians, the testis produces an SHBG homolog known as the testicular androgen-binding protein (ABP) which is produced and secreted by Sertoli cells primarily under the influence of follicle-stimulating hormone and then secreted into the seminiferous tubular lumen to regulate androgen availability in the male reproductive tract [92]. Studies of the SHBG steroid-binding characteristics in several fish species [93,94] have shown that its affinity for endogenous sex steroids (testosterone and estradiol), and xenobiotics [95–97] varies between species; moreover, some evidence showed that plasma SHBG levels fluctuate in fish during the reproductive cycle [98,99]. In zebrafish, the tissue distribution of SHBG transcript was detected in the digestive tract and hepatopancreas; in addition, a low expression was detected in the testis using RT-PCR [68]. However, several data indicate, in agreement with the literature about mammalians, that the liver is the main expression site of SHBG in teleost fish [100,101]. In some species, it was also found that there was a significant expression of SHBG in several other tissues, which suggests that the circulating SHBG could have an extra-hepatic origin. In mammals, a local expression of SHBG in several target organs has also been evidenced and associated with a modulation of the steroidogenic signal [102]. Similarly, non-hepatic expression sites of SHBG in teleost could also be associated with local action in target organs. For this reason, the transcription of SHBG in fish could depend on different variants which have to still be identified. It is not known how the expression of the SHBG gene is regulated [68]. Our results evidence that an expression of SHBG could be increased in the presence of NPs. This was supported by the result of gene expression. Unfortunately, the biological importance of SHBG in fish is not studied as well as it is in mammals. It may be reasonably hypothesized that the function of SHBG proteins expressed locally in target organs and tissues could be different from the circulation of SHBG. As in mammals, fish SHBG protein is involved in sex steroid transport, regulation, and action [103]. Considering our evidence, SHBG could improve the spermatogenesis process because their localization in the testis could bring a higher intake of androgen hormone or TiO_2-NPs could act similarly to androgen hormones. It could be supported by the result of the crystal structure of the SHBG, as shown in Figure 14, additional in support of the action of TiO_2-NPs like androgen hormones are the result of gene expression on the SRD5A2 gene. This gene, in all vertebrates, encodes for the steroid enzyme 5-αreductase α-polypeptide 2 (SRD5α2), an enzyme of spermatogenesis [104], and it works by converting testosterone (T) to dihydrotestosterone (DHT). 5α-dihydrotestosterone (DHT) plays a physiologically important role in some fish species and it is associated with spermatogenesis. Several pieces of evidence have documented the consequences of a lack of SRD5α2 activity that brought a decrease of 5α-DHT levels in those tissues, but also negative effects on spermatogenesis and the structure of seminiferous tubules [105,106]. The lack of 5α DHT mainly affects the Sertoli cells because they are involved in spermatogenesis by supporting germ cell development, as well as playing an important role in the structure of the seminiferous tubules and the maintenance of the blood-testis barrier. Nevertheless, our results differ from the negative effects due to the lack of enzyme activity. The expression of the SRD5α2 was increased in the gonads exposed to nanoparticles, and this consequently leads to its greater activity and production of DH. Then, exposure to TiO_2-NPs does not alter spermatogenesis because they encourage the activity of enzyme SRD5α2. Similar data were observed in fish treated with 5α-reductase inhibitors (5ARIs) in which spermatogenesis is unaltered or even increased because levels of T and 11KT are increased [107,108]. Since the synthesis of SRD5αs is regulated by the androgens they produce, and considering our results, it can be hypothesized that nanoparticles may behave similarly to androgens. Unfortunately, no unified theory about how hormone levels change after exposure to NPs exists. Some studies showed that low-dose (1 mg/kg/dose) AgNPs intravenously injected into male CD1 mouse serum caused a significative increase of the intratesticular testosterone (T) [109]; whereas, it was discovered in another study that CeO_2-NPs treatment caused decreases in T, FSH, LH, and prolactin (PRL) [110]. Indeed, these changes might be influenced by different factors such as particle type, size, and time of exposure. Therefore, research on how NPs affect hormones should be conducted and the lack of in vitro studies is still a problem. The cytochrome P540 (Cyp19b) confirmed the ability of nanoparticles to induce oxidative stress. CYPs are a large superfamily of enzymes capable of metabolizing several substances including steroids, pharmaceuticals, and xenobiotic compounds. They catalyze mixed-function oxidation reactions and the induction of their catalytic activity, measured as ethoxyresorufin-O-deethylase (EROD) activity, and expression (protein and transcript) is a useful biomarker of exposure to xenobiotics. Their catalyzing activity leads to the activation or inactivation of many endogenous and exogenous chemicals, with consequences for normal physiology and disease processes such as oxidative stress [111]. Nanomaterials of varying chemical compositions such as fullerenes, CNT, and metal oxides are able to induce oxidative stress [112]. Particularly in fish, TiO_2-NPs cause oxidative stress with the production of reactive oxygen species (ROS) [85,112], and titanium dioxide nanoparticle aggregates (TiO_2-NMs) cause oxidative stress in zebrafish embryos [48]. However, the investigations about oxidative stress by TiO_2-NPs on testis are insufficient. Oxidative stress may be induced through oxidizing (e.g., hydrogen peroxide, H_2O_2) or photo-oxidizing (e.g., fluoranthene) agents that react with oxygen-producing reactive oxygen species (ROS) [113,114] Since the reactive oxygen species (ROS) are able to affect the physiology, growth, and survival in aquatic organisms [115,116], they, similar to mammals, developed an antioxidant defense system for neutralizing the toxic effects of ROS [117]. Antioxidant enzymes such as SOD, CAT, GPX, POD, and low-molecular-weight, nonenzymatic antioxidants (e.g., GSH) are important components of the antioxidant defense system in animals [118,119]. In our study, it is clear that TiO_2-NPs are able to induce oxidative stress in the testis involving antioxidant defense [120–123]. The exposure to the lower concentration (1 mg/L) motivated the SOD to eliminate generated ROS as a protection mechanism against oxidative stress and a major increase was observed at the higher concentration (4 mg/L). Additionally, induction in the mRNA level of the GPX gene was higher at 1 mg/L concentration than at 4 mg/L. Fortunately, our results showed that the antioxidant defense of fish is induced by mild oxidative stress due to TiO_2-NPs, and it does not overwhelm the detoxifying or antioxidant mechanisms.

5. Conclusions

In conclusion, through acute and long-term TiO_2-NP exposure experiments on zebrafish, respectively, on embryos and adults, this research has improved knowledge of the action of TiO_2-NPs on embryonic development and the male reproductive system.

We provide evidence that TiO_2-NPs do not interfere with the development of zebrafish embryos, nor cause premature death in embryos; however, the embryos did show alterations in heartbeat, body length, and, above all, an increase in oxidative stress with the production of ROS and the expression of biomarkers associated to endocrine disruptors. The expression of the SHBG protein corresponding to the androgen binding protein (ABP) is increased in the presence of TiO_2-NPs, both in zebrafish larvae and in male gonads. This result is important because ABP is known to increase with xenobiotics. An increase in gene expression has been recorded in the testes. Consequently, it can be assumed that the TiO_2-NPs have an androgenic-like effect, as suggested by the increase in gene expression of SRD5A2, an enzyme that converts testosterone (T) to dihydrotestosterone (DHT). There is a greater need for knowledge of NP-induced reproductive toxicity since the production of engineered nanoparticles is continuously increasing with the nano revolution; therefore, the risk of exposure to nanoparticles is very common.

Supplementary Materials: The following supporting information can be downloaded at https://www.mdpi.com/article/10.3390/nano13111783/s1, Text S1: Evaluation of toxicological endpoints and DanioScope™ analysis; Text S2: Immunohistochemical analysis on zebrafish larvae; Text S3: TiO_2-NPs accumulation; Text S4: Histological Examination; Text S5: Protocol for preparation of semithin sections and Electron microscopy study; Text S6: RNA extraction and qRT-PCR; Text S7: Crystal structure of sex hormone-binding globulin (SHBG); Figure S1: Coagulated eggs rate; Figure S2: Hatching rate; Figure S3: Body length and beats per minute of larvae.

Author Contributions: Conceptualization, E.M.S. and M.V.B.; methodology, E.M.S. and R.P.; software, C.G.F., M.Z., R.P. and E.M.S.; validation, M.V.B. and G.I.; investigation, E.M.S. and R.P.; resources, F.C. and G.I; data curation, E.M.S. and R.P.; writing—original draft preparation, E.M.S.; writing—review and editing, E.M.S. and R.P; visualization, M.V.B., C.G.F. and A.S.; supervision, M.V.B. All authors have read and agreed to the published version of the manuscript.

Funding: This research was funded by "fondi di ateneo 2020–2022, Università di Catania, linea Open Access".

Institutional Review Board Statement: The animal study protocol was approved by the Institutional Ethics Committee of the Italian Health Ministry (authorization n°1244/2015-PR).

Informed Consent Statement: Not applicable.

Data Availability Statement: Original data are available on request.

Acknowledgments: E.M.S. thanks the University of Catania, Italy, MIUR XXXIV cycle.

Conflicts of Interest: The authors declare no conflict of interest.

References

1. Wilson, N. Nanoparticles: Environmental problems or problem solvers? *BioScience* **2018**, *68*, 241–246. [CrossRef]
2. Chiavola, A.; Amato, E.D.; Stoller, M.; Chianese, A.; Boni, M.R. Application of iron based nanoparticles as adsorbents for arsenic removal from water. *Chem. Eng. Trans.* **2016**, *47*, 325–330.
3. Vilardi, G.; Ochando-Pulido, J.M.; Stoller, M.; Verdone, N.; Di Palma, L. Fenton oxidation and chromium recovery from tannery wastewater by means of iron-based coated biomass as heterogeneous catalyst in fixed-bed columns. *Chem. Eng. J.* **2018**, *351*, 1–11. [CrossRef]
4. Piccinno, F.; Gottschalk, F.; Seeger, S.; Nowack, B. Industrial production quantities and uses of ten engineered nanomaterials in Europe and the world. *J. Nanoparticle Res.* **2012**, *14*, 1–11. [CrossRef]
5. Sharma, S.; Sharma, R.K.; Gaur, K.; Cátala Torres, J.F.; Loza-Rosas, S.A.; Torres, A.; Saxena, M.; Julin, M.; Tinoco, A.D. Fueling a hot debate on the application of TiO_2 nanoparticles in sunscreen. *Materials* **2019**, *12*, 2317. [CrossRef]
6. Tan, W.; Peralta-Videa, J.R.; Gardea-Torresdey, J.L. Interaction of titanium dioxide nanoparticles with soil components and plants: Current knowledge and future research needs-A critical review. *Environ. Sci. Nano* **2018**, *5*, 257–278. [CrossRef]
7. Scuderi, V.; Impellizzeri, G.; Romano, L.; Scuderi, M.; Nicotra, G.; Bergum, K.; Irrera, A.; Svensson, B.; Privitera, V. TiO_2-coated nanostructures for dye photo-degradation in water. *Nanoscale Res. Lett.* **2014**, *9*, 1–7. [CrossRef] [PubMed]
8. Zimbone, M.; Cacciato, G.; Spitaleri, L.; Egdell, R.G.; Grimaldi, M.G.; Gulino, A. Sb-doped titanium oxide: A rationale for its photocatalytic activity for environmental remediation. *ACS Omega* **2018**, *3*, 11270–11277. [CrossRef]
9. Guo, M.; Song, W.; Wang, T.; Li, Y.; Wang, X.; Du, X. Phenyl-functionalization of titanium dioxide-nanosheets coating fabricated on a titanium wire for selective solid-phase microextraction of polycyclic aromatic hydrocarbons from environment water samples. *Talanta* **2015**, *144*, 998–1006. [CrossRef]
10. Ohsaka, T.; Shinozaki, K.; Tsuruta, K.; Hirano, K. Photo-electrochemical degradation of some chlorinated organic compounds on n-TiO_2 electrode. *Chemosphere* **2008**, *73*, 1279–1283. [CrossRef]
11. Lee, Y.S.; Kim, S.J.; Venkateswaran, P.; Jang, J.S.; Kim, H.; Kim, J.G. Anion co-doped Titania for solar photocatalytic degradation of dyes. *Carbon Lett.* **2008**, *9*, 131–136. [CrossRef]
12. Kim, S.H.; Lee, S.W.; Lee, G.M.; Lee, B.T.; Yun, S.T.; Kim, S.O. Monitoring of TiO_2-catalytic UV-LED photo-oxidation of cyanide contained in mine wastewater and leachate. *Chemosphere* **2016**, *143*, 106–114. [CrossRef] [PubMed]
13. Nguyen, A.T.; Hsieh, C.T.; Juang, R.S. Substituent effects on photodegradation of phenols in binary mixtures by hybrid H_2O_2 and TiO_2 suspensions under UV irradiation. *J. Taiwan Inst. Chem. Eng.* **2016**, *62*, 68–75. [CrossRef]
14. Moon, G.H.; Kim, D.H.; Kim, H.I.; Bokare, A.D.; Choi, W. Platinum-like behavior of reduced graphene oxide as a cocatalyst on TiO_2 for the efficient photocatalytic oxidation of arsenite. *Environ. Sci. Technol. Lett.* **2014**, *1*, 185–190. [CrossRef]
15. Chen, Z.; Li, Y.; Guo, M.; Xu, F.; Wang, P.; Du, Y.; Na, P. One-pot synthesis of Mn-doped TiO_2 grown on graphene and the mechanism for removal of Cr (VI) and Cr (III). *J. Hazard. Mater.* **2016**, *310*, 188–198. [CrossRef] [PubMed]
16. Weir, A.; Westerhoff, P.; Fabricius, L.; Hristovski, K.; Von Goetz, N. Titanium dioxide nanoparticles in food and personal care products. *Environ. Sci. Technol.* **2012**, *46*, 2242–2250. [CrossRef] [PubMed]
17. Wu, F.; Hicks, A.L. Estimating human exposure to titanium dioxide from personal care products through a social survey approach. *Integr. Environ. Assess. Manag.* **2020**, *16*, 10–16. [CrossRef] [PubMed]
18. Larue, C.; Castillo-Michel, H.; Sobanska, S.; Trcera, N.; Sorieul, S.; Cecillon, L.; Ouerdane, L.; Legros, S.; Sarret, G. Fate of pristine TiO_2 nanoparticles and aged paint-containing TiO_2 nanoparticles in lettuce crop after foliar exposure. *J. Harzard. Mater.* **2014**, *273*, 17–26. [CrossRef]
19. Klaine, S.J.; Alvarez, P.J.; Batley, G.E.; Fernandes, T.F.; Handy, R.D.; Lyon, D.Y.; Mahendra, S.; McLaughlin, M.J.; Lead, J.R. Nanomaterials in the environment: Behavior, fate, bioavailability, and effects. *Environ. Toxicol. Chem.* **2008**, *27*, 1825–1851. [CrossRef]
20. Handy, R.D.; Owen, R.; Valsami-Jones, E. The ecotoxicology of nanoparticles and nanomaterials: Current status, knowledge gaps, challenges, and future needs. *Ecotoxicology* **2008**, *17*, 315–325. [CrossRef]
21. Wu, T.; Tang, M. Review of the effects of manufactured nanoparticles on mammalian target organs. *J. Appl. Toxicol.* **2018**, *38*, 25–40. [CrossRef] [PubMed]
22. Nemmar, A.; Hoet, P.M.; Vanquickenborne, B.; Dinsdale, D.; Thomeer, M.; Hoylaerts, M.F.; Vanbilloen, H.; Mortelmans, L.; Nemery, B. Passage of inhaled particles into the blood circulation in humans. *Circulation* **2002**, *105*, 411–414. [CrossRef] [PubMed]
23. Nel, A.; Xia, T.; Madler, L.; Li, N. Toxic potential of materials at the nanolevel. *Science* **2006**, *311*, 622–627. [CrossRef]
24. Mu, Q.; Jiang, G.; Chen, L.; Zhou, H.; Fourches, D.; Tropsha, A.; Yan, B. Chemical basis of interactions between engineered nanoparticles and biological systems. *Chem. Rev.* **2014**, *114*, 7740–7781. [CrossRef]
25. Sajid, M.; Ilyas, M.; Basheer, C.; Tariq, M.; Daud, M.; Baig, N.; Shehzad, F. Impact of nanoparticles on human and environment: Review of toxicity factors, exposures, control strategies, and future prospects. *Environ. Sci. Pollut. Res.* **2015**, *22*, 4122–4143. [CrossRef]
26. Behzadi, S.; Serpooshan, V.; Tao, W.; Hamaly, M.A.; Alkawareek, M.Y.; Dreaden, E.C.; Brown, D.; Alkilany, A.M.; Farokhzad, O.C.; Mahmoudi, M. Cellular uptake of nanoparticles: Journey inside the cell. *Chem. Soc. Rev.* **2017**, *46*, 4218–4244. [CrossRef] [PubMed]
27. Iavicoli, I.; Fontana, L.; Leso, V.; Bergamaschi, A. The effects of nanomaterials as endocrine disruptors. *Int. J. Mol. Sci.* **2013**, *14*, 16732–16801. [CrossRef]

28. Matthiessen, P.; Johnson, I. Implications of research on endocrine disruption for the environmental risk assessment, regulation and monitoring of chemicals in the European. *Union. Environ. Pollut.* **2007**, *146*, 9–18. [CrossRef]
29. Jeong, T.Y.; Simpson, M.J. Endocrine disruptor exposure causes infochemical dysregulation and an ecological cascade from zooplankton to algae. *Environ. Sci. Technol.* **2021**, *55*, 3845–3854. [CrossRef]
30. Iavicoli, I.; Fontana, L.; Bergamaschi, A. The effects of metals as endocrine disruptors. *J. Toxicol. Environ. Health B* **2009**, *12*, 206–223. [CrossRef]
31. Fang, Q.; Shi, Q.; Guo, Y.; Hua, J.; Wang, X.; Zhou, B. Enhanced bioconcentration of bisphenol A in the presence of nano-TiO_2 can lead to adverse reproductive outcomes in zebrafish. *Environ. Sci. Technol.* **2016**, *50*, 1005–1013. [CrossRef]
32. Kotil, T.; Akbulut, C.; Yön, N.D. The effects of titanium dioxide nanoparticles on ultrastructure of zebrafish testis (*Danio rerio*). *Micron* **2017**, *100*, 38–44. [CrossRef]
33. Nüsselin-Volhard, C.; Dahm, R. *Zebrafish a Practial Approch*; Oxford University Press: New York, NY, USA, 2000.
34. Pecoraro, R.; Marino, F.; Salvaggio, A.; Capparucci, F.; Di Caro, G.; Iaria, C.; Salvo, A.; Rotondo, A.; Tibullo, D.; Guerriero, G.; et al. Evaluation of Chronic Nanosilver Toxicity to Adult Zebrafish. *Front. Physiol.* **2017**, *8*, 1011. [CrossRef]
35. OECD. *Test No. 236: Fish Embryo Acute Toxicity (FET) Test, OECD Guidelines for the Testing of Chemicals*; Section 2; OECD Publishing: Paris, France, 2013.
36. Sobanska, M.; Scholz, S.; Nyman, A.M.; Cesnaitis, R.; Gutierrez, A.S.; Klüver, N.; De Coen, W. Applicability of the fish embryo acute toxicity (FET) test (OECD 236) in the regulatory context of registration, evaluation, authorisation, and restriction of chemicals (REACH). *Environ. Toxicol. Chem.* **2018**, *37*, 657–670. [CrossRef]
37. Pecoraro, R.; Salvaggio, A.; Marino, F.; Di Caro, G.; Capparucci, F.; Lombardo, B.M.; Messina, G.; Scalisi, E.M.; Tummino, M.; Loreto, F.; et al. Metallic nano-composite toxicity evaluation by zebrafish embryo toxicity test with identification of specific exposure biomarkers. *Curr. Protoc. Toxicol.* **2017**, *74*, 1–14.
38. Kimmel, C.B.; Ballard, W.W.; Kimmel, S.R.; Ullmann, B.; Schilling, T.F. Stages of embryonic development of the zebrafish. *Dev. Dynam.* **1995**, *203*, 253e310. [CrossRef]
39. Giannaccini, M.; Cushieri, A.; Dente, L.; Raffa, V. Non-mammalian vertebrate embryos as models in nanomedicine. *Nanomed. Nanotechnol. Biol. Med.* **2014**, *10*, 703–719. [CrossRef] [PubMed]
40. Zimbone, M.; Calcagno, L.; Messina, G.; Baeri, P.; Compagnini, G. Dynamic light scattering and UV–vis spectroscopy of gold nanoparticles solution. *Mater. Lett.* **2011**, *65*, 2906–2909. [CrossRef]
41. Zimbone, M.; Musumeci, P.; Baeri, P.; Messina, E.; Boninelli, S.; Compagnini, G.; Calcagno, L. Rotational dynamics of gold nanoparticle chains in water solution. *J. Nanoparticle Res.* **2012**, *14*, 1308. [CrossRef]
42. Kristofco, L.A.; Haddad, S.P.; Chambliss, C.K.; Brooks, B.W. Differential uptake of and sensitivity to diphenhydramine in embryonic and larval zebrafish. *Environ. Toxicol. Chem.* **2018**, *37*, 1175–1181. [CrossRef]
43. Rawson, D.M.; Zhang, T.; Kalicharan, D.; Jogebloed, W.L. Field emission scanning electron microscopy and transmission electron microscopy studies of the chorion, plasma membrane and syncytial layers of the gastrula-stage embryo of the zebrafish *Brachy Danio rerio*: A consideration of the structural and functional relationships with respect to cryoprotectant penetration. *Aquacult. Res.* **2001**, *31*, 325–336.
44. Fent, K.; Weisbrod, C.J.; Wirth-Heller, A.; Pieles, U. Assessment of uptake and toxicity of fluorescent silica nanoparticles in zebrafish (*Danio rerio*) early lifestages. *Aquat. Toxicol.* **2010**, *100*, 218–228. [CrossRef] [PubMed]
45. Cheng, J.; Flahaut, E.; Cheng, S.H. Effect of carbon nanotubes on developing zebrafish (*Danio rerio*) embryos. *Environ. Toxicol. Chem.* **2007**, *26*, 708–716. [CrossRef] [PubMed]
46. Pereira, A.C.; Gomes, T.; Machado, M.R.F.; Rocha, T.L. The zebrafish embryotoxicity test (ZET) for nanotoxicity assessment: From morphological to molecular approach. *Environ. Pollut.* **2019**, *252*, 1841–1853. [CrossRef] [PubMed]
47. Federici, G.; Shaw, B.J.; Handy, R.D. Toxicity of titanium dioxide nanoparticles to rainbow trout (*Oncorhynchus mykiss*): Gill injury, oxidative stress, and other physiological effects. *Aquat. Toxicol.* **2007**, *84*, 415–430. [CrossRef]
48. Chen, T.H.; Lin, C.Y.; Tseng, M.C. Behavioral effects of titanium dioxide nanoparticles on larval zebrafish (Danio rerio). *Mar. Pollut. Bull.* **2011**, *63*, 303–308. [CrossRef]
49. Wang, Y.J.; He, Z.Z.; Fang, Y.W.; Xu, Y.; Chen, Y.N.; Wang, G.Q.; Yang, Y.Q.; Yang, Z.; Li, Y.H. Effect of titanium dioxide nanoparticles on zebrafish embryos and developing retina. *Int. J. Ophthalmol.* **2014**, *7*, 917.
50. Faria, M.; Navas, J.M.; Soares, A.M.; Barata, C. Oxidative stress effects of titanium dioxide nanoparticle aggregates in zebrafish embryos. *Sci. Total Environ.* **2014**, *470*, 379–389. [CrossRef]
51. Bai, W.; Zhang, Z.; Tian, W.; He, X.; Ma, Y.; Zhao, Y.; Chai, Z. Toxicity of zinc oxide nanoparticles to zebrafish embryo: A physicochemical study of toxicity mechanism. *J. Nanopart. Res.* **2010**, *12*, 1645–1654. [CrossRef]
52. Johnson, A.; Carew, E.; Sloman, K.A. The effects of copper on the morphological and functional development of zebrafish embryos. *Aquat. Toxicol.* **2007**, *84*, 431–438. [CrossRef]
53. Asharani, P.V.; Lianwu, Y.I.; Gong, Z.; Valiyaveettil, S. Comparison of the toxicity of silver, gold and platinum nanoparticles in developing zebrafish embryos. *Nanotoxicology* **2011**, *5*, 43–54. [CrossRef]
54. Chakraborty, C.; Sharma, A.R.; Sharma, G.; Lee, S.S. Zebrafish: A complete animal model to enumerate the nanoparticle toxicity. *J. Nanobiotechnol.* **2016**, *14*, 1–13. [CrossRef]
55. Cerbinskaite, A.; Mukhopadhyay, A.; Plummer, E.R.; Curtin, N.J.; Edmondson, R.J. Defective homologous recombination in human cancers. *Cancer Treat. Rev.* **2012**, *38*, 89–100. [CrossRef] [PubMed]

56. De Murcia, J.M.; Niedergang, C.; Trucco, C.; Ricoul, M.; Dutrillaux, B.; Mark, M.; Oliver, F.J.; Masson, M.; Dierich, A.; LeMeur, M.; et al. Requirement of poly (ADP-ribose) polymerase in recovery from DNA damage in mice and in cells. *Proc. Natl. Acad. Sci. USA* **1997**, *94*, 7303–7307. [CrossRef] [PubMed]
57. Heacock, M.L.; Stefanick, D.F.; Horton, J.K.; Wilson, S.H. Alkylation DNA damage in combination with PARP inhibition results in formation of S-phase-dependent double-strand breaks. *DNA Repair* **2010**, *9*, 929–936. [CrossRef]
58. Kondo, N.; Takahashi, A.; Ono, K.; Ohnishi, T. DNA damage induced by alkylating agents and repair pathways. *J. Nucleic Acids.* **2010**, *21*, 1–7. [CrossRef]
59. Lindquist, S.; Craig, E.A. The heat-shock proteins. *Ann. Rev. Genet.* **1988**, *22*, 631–677. [CrossRef] [PubMed]
60. Hu, Y.L.; Qi, W.; Han, F.; Shao, J.Z.; Gao, J.Q. Toxicity evaluation of biodegradable chitosan nanoparticles using a zebrafish embryo model. *Int. J. Nanomed.* **2011**, *6*, 3351.
61. Lin, S.; Zhao, Y.; Xia, T.; Meng, H.; Ji, Z.; Liu, R.; George, S.; Xiong, S.; Wang, X.; Zhang, H.; et al. High content screening in zebrafish speeds up hazard ranking of transition metal oxide nanoparticles. *ACS Nano* **2011**, *5*, 7284–7295. [CrossRef] [PubMed]
62. Dziegiel, P. Expression of metallothioneins in tumor cells. *Pol. J. Pathol.* **2004**, *55*, 3–12. [PubMed]
63. Brundo, M.V.; Pecoraro, R.; Marino, F.; Salvaggio, A.; Tibullo, D.; Saccone, S.; Bramanti, V.; Buccheri, M.A.; Impellizzeri, G.; Scuderi, V.; et al. Toxicity evaluation of new engineered nanomaterials in zebrafish. *Front. Physiol.* **2016**, *7*, 130. [CrossRef] [PubMed]
64. Lin, H.Y.; Muller, Y.A.; Hammond, G.L. Molecular and structural basis of steroid hormone binding and release from corticosteroid-binding globulin. *Mol. Cell. Endocrinol.* **2010**, *316*, 3–12. [CrossRef]
65. Hammond, G.L. Diverse roles for sex hormone-binding globulin in reproduction. *Biol. Reprod.* **2011**, *85*, 431–441. [CrossRef] [PubMed]
66. Miguel-Queralt, S.; Hammond, G.L. Sex hormone-binding globulin in fish gills is a portal for sex steroids breached by xenobiotics. *Endocrinology* **2008**, *149*, 4269–4275. [CrossRef] [PubMed]
67. Dube, J.; Guiguen, Y.; Fostier, A. Diversity and biological significance of sexhormone-binding globulin in fish, an evolutionary perspective. *Mol. Cell. Endocrinol.* **2010**, *316*, 66–78. [CrossRef] [PubMed]
68. Miguel-Queralt, S.; Knowlton, M.; Avvakumov, G.V.; Al-Nouno, R.; Kelly, G.M.; Hammond, G.L. Molecular and functional characterization of sex hormone binding globulin in zebrafish. *Endocrinology* **2004**, *145*, 5221–5230. [CrossRef]
69. Dechaud, H.; Ravard, C.; Claustrat, F.; de la Perrière, A.B.; Pugeat, M. Xenoestrogen interaction with human sex hormone-binding globulin (hSHBG). *Steroids* **1999**, *64*, 328–334. [CrossRef] [PubMed]
70. Hodgert Jury, H.; Zacharewski, T.R.; Hammond, G.L. Interactions between human plasma sex hormone-binding globulin and xenobiotic ligands. *J. Steroid Biochem. Mol. Biol.* **2000**, *75*, 167–176. [CrossRef]
71. Hong, H.; Branham, W.S.; Ng, H.W.; Moland, C.L.; Dial, S.L.; Fang, H.; Perkins, R.; Sheehan, D.; Tong, W. Human sex hormone-binding globulin binding affinities of 125 structurally diverse chemicals and comparison with their binding to androgen receptor, estrogen receptor, and α-fetoprotein. *Toxicol. Sci.* **2015**, *143*, 333–348. [CrossRef]
72. Desvergne, B.; Feige, J.N.; Casals-Casas, C. PPAR-mediated activity of phthalates: A link to the obesity epidemic? *Mol. Cell. Endocrinol.* **2009**, *304*, 43–48. [CrossRef]
73. Chen, P.; Wang, Q.; Chen, M.; Yang, J.; Wang, R.; Zhong, W.; Zhu, L.; Yang, L. Antagonistic estrogenic effects displayed by bisphenol AF and perfluorooctanoic acid on zebrafish (*Danio rerio*) at an early developmental stage. *Environ. Sci. Technol. Lett.* **2018**, *5*, 655–661. [CrossRef]
74. Haritos, A.A.; Goodall, G.J.; Horecker, B.L. Prothymosin alpha: Isolation and properties of the major immunoreactive form of thymosin alpha 1 in rat thymus. *Proc. Natl. Acad. Sci. USA* **1984**, *81*, 1008–1011. [CrossRef] [PubMed]
75. Dosil, M.; Freire, M.; Gomez-Marquez, J. Tissue-specific and differential expression of prothymosin alpha gene during rat development. *FEBS Lett.* **1990**, *269*, 373–376. [CrossRef] [PubMed]
76. Yoshida, S.; Ono, N.; Tsukue, N.; Oshio, S.; Umeda, T.; Takano, H.; Takeda, K. *In utero* exposure to diesel exhaust increased accessory reproductive gland weight and serum testosterone concentration in male mice. *Environ. Sci.* **2006**, *13*, 139–147.
77. Asare, N.; Instanesa, C.; Sandberga, W.J.; Refsnesa, M.; Schwarzea, P.; Kruszewskib, M.; Brunborg, G. Cytotoxic and genotoxic effects of silver nanoparticles in testicular cells. *Toxicology* **2012**, *291*, 65–72. [CrossRef]
78. Gao, G.; Ze, Y.; Zhao, X.; Sang, X.; Zheng, L.; Ze, X.; Gui, S.; Sheng, L.; Sun, Q.; Hong, J.; et al. Titanium dioxide nanoparticle-induced testicular damage, spermatogenesis suppression, and gene expression alterations in male mice. *J. Hazard. Mater.* **2013**, *258*, 133–143. [CrossRef] [PubMed]
79. De Jong, W.H.; Hagens, W.I.; Krystek, P.; Burger, M.C.; Sips, A.J.; Geertsma, R.E. Particle size-dependent organ distribution of gold nanoparticles after intravenous administration. *Biomaterials* **2008**, *29*, 1912–1919. [CrossRef]
80. Sadauskas, E.; Jacobsen, N.R.; Danscher, G.; Stoltenberg, M.; Vogel, U.; Larsen, A.; Kreyling, W.; Wallin, H. Biodistribution of gold nanoparticles in mouse lung following intratracheal instillation. *Chem. Cent. J.* **2009**, *3*, 16. [CrossRef]
81. Lankveld, D.P.; Oomen, A.G.; Krystek, P.; Neigh, A.; Troostde Jong, A.; Noorlander, C.W.; Van Eijkeren, J.C.; Geertsma, R.E.; De Jong, W.H. The kinetics of the tissue distribution of silver nanoparticles of different sizes. *Biomaterials* **2010**, *31*, 8350–8361. [CrossRef]
82. Lankveld, D.P.; Rayavarapu, R.G.; Krystek, P.; Oomen, A.G.; Verharen, H.W.; van Leeuwen, T.G.; De Jong, W.H.; Manohar, S. Blood clearance and tissue distribution of PEGylated and non-PEGylated gold nanorods after intravenous administration in rats. *Nanomedicine* **2011**, *6*, 339–349. [CrossRef]

83. Lead, J.R.; Batley, G.E.; Alvarez, P.J.J.; Croteau, M.N.; Handy, R.D.; Mclaughlin, M.J.; Judy, J.D.; Schrimer, K. Nanomaterials in the environment: Behavior, bioavailability, and effects—An updated review. *Environ. Toxicol. Chem.* **2018**, *37*, 2029–2063. [CrossRef] [PubMed]
84. Griffitt, R.J.; Lavelle, C.M.; Kane, A.S.; Denslow, N.D.; Barber, D.S. Chronic nanoparticulate silver exposure results in tissue accumulation and transcriptomic changes in zebrafish. *Aquat. Toxicol.* **2013**, *130*, 192–200. [CrossRef] [PubMed]
85. Xiong, D.; Fang, T.; Yu, L.; Sima, X.; Zhu, W. Effects of nano-scale TiO_2, zno and their bulk counterparts on zebrafish: Acute toxicity, oxidative stress and oxidative damage. *Sci. Total Environ.* **2011**, *409*, 1444–1452. [CrossRef]
86. Griffitt, R.J.; Hyndman, K.; Denslow, N.D.; Barber, D.S. Comparison of molecular and histological changes in zebrafish gills exposed to metallic nanoparticles. *Toxicol. Sci.* **2009**, *107*, 404–415. [CrossRef]
87. Song, G.; Lin, L.; Liu, L.; Wang, K.; Ding, Y.; Niu, Q.; Mu, L.; Wang, H.; Shen, H.; Guo, S. Toxic effects of anatase titanium dioxide nanoparticles on spermatogenesis and testicles in male mice. *Pol. J. Environ. Stud.* **2017**, *26*, 2739–2745. [CrossRef] [PubMed]
88. Baker, B.B.; Yee, J.S.; Meyer, D.N.; Yang, D.; Baker, T.R. Histological and transcriptomic changes in male zebrafish testes due to early life exposure to low level 2, 3, 7, 8-tetrachlorodibenzo-p-dioxin. *Zebrafish* **2016**, *13*, 413–423. [CrossRef] [PubMed]
89. Nittoli, V.; Colella, M.; Porciello, A.; Reale, C.; Roberto, L.; Russo, F.; Russo, N.A.; Porreca, I.; De Felice, M.; Mallardo, M.; et al. Multi Species Analyses Reveal Testicular T3 Metabolism and Signalling as a Target of Environmental Pesticides. *Cells* **2021**, *10*, 2187. [CrossRef]
90. Lora, A.J.; Molina, A.M.; Bellido, C.; Blanco, A.; Monterde, J.G.; Moyano, M.R. Adverse effects of bisphenol A on the testicular parenchyma of zebrafish revealed using histomorphological methods. *Vet. Med.* **2016**, *61*, 577–589. [CrossRef]
91. Siiteri, P.K.; Murai, J.T.; Raymoure, W.J.; Kuhn, R.W.; Hammond, G.L.; Nisker, J.A. The Serum Transport of Steroid Hormones. In Proceedings of the 1981 Laurentian Hormone Conference; Academic Press: Cambridge, MA, USA, 1982; pp. 457–510.
92. Joseph, D.R. Structure, function, and regulation of androgen-binding protein/sex hormone-binding globulin. *Vitam. Horm.* **1994**, *49*, 197–280.
93. Mak, P.; Callard, G.V. A novel steroid-binding protein in the testis of the dogfish Squalus acanthias. *Gen. Comp. Endocrinol.* **1987**, *68*, 104–112. [CrossRef]
94. Ovrevik, J.; Stenersen, J.; Nilssen, K.; Tollefsen, K.E. Partial characterization of a sex steroid-binding protein in plasma from arctic charr (*Salvelinus alpinus* L.). *Gen. Comp. Endocrinol.* **2001**, *122*, 31–39. [CrossRef] [PubMed]
95. Milligan, S.R.; Khan, O.; Nash, M. Competitive binding of xenobiotic oestrogens to rat fetoprotein and to sex steroid binding proteins in human and rainbow trout (*Oncorhynchus mykiss*) plasma. *Gen. Comp. Endocrinol.* **1998**, *112*, 89–95. [CrossRef] [PubMed]
96. Tollefsen, K.E. Interaction of estrogen mimics, singly and in combination, with plasma sex steroid-binding proteins in rainbow trout (*Oncorhynchus mykiss*). *Aquat. Toxicol.* **2002**, *56*, 215–225. [CrossRef] [PubMed]
97. Pryce-Hobby, A.C.; McMaster, M.E.; Hewitt, L.M.; Van Der Kraak, G. The effects of pulp mill effluent on the sex steroid binding protein in white sucker (*Catostomus commersoni*) and longnose sucker (*C. catostomus*). *Comp. Biochem. Physiol.* **2003**, *134*, 241–250. [CrossRef]
98. Foucher, J.L.; Le Bail, P.Y.; Le Gac, F. Influence of hypophysectomy, castration, fasting, and spermiation on SBP concentration in male rainbow trout (*Oncorhynchus mykiss*). *Gen. Comp. Endocrinol.* **1992**, *85*, 101–110. [CrossRef]
99. Hobby, A.C.; Geraghty, D.P.; Pankhurst, N.W. Differences in binding characteristics of sex steroid binding protein in reproductive and non reproductive female rainbow trout (*Oncorhynchus mykiss*), black bream (*Acanthopagrus butcheri*), and greenback flounder (*Rhombosolea tapirina*). *Gen. Comp. Endocrinol.* **2000**, *120*, 249–259. [CrossRef]
100. Bobe, J.; Mahé, S.; Nguyen, T.; Rime, H.; Vizziano, D.; Fostier, A.; Guiguen, Y. A novel, functional, and highly divergent sex hormone-binding globulin that may participate in the local control of ovarian functions in salmonids. *Endocrinology* **2008**, *149*, 2980–2989. [CrossRef]
101. Miguel-Queralt, S.; Underhill, C.; Devlin, R.H.; Hammond, G.L. Characterization and Measurement of the Plasma α-and β-Sex Hormone-Binding Globulin Paralogs in Salmon. *Endocrinology* **2009**, *150*, 366–375. [CrossRef]
102. Hryb, D.J.; Nakhla, A.M.; Kahn, S.M.; St George, J.; Levy, N.C.; Romas, N.A.; Rosner, W. Sex hormone-binding globulin in the human prostate is locally synthesized and may act as an autocrine/paracrine effector. *J. Biol. Chem.* **2002**, *277*, 26618–26622. [CrossRef]
103. Tokarz, J.; Möller, G.; de Angelis, M.H.; Adamski, J. Zebrafish and steroids: What do we know and what do we need to know? *J. Steroid Biochem. Mol. Biol.* **2013**, *137*, 165–173. [CrossRef]
104. Hering, D.M.; Olenski, K.; Kaminski, S. Genome-wide association study for poor sperm motility in Holstein-Friesian bulls. *Anim. Reprod. Sci.* **2014**, *146*, 89–97. [CrossRef]
105. Urbatzka, R.; Watermann, B.; Lutz, I.; Kloas, W. Exposure of Xenopus laevis tadpoles to finasteride, an inhibitor of 5-alpha reductase activity, impairs spermatogenesis and alters hypophyseal feedback mechanisms. *J. Mol. Endocrinol.* **2009**, *43*, 209–219. [CrossRef]
106. Kang, H.J.; Imperato-McGinley, J.; Zhu, Y.S.; Rosenwaks, Z. The effect of 5α-reductase-2 deficiency on human fertility. *Fertil. Steril.* **2014**, *101*, 310–316. [CrossRef] [PubMed]
107. Margiotta-Casaluci, L.; Hannah, R.E.; Sumpter, J.P. Mode of action of human pharmaceuticals in fish: The effects of the 5-alpha-reductase inhibitor, dutasteride, on reproduction as a case study. *Aquat. Toxicol.* **2013**, *128*, 113–123. [CrossRef] [PubMed]

108. Garcìa-Garcìa, M.; Sànchez-Hernàndez, M.; Garcìa-Hernàndez, M.P.; Garcìa-Ayala, A.; Chaves-Pozo, E. Role of 5α-dihydrotestosterone in testicular development of gilthead seabream following finasteride administration. *J. Steroid Biochem. Mol. Biol.* **2017**, *174*, 48–55. [CrossRef] [PubMed]
109. Garcia, T.X.; Costa, G.M.; França, L.R.; Hofmann, M.C. Sub-acute intravenous administration of silver nanoparticles in male mice alters Leydig cell function and testosterone levels. *Reprod. Toxicol.* **2014**, *45*, 59–70. [CrossRef]
110. Adebayo, O.A.; Akinloye, O.; Adaramoye, O.A. Cerium oxide nanoparticle elicits oxidative stress, endocrine imbalance and lowers sperm characteristics in testes of balb/c mice. *Andrologia* **2018**, *50*, e12920. [CrossRef]
111. Goldstone, J.V.; McArthur, A.G.; Kubota, A.; Zanette, J.; Parente, T.; Jönsson, M.E.; Nelson, D.R.; Stegeman, J.J. Identification and developmental expression of the full complement of Cytochrome P450 genes in Zebrafish. *BMC Genom.* **2010**, *11*, 1–21. [CrossRef]
112. Vallyathan, V.; Shi, X. The role of oxygen free radicals in occupational and environmental lung diseases. *Environ. Health Perspect.* **1997**, *105*, 165–177.
113. Bar-Ilan, O.; Louis, K.M.; Yang, S.P.; Pedersen, J.A.; Hamers, R.J.; Peterson, R.E.; Heideman, W. Titanium dioxide nanoparticles produce phototoxicity in the developing zebrafish. *Nanotoxicology* **2012**, *6*, 670–679. [CrossRef]
114. Fu, P.P.; Xia, Q.; Sun, X.; Yu, H. Phototoxicity and environmental transformation of polycyclic aromatic hydrocarbons (PAHs)-light-induced reactive oxygen species, lipid peroxidation, and DNA damage. *J. Environ. Sci. Health Part C Environ. Carcinog. Ecotoxicol. Rev.* **2012**, *30*, 1–41. [CrossRef] [PubMed]
115. Filho, D.W. Fish antioxidant defenses—A comparative approach. *Braz. J. Med. Biol. Res.* **1996**, *29*, 1735–1742.
116. Pandey, S.; Parvez, S.; Sayeed1, I.; Haque, R.; Bin-Hafeez, B.; Raisuddin, S. Biomarkers of oxidative stress: A comparative study of river Yamuna fish Wallago attu (Bl. & Schn.). *Sci. Total Environ.* **2003**, *309*, 105–115.
117. Zhu, X.; Zhou, J.; Cai, Z. The toxicity and oxidative stress of TiO_2 nanoparticles in marine abalone (*Haliotis diversicolor supertexta*). *Mar. Pollut. Bull.* **2011**, *63*, 334–338. [CrossRef]
118. Van der Oost, R.; Beyer, J.; Vermeulen, N.P. Fish bioaccumulation and biomarkers in environmental risk assessment: A review. *Environ. Toxicol. Pharmacol.* **2003**, *13*, 57–149. [CrossRef]
119. Zhang, J.; Shen, H.; Wang, X.; Wu, J.; Xue, Y. Effects of chronic exposure of 2, 4-dichlorophenol on the antioxidant system in liver of freshwater fish Carassius auratus. *Chemosphere* **2004**, *55*, 167–174. [CrossRef] [PubMed]
120. Lin, T.; Zhou, D.; Dong, J.; Jiang, F.; Chen, W. Acute toxicity of dichloroacetonitrile (DCAN), a typical nitrogenous disinfection by-product (N-DBP), on zebrafish (*Danio rerio*). *Ecotox. Environ. Safe* **2016**, *133*, 97–104. [CrossRef]
121. Asakura, H.; Kitahora, T. Antioxidants and Polyphenols in Inflammatory Bowel Disease: Ulcerative Colitis and Crohn Disease. In *Polyphenols: Prevention and Treatment of Human Disease*; Ronald, R.W., Victor, R.P., Sherma, Z., Eds.; Academic Press: Cambridge, MA, USA, 2018; pp. 279–292.
122. Celardo, I.; Pedersen, J.Z.; Traversa, E.; Ghibelli, L. Pharmacological potential of cerium oxide nanoparticles. *Nanoscale* **2011**, *3*, 1411–1420. [CrossRef]
123. Nash, K.M.; Ahmed, S. Nanomedicine in the ROS-mediated pathophysiology: Applications and clinical advances. *Nanomed. Nanotechnol. Biol. Med.* **2015**, *11*, 2033–2040. [CrossRef]

nanomaterials

MDPI

Article

Interaction of Cerium Oxide Nanoparticles and Ionic Cerium with Duckweed (*Lemna minor* L.): Uptake, Distribution, and Phytotoxicity

Yang Liu [1], Xuepeng Zhao [1], Yuhui Ma [2,*], Wanqin Dai [2], Zhuda Song [2], Yun Wang [2], Jiaqi Shen [2], Xiao He [2], Fang Yang [1,*] and Zhiyong Zhang [2,3,*]

1 Hebei Provincial Key Laboratory of Green Chemical Technology & High Efficient Energy Saving, School of Chemical Engineering and Technology, Hebei University of Technology, Tianjin 300130, China; liuy99@ihep.ac.cn (Y.L.); zhaoxuepeng@ihep.ac.cn (X.Z.)

2 Key Laboratory for Biomedical Effects of Nanomaterials and Nanosafety, Institute of High Energy Physics, Chinese Academy of Sciences, Beijing 100049, China; daiwanqin@ihep.ac.cn (W.D.); songzd@ihep.ac.cn (Z.S.); yunwang@ihep.ac.cn (Y.W.); shenjq@ihep.ac.cn (J.S.); hexiao@ihep.ac.cn (X.H.)

3 School of Nuclear Science and Technology, University of the Chinese Academy of Sciences, Beijing 100049, China

* Correspondence: mayh@ihep.ac.cn (Y.M.); yangfang@hebut.edu.cn (F.Y.); zhangzhy@ihep.ac.cn (Z.Z.)

Abstract: As one of the most widely used nanomaterials, CeO_2 nanoparticles (NPs) might be released into the aquatic environment. In this paper, the interaction of CeO_2 NPs and Ce^{3+} ions (0~10 mg/L) with duckweed (*Lemna minor* L.) was investigated. CeO_2 NPs significantly inhibited the root elongation of duckweed at concentrations higher than 0.1 mg/L, while the inhibition threshold of Ce^{3+} ions was 0.02 mg/L. At high doses, both reduced photosynthetic pigment contents led to cell death and induced stomatal deformation, but the toxicity of Ce^{3+} ions was greater than that of CeO_2 NPs at the same concentration. According to the in situ distribution of Ce in plant tissues by μ-XRF, the intensity of Ce signal was in the order of root > old frond > new frond, suggesting that roots play a major role in the uptake of Ce. The result of XANES showed that 27.6% of Ce(IV) was reduced to Ce(III) in duckweed treated with CeO_2 NPs. We speculated that the toxicity of CeO_2 NPs to duckweed was mainly due to its high sensitivity to the released Ce^{3+} ions. To our knowledge, this is the first study on the toxicity of CeO_2 NPs to an aquatic higher plant.

Keywords: CeO_2 nanoparticles; duckweed; uptake; distribution; toxicity; transformation

1. Introduction

Due to their unique physicochemical properties, the applications of nanoparticles (NPs) are growing rapidly. Among various NPs, cerium oxide (CeO_2) NPs have been widely used in a variety of industry applications, such as in catalysts, polishing agents, fuel additives, etc. [1–3]. The global production of CeO_2 NPs was estimated to be about 10,000 tons per year [4]. Along with manufacture, transportation, use, and disposal, NPs (including CeO_2 NPs) are inevitably released and accumulated in the environment. Therefore, there is growing interest in studying the fate of NPs in the environment and assessing their impact on the environment and biota [5–8].

The aquatic environment is an important sink of NPs, including CeO_2 NPs. The predicted environmental concentrations (PEC) of CeO_2 NPs in fresh water and sediments were 0.51–25.59 ng/L and 0.34–253.49 μg/kg in 2017, respectively [9]. Moreover, the maximum concentration of CeO_2 NPs in sewage treatment effluent was predicted to be 1.87 mg/L [9]. Aquatic plants are an important part of the aquatic ecosystem. They are not only the habitat and food for aquatic animals but also a source of nutrients. Currently, there are few studies on the effects of CeO_2 NPs on aquatic plants, mainly focusing on unicellular algae. The EC_{50} (50% effective concentration) values of CeO_2 NPs to the freshwater microalgae ranged from 4.1 to 29.6 mg/L according to several studies, and the inhibitory effects were mainly caused by membrane damage due to direct contact or indirectly induced reactive oxygen species (ROS) generation [10,11].

Duckweed (e.g., *Lemna minor* L.) is a small floating aquatic plant which has the physiological characteristics such as small size, high reproduction rate, asexual reproduction and sensitivity to certain chemicals, etc. It has been widely used as a model for environmental monitoring and toxicity testing [12–15]. There are currently several studies on the interaction between some NPs and duckweed [16–19]. For instance, Ag NPs (6.84 mg/L) could change cellular ultrastructure, reduce photosynthetic pigments and starch grains along with the Ag accumulation in duckweed (*Landoltia punctata*) [16]. After exposure to Ag/Ag_2S NPs and $AgNO_3$, Ag_2S NPs remained as Ag_2S in duckweed (*Landoltia punctata*), while Ag NPs was transformed to Ag_2S and silver thiol species and $AgNO_3$ was transformed to Ag and sulfur-associated Ag species in tissues [17]. The growth of the duckweed species *Landoltia punctata* was not strongly affected by 1000 mg/L CuO NPs, but the reduction in photosynthetic pigments and the destruction of cellular structure were observed [18]. In another study, 150 µg/L CuO NPs altered the frond histomorphology and root cell integrity of duckweed (*Lemna minor*). Moreover, the authors confirmed that root uptake was the main pathway for the internalization of CuO NPs and chloroplasts were the sites of ROS production [19]. Unfortunately, there is no report on the biological effects of CeO_2 NPs on duckweed. Given the possible negative impacts of CeO_2 NPs on environmental species, it is very important to address this information gap in aquatic ecotoxicology.

In the current study, we aimed to study the uptake, distribution, and toxicity of CeO_2 NPs and Ce^{3+} ions to duckweed (*Lemna minor* L.) and to discuss the possible mechanisms. A range of endpoints such as plant growth, photosynthetic pigment contents, and cell death were assessed. The distribution of Ce in duckweed tissues was analyzed using micro-X-ray fluorescence (μ-XRF). The transformation of CeO_2 NPs was investigated using X-ray absorption near-edge spectroscopy (XANES), and the contribution of NP-dissolution to the toxicity of CeO_2 NPs to duckweed was estimated. Our results will provide further insight on the mechanisms underlying the impacts of metal-based NPs on aquatic species.

2. Materials and Methods

2.1. Material Preparation and Characterization

CeO_2 NPs were synthesized via a precipitation method as previously described [20]. All chemicals were analytical grade and obtained from Aladdin Reagent Co., Ltd. (Shanghai, China). The synthesized CeO_2 NPs were observed using scanning electron microscopy (SEM, Hitachi S-4800, Tokyo, Japan) and transmission electron microscopy (TEM, Tecnai G2 20 S-Twin, FEI, Tokyo, Japan). The hydrodynamic diameter and zeta potential of CeO_2 NPs were determined using dynamic light scattering (DLS, ZetaSizer, Malvern Instruments, Malvern, UK) analysis.

2.2. Plant Cultivation and Exposure

Duckweed (*Lemna minor* L.) was collected from a wetland (40 N latitude, 116 E longitude) in the southwest of Beijing, China and then washed with deionized water and submerged in 0.025% sodium hypochlorite solution for 1 to 3 min to remove algae and other impurities. The plants were pre-cultured in 1/10 modified Steinberg's medium for one month under laboratory conditions according to the Organization for Economic Co-operation and Development (OECD) 2006 protocol [21] with a 16 h photoperiod in fluorescent light (8000 lux) and 8 h in the dark at 24 ± 2 °C. For exposure, the two-fronded duckweed with similar size were selected and incubated in CeO_2 NP suspensions containing 1/10 modified Steinberg nutrition medium. In each treatment, 15 duckweed plants with their roots cut off were placed on a Petri dish. The duckweed was exposed to 0, 0.02, 0.1, 0.5, 2, and 10 mg/L of CeO_2 NPs and their ionic counterpart Ce^{3+} ions (obtained by dissolving $Ce(NO_3)_3$ in water), respectively. Each treatment was replicated at least three times. All analyses were performed after 3 days of exposure.

2.3. Growth Related Parameters

At the end of experiment, the plants were thoroughly washed with deionized water and blotted with paper towels. The root lengths and fresh weight (Fw) of the plants were measured for each Petri dish. After weighing the Fw, the duckweed was oven-dried at 60 °C to a constant weight, and the dry weight (Dw) was determined. Photosynthetic pigment (chlorophyll a/b, total chlorophyll, and total carotenoid) contents were measured, and the detailed methods were described in Supporting Information (SI). The average specific growth rate (r) was calculated according to the method shown in the SI.

2.4. Cell Death and SEM/TEM Observation

After treatment with 10 mg/L CeO_2 NPs and Ce^{3+} ions for 3 days, the cell death in the frond and root of duckweed under different treatments was visualized using Evans blue (Eb) staining according to the previous methods with some modification [22–24]. The detailed methods were described in SI.

After treatment with 10 mg/L CeO_2 NPs and Ce^{3+} ions for 3 days, the morphology changes of duckweed were observed using SEM (S4800, Hitachi, Tokyo, Japan), and the distribution of Ce-containing materials in tissues was observed using TEM (JEM-1230, JEOL, Tokyo, Japan). The samples were prepared as described in our previous report [11].

2.5. Ce and Mineral Elements Analysis

The bioaccumulation of Ce and mineral elements by duckweed under different treatments was determined. The lyophilized dried samples were ground into powders and digested with a 4:1 (v/v) mixture of concentrated HNO_3 and H_2O_2. The elemental content of Ce was analyzed by inductively coupled plasma mass spectrometry (ICP-MS, Thermo Elemental X7, Waltham, MA, USA). Mineral elements (Ca, Mg, Fe, Mn, Zn) were analyzed using ICP-optical emission spectrometry (OES, Perkin Elmer Optima 8000, Waltham, MA, USA).

2.6. Elemental Mapping by μ-XRF

Duckweed treated with 10 mg/L CeO_2 NPs and Ce^{3+} ions were washed thoroughly with deionized water and were lyophilized. The dried samples were fixed on 3M tape and the μ-XRF experiments were performed at 4W1B beamline of Beijing Synchrotron Radiation Facility (BSRF, Beijing, China). The energy of the storage ring was 2.5 GeV, and the current intensity was 150 to 250 mA. The spot was focused to 20 × 50 μm by the polycapillary lens. The sample was held on a precision motor-driven stage, and the elemental 2D mapping was acquired in a stepwise manner with 50/100 μm steps. The data were analyzed using the PyMca package, and the elemental mappings were created using Origin.

2.7. Speciation of Ce in Duckweed

To explore the speciation of Ce in duckweed treated with 10 mg/L CeO_2 NPs and Ce^{3+} ions, XANES analysis was performed, and samples were prepared as previously described [25]. Ce L_{III}-edge spectra were collected at the 1W1B beamline of BSRF (Beijing, China). CeO_2 and $CePO_4$ were used as standard compounds, and their spectra were collected using transmission mode. The spectra of samples were collected using fluorescence mode. XANES spectra were normalized using Athena software and the speciation of Ce in samples was analyzed by linear combination fitting (LCF).

2.8. Statistical Analysis

Data were given as the mean ± standard deviation (SD). All statistical analyses were performed on the Statistical Package for Social Sciences (SPSS) version 22.0. One-way analysis of variance (ANOVA) and Tukey's HSD post hoc test were used to check for statistically significant differences at $p < 0.05$.

3. Results

3.1. Characterization of CeO_2 NPs

The TEM and SEM images of CeO_2 NPs are shown in Figure S1. The average size of the NPs was calculated to be 7.3 ± 1.1 nm. XRD spectrum showed that CeO_2 NPs have a cubic fluorite structure. The hydrodynamic particle sizes of CeO_2 NPs (10 mg/L) in deionized water and Steinberg's medium were 71.7 ± 0.5 nm and 306.1 ± 15.6 nm (Figure S2), respectively, indicating that CeO_2 NPs was easier to agglomerate in the culture medium. After exposure, the hydrodynamic particle size of CeO_2 NPs increased to 723.8 ± 106.1 nm (Figure S3). The zeta potential of CeO_2 NPs suspensions changed from 32.4 ± 1.5 mV in deionized water to -14.7 ± 1.8 mV in Steinberg's medium.

3.2. Effect of CeO_2 NPs and Ce^{3+} Ions on the Growth of Duckweed

After exposure to CeO_2 NPs or Ce^{3+} ions, the root length, frond area, and biomass of duckweed are shown in Figure 1A. Compared with the control, the root elongation was significantly inhibited under the two treatments (except 0.02 mg/L CeO_2 NPs). Moreover, the root length of Ce^{3+} ions treatment was significantly shorter than that of CeO_2 NPs treatment at each exposure concentration. At 10 mg/L, the inhibition rate of Ce^{3+} ions to duckweed was 96%, and the root hardly developed. The average specific growth rate was calculated using the frond area of duckweed (Figure 1B). Similar to the result of root elongation, the average specific growth rate of duckweed fronds decreased with increasing concentrations of exposure materials. At the highest concentration, the average specific growth rates of duckweed in the NP group and the ionic group were 21.0% and 6.3%, respectively. The EC_{20} values of CeO_2 NPs and Ce^{3+} ions for duckweed were calculated to be 1.78 mg/L and 0.39 mg/L, respectively, indicating that the latter was more toxic than the former.

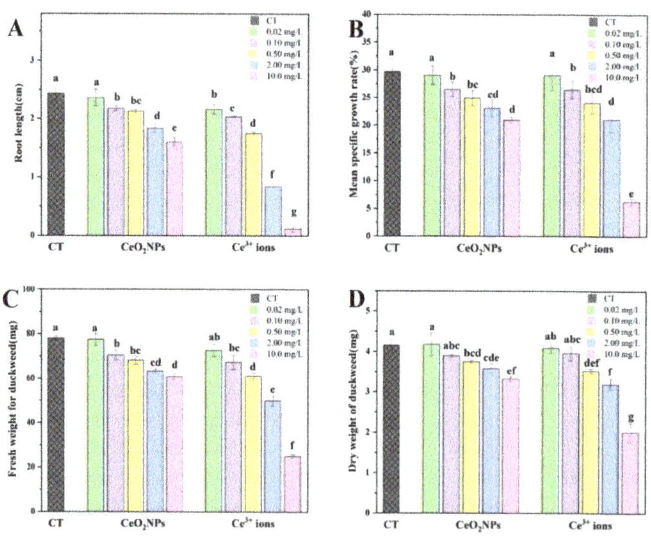

Figure 1. Root length (**A**), mean specific growth rate (**B**), fresh weight (**C**), and dry weight (**D**) of the duckweed treated with CeO_2 NPs and Ce^{3+} ions. Different letters indicate significant differences ($p < 0.05$).

Biomass is an important parameter related to photosynthetic efficiency. Fw and Dw as functions of concentration are shown in Figure 1C,D. Similarly to the results of root length and frond area, Fw and Dw of duckweed treated with CeO_2 NPs and Ce^{3+} ions were not affected at lower concentrations but significantly decreased at concentrations higher than

0.1 mg/L. By comparing the apparent toxicity indexes, it was found that the effect of the highest concentration of 10 mg/L CeO_2 NPs on duckweed was approximately comparable to that of Ce^{3+} ions at 0.5–2 mg/L.

3.3. Pigments and Mineral Elements Analyses

The photosynthetic damage caused by CeO_2 NPs and Ce^{3+} ions was analyzed based on the contents of photosynthetic pigments (Figure S3). At concentrations higher than 0.5 mg/L, the contents of chlorophyll a, chlorophyll b, and carotenoids in the fronds of duckweed under the two treatments were significantly reduced compared to the control, which was consistent with the result of biomass. At the end of the experiment, the fronds of duckweed treated with CeO_2 NPs (10 mg/L) showed chlorosis, while the fronds of the ion-treated group grew more slowly and showed severe chlorosis, and the edges of the fronds became transparent (Figure S5).

The contents of mineral elements, including macro elements (Ca and Mg) and trace elements (Fe, Mn, and Zn), in duckweed under different treatments are shown in Table 1. With the increase in exposure concentration, Ca contents in duckweed decreased significantly in both treatments, while Mg contents were almost unchanged (except 10 mg/L Ce^{3+} ions). In the NP group. Fe contents in duckweed gradually decreased with the increase in concentration. Different from this, Fe contents in the ionic group were significantly reduced compared to the control, but there was no concentration-dependent effect. Mn contents in duckweed exposed to different concentrations of CeO_2 NPs showed a similar decreasing trend to the Fe contents, while Mn contents in the ionic group did not change significantly (except 10 mg/L Ce^{3+} ions). The change in Zn contents in duckweed as similar to that of Fe.

Table 1. Mineral element concentrations in duckweed after exposure to CeO_2 NPs and Ce^{3+} ion [1].

Treatments (mg/L)	Macroelements (μg/g·Dw)		Microelements (μg/g·Dw)		
	Ca	**Mg**	**Fe**	**Mn**	**Zn**
0 (CT)	6209.1 ± 72.7 a	2249.91 ± 70.6 a	168.3 ± 14.3 a	98.8 ± 3.9 a	30.4 ± 5.3 a
CeO_2 NPs					
0.02	5464.7 ± 70.2 b	2312.4 ± 143.3 a	135.48 ± 5.5 b	92.4± 0.7 b	24.1 ± 9.6 ab
0.1	5191.4 ± 114.7 bc	2227.3 ± 78.2 a	143.0 ± 35.4 ab	85.5 ± 1.5 c	19.8 ± 5.9 abc
0.5	5234.3 ± 295.2 bc	2343.4 ± 93.1 a	96.1 ± 1.444 c	83.4 ± 1.7 c	11.4 ± 0.7 bc
2	4948.5 ± 89.6 c	2401.45 ± 17.1 a	83.3 ± 6.8 c	84.6 ± 1.7 c	15.4 ± 4.3 bc
10	4970.2 ± 172.0 c	2265.7 ± 47.8 a	83.8 ± 1.4 c	86.2 ± 2.0 c	7.3 ± 0.7 c
Ce^{3+} ions					
0.02	5606.9 ± 146.3 b	2188.8 ± 93.7 a	125.3 ± 16.6 b	87.8 ± 1.6 a	12.3 ± 5.0 b
0.1	5128.4 ± 211.8 c	2254.2 ± 196.3 a	99.7 ± 0.5 b	88.4 ± 3.8 a	22.9 ± 3.0 ab
0.5	5262.0 ± 221.0 bc	2273.8 ± 69.5 a	108.6 ± 2.3 b	82.6 ± 2.0 a	13.2 ± 4.3 b
2	5037.6 ± 132.1 c	2652.4 ± 542.3 a	124.6 ± 29.4 b	100.6 ± 25.4 a	21.3 ± 4.8 ab
10	3502.1 ± 153.6 d	929.2 ± 24.4 b	133.9 ± 11.7 b	51.5 ± 1.3 b	17.2 ± 4.2 b

[1] Data are average of three replicates ± SD. Different letters among columns indicate significant difference at $p < 0.05$.

3.4. Visualization and Quantification of Plant Cell Activity

The cell membranes of plants are selective, and the dye can be excreted by living cells when plant tissues are immersed in the Eb dye. Therefore, Eb staining can show cell death and assess the degree of cell membrane damage. As shown in Figure 2A, with the increase of concentrations of CeO_2 NPs and Ce^{3+} ions, the fronds of duckweed were gradually stained with Eb, indicating that the cell activity decreased and oxidative stress increased. The quantitative results showed that the amount of Eb in the ionic group was about twice that in the NPs group at the highest concentration (Figure 2B), suggesting that the former caused more severe oxidative damage to duckweed. The staining of duckweed roots also showed concentration dependence, but there was no significant difference between the two treatments when the concentrations was lower than 0.5 mg/L (Figure 2C). At 10 mg/L Ce^{3+} ions, the root elongation was too short to be collected and measured.

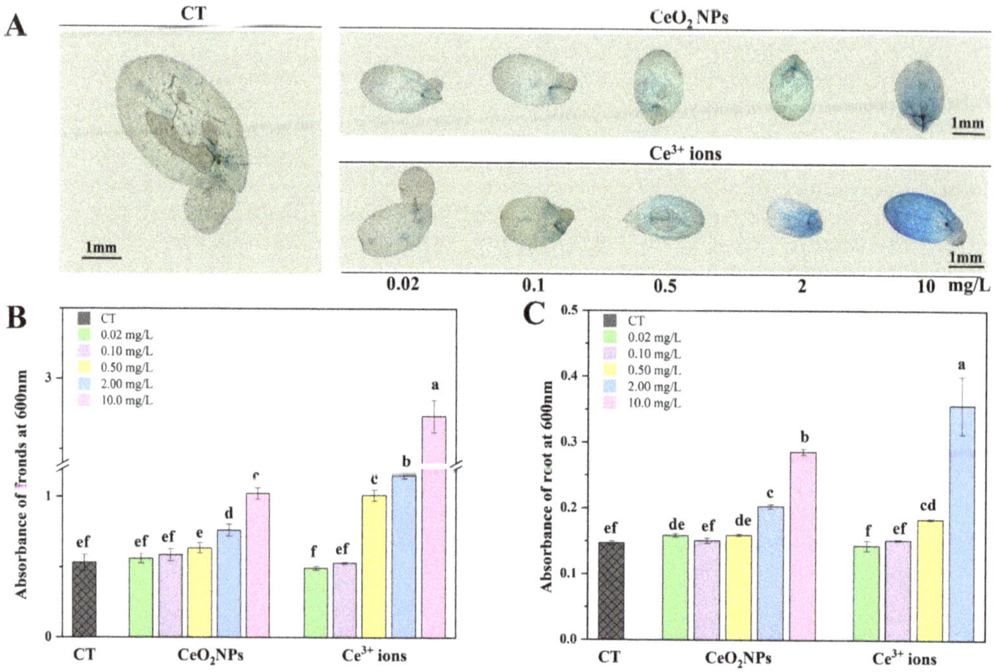

Figure 2. Visualization of cell death in fronds stained with Eb (**A**). The absorbance values of Eb in fronds (**B**) and roots (**C**) of duckweed under different treatments. Different letters indicate significant differences ($p < 0.05$).

3.5. SEM and TEM Observations

After exposure to 10 mg/L CeO_2 NPs and Ce^{3+} ions, the morphological changes of duckweed fronds were observed via SEM. Compared to the control (Figure 3A), the frond area of the NP group was almost unchanged (Figure 3D), whereas that of the ionic group significantly decreased (Figure 3G). In addition, the stomata of the treated fronds were significantly deformed. The stomata of the frond in the control were intact and elliptical, with larger stomatal opening apertures. In contrast, the stomata in the NP group became narrower and the aperture became smaller (Figure 3F). The stomata in the ionic group were changed to be diamond-like, and the morphology of the guard cells was altered (Figure 3I). The aperture width and area of the duckweed stomata under the different treatments were calculated (at least 20 stomata), and the results are shown in Figure S6. The stomatal width and area in the NPs group were significantly lower than those of the control group, while the stomatal width and area in the ionic group were significantly higher than those of the control, which might be caused by the stomatal deformation under stress.

TEM images of duckweed root sections under different treatments are shown in Figure S7. The epidermis of the roots in the control group was smooth, the internal organelles were intact (Figure S8A,B), and no deposits were found inside or outside the roots. In contrast, some particles were adsorbed on the root surface of duckweed in the CeO_2 NP group, and some flocculent deposits were found in the internal intercellular spaces (Figure S8C,D). Surprisingly, there were many needle-like deposits on the root surface of duckweed in the ionic group (Figure S8E), and the morphology of the deposits was very similar to that of $CePO_4$, which may be caused by the reaction of Ce^{3+} ions with phosphate in the culture medium. At the same time, high-electron-density substances were also found inside the cell (Figure S8F), indicating that Ce^{3+} ions could be uptaken by duckweed roots and transformed into precipitates.

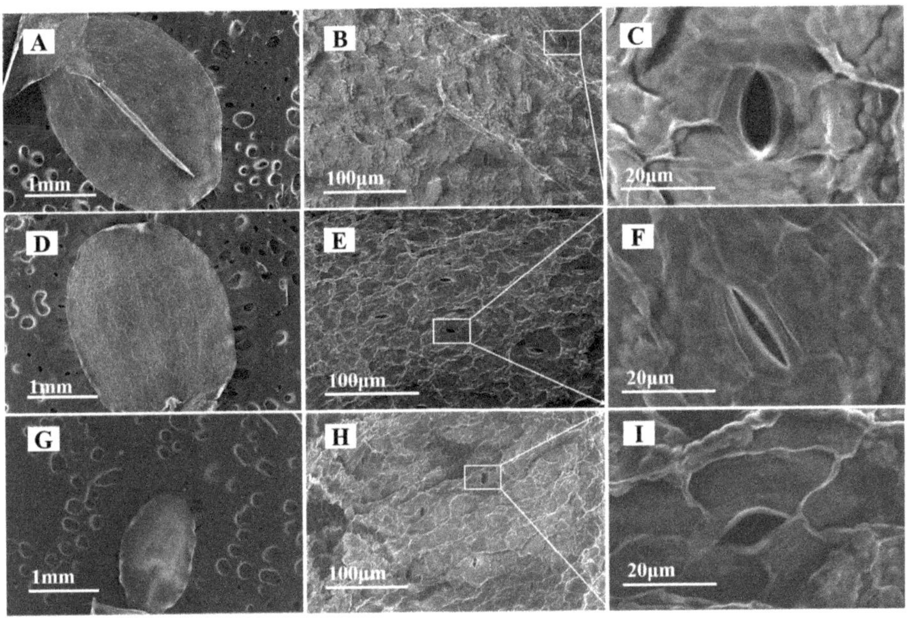

Figure 3. SEM observations of fronds and stomatal morphology of duckweed unexposed (**A–C**), exposed to 10 mg/L CeO_2 NPs (**D–F**) and exposed to10 mg/L to Ce^{3+} ions (**G–I**). Panels (**C,F,I**) are enlarged from panels (**B,E,H**), respectively.

3.6. Elemental Mapping by μ-XRF

The distribution and localization of Ce, Ca, and Fe in the tissues are shown in Figure 4. After normalization, the minimum (blue) to maximum (red) values of fluorescence intensity were represented by color-coding diagrams. The high concentration of Ce in the NPs group was mainly accumulated at the edge of the fronds (Figure 4A), while that in the ionic group was mainly concentrated along the leaf veins (Figure 4B). Notably, strong Ce signals were observed at the root–frond junctions under both treatments, whereas no Ce signal was detected in the new frond of the ionic group (the upper-right corner of Figure 4B). In the roots, Ce signals in the NP-treated duckweed was mainly concentrated at the root tip (Figure 4C), while those in the ion-treated duckweed were detected mainly in the elongation region (Figure 4D). In addition, the intensity of Ce in the roots was much higher that in the fronds upon exposure to CeO_2 NPs.

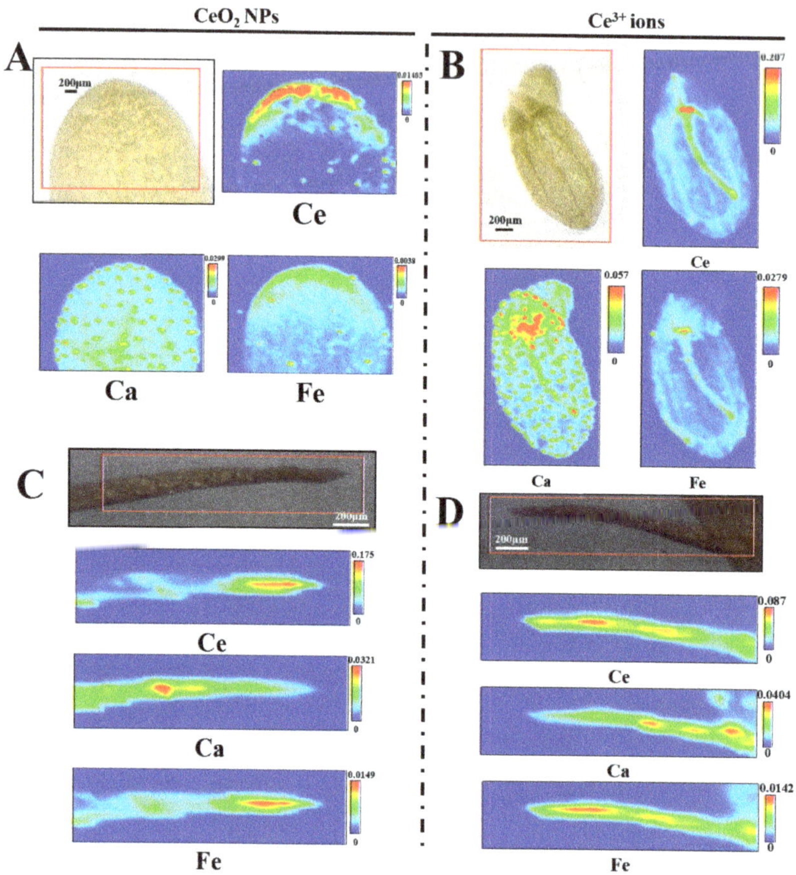

Figure 4. μ-XRF images of Ce, Ca, and Fe in the fronds (**A**,**B**) and roots (**C**,**D**) of duckweed exposed to 10 mg/L CeO_2 NPs (**A**,**C**) and Ce^{3+} ions (**B**,**D**). The red color in each image corresponds to the maximum concentration of the element.

3.7. Contents and Speciation of Ce in Duckweed Tissues

Ce contents in duckweed exhibited dose-dependent manner under both treatments and increased with the increase of exposure concentration (Figure S7). Moreover, the Ce content in the ionic group was significantly higher than that of the NPs group at concentrations higher than 0.5 mg/L. The BAF values of Ce (defined as the ratio of Ce concentration in organisms to those in water) in duckweed decreased with increasing exposure concentrations in both groups (except 10 mg/L CeO_2 NPs and 0.1 mg/L Ce^{3+} ions) (Table S1). At each concentration, the BAF value of the ionic group was higher than that of the NP group.

The speciation of Ce in the duckweed treated with 10 mg/L CeO_2 NPs and Ce^{3+} ions was analyzed using XANES. In Figure 5, line a represents the characteristic peak of Ce(III), and lines b and c represent the characteristic peaks of Ce(IV). The spectrum of Ce in the sample of the NP group seemed to be a mixture of Ce(III) and Ce(IV) standard reference spectra. In contrast, the peak of Ce in the ionic group sample completely coincided with line a, suggesting that only Ce(III) existed. LCF results revealed that about 27.6% of Ce(IV) was reduced to Ce(III) in the NPs group, whereas 100% of Ce was present as Ce(III) in the ions group.

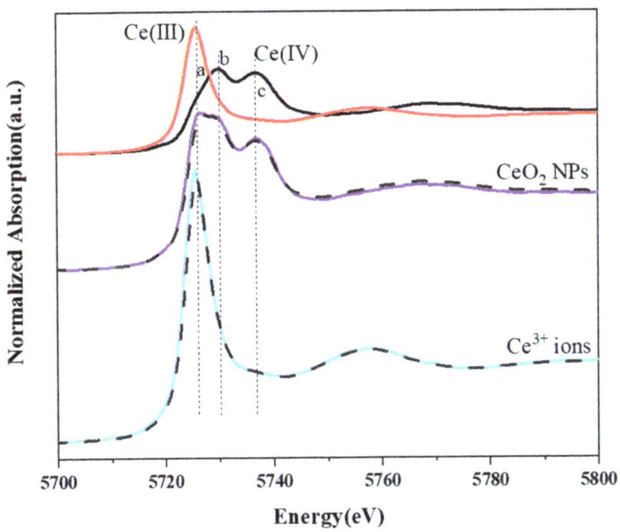

Figure 5. XANES spectra of Ce L_{III}-edge in duckweed treated with 10 mg/L CeO_2 NPs and Ce^{3+} ions.

4. Discussion

It has been reported that CeO_2 NPs are non-toxic to several terrestrial plant species but specifically toxic to *Lactuca* plants [26], with the degree of inhibition being related to the species and the type of culture medium [27]. However, there are few reports about the effects of CeO_2 NPs on the growth of aquatic plants. In the present study, we compared the toxicity of CeO_2 NPs and Ce^{3+} ions to the aquatic plant duckweed at concentrations ranging from 0 to 10 mg/L. The significant inhibitory effect of CeO_2 NPs on the growth of duckweed at such a low concentration (0.1 mg/L) was surprising, indicating that duckweed is very sensitive to non-environmental stresses (such as exogenous NPs) and can be used as an indicator of environmental pollution. The toxicity of Ce^{3+} ions to duckweed was greater than that of CeO_2 NPs at the same concentration. This was similar to the results of a previous study, in which the authors found that ionic cerium (Ce^{3+} ions) at 10 mg/L had a negative effect on the growth of radish, while bulk and nanoparticulate CeO_2 at the same concentration had no effect on radish growth and increased plant biomass, respectively [28].

Many physiological and biochemical activities during plant growth, such as chloroplast synthesis and photosynthesis are associated with the uptake and utilization of essential elements [29,30]. Deficiencies of certain mineral elements may affect plant respiration and photosynthesis, resulting in growth inhibition [31]. The mineral element contents of duckweed were disturbed under both CeO_2 NPs and Ce^{3+} ions treatments (Table 1). The effective ion radius of REE ions are close to that of Ca^{2+}, they may compete with Ca for organic ligands or replace it in the cell walls [32]. The decrease in Ca content may be due to the replacement of Ca^{2+} by Ce^{3+} ions in the cell wall. Moreover, the decrease in Ca content may affect membrane stability, leading to a decrease in chlorophyll content [33]. Therefore, the reduction of photosynthetic pigments induced by CeO_2 NPs and Ce^{3+} ions in this study might be related to the decrease of Ca content. Fe is an activator of coenzymes related to the synthesis of chlorophyll, and the decrease in Fe will reduce the content of chlorophyll [34,35], which might also be one of the reasons for the reduction in photosynthetic pigments in the treated duckweed. It has been reported that rare earth elements (REEs) can regulate plant growth by influencing the mineral elements in the plant tissues [36], and the uptake of Ca, Na, Zn and Mn by corn and mung bean decreased with increases in La^{3+} and Ce^{3+} ion concentrations (0–5 μM) in the exposed medium [37]. Similarly, reductions in the content of related elements and corresponding growth inhibition were observed in duckweed treated with CeO_2 NPs and Ce^{3+} ions.

The state of stomata can reflect the growth of plant leaves, and photosynthesis is sensitive to the disturbance of gas exchange through stomata. The decrease of stomatal aperture may result in the reduction in net photosynthesis and transpiration rate [38]. In this study, duckweed stomata were deformed when exposed to both CeO_2 NPs and Ce^{3+} ions, indicating that the morphology and function of stomata were severely damaged. Similar stomatal closure was also observed in the fronds of purple-backed duckweed exposed to Ag NPs [16] and *Eichhornia crassipes* exposed to CuO NPs [39]. The authors attributed this alteration to the overproduction of ROS. In this study, the results of Eb staining indicated that both CeO_2 NPs and Ce^{3+} ions produced excessive ROS in duckweed at high concentrations, which led to cell death and affected plant physiological characteristics, such as root elongation, frond area, and morphology.

The different in situ distributions of Ce in duckweed tissues under the two treatments implied that the uptake and transport of the two materials might be different. The distribution of Ce in duckweed fronds was similar to that previously reported in cucumber—namely, Ce mainly concentrated at the edge of CeO_2 NPs-treated leaves, whereas Ce in the Ce^{3+}-ion-treated leaf mainly concentrated along the veins [20]. Although both roots and fronds of duckweed were in contact with CeO_2 NPs (or Ce^{3+} ions) upon exposure, the distribution of Ce in duckweed fronds was similar to that in cucumber leaves under the same treatment, indicating that roots play an important role in the accumulation of Ce in duckweed. In the roots, the Ce signal in the NPs group was concentrated at the root tip and elongation region, while that in the ionic group was mainly detected in the elongation region. Plant roots can secrete mucilage containing organic acids, which can promote the adsorption and internalization of NPs. One of the possible ways by which NPs enter plants is from the root meristem entering into the vascular bundle through the apoplast; they are then transported to shoots through transpiration [20]. The accumulation of Ce in the root tips of duckweed exposed to CeO_2 NPs in this study also suggested this possibility. In addition, high-Ce signals were observed at the root and frond junctions under both treatments, while no Ce signals were detected in the new fronds. This suggested that Ce might be taken up by the roots and then transported upward to the fronds by transpiration with the water flow rather than being absorbed directly through the fronds. This result was consistent with the accumulation of Cu in duckweed exposed to CuO NPs being mainly attributed to the roots [19].

Due to the highly reactive and dynamic properties of NPs, they may undergo transformation upon interaction with plants [40]. Since the "acquired" and "original" characteristics of NPs coexist in the real environment, relevant factors must be considered when assessing their environmental fate and biological effects [41]. CeO_2 NPs were originally thought to be stable and insoluble compounds in the environment. However, later studies proved that dissolution of CeO_2 NPs did occur under certain circumstances [42]. For example, when CeO_2 NPs interact with terrestrial plants, they can be transformed and release Ce(III) ions with the assistant of root exudates [43,44]. Root exudates contain a variety of low molecular compounds, such as organic acids, amino acids, carbohydrates, polysaccharides, and phenolic compounds, which were critical for the dissolution and uptake of CeO_2 NPs by plants. Similar to terrestrial plants, aquatic plants have a special root-affected area, where root exudates interact with NPs [45]. Root exudates of aquatic plants mainly consist of carbon-containing compounds, such as organic acids, amino acids, carbohydrates, polysaccharides, proteins, tannins, phenolic compounds, etc. [46]. As we demonstrated, CeO_2 NPs could be reduced to Ce(III) in aquatic plant duckweed with the assistant of organic acids and reducing substances (such as phenols). The degree of transformation of CeO_2 NPs (the amount of ions released) upon interacting with plants varied from species to species, and the phytotoxicity of CeO_2 NPs was mainly plant-species-dependent. For instance, the phytotoxicity of CeO_2 NPs to *Lactuca* plants was due to the high sensitivity of this species to Ce^{3+} ions produced through dissolution after the interaction between them [26,27]. According to the proportion of Ce(III) and the total Ce content in duckweed treated with CeO_2 NPs (10 mg/L), the absolute content of Ce(III) was calculated to be 948.0 mg/kg, which was comparable to that in duckweed treated with Ce^{3+} ions of 2 mg/L. By retrieving the apparent toxicity parameters (Figure 1), it was found that the inhibition degree of 10 mg/L CeO_2 NPs on duckweed was exactly between that of 0.5 and 2 mg/L Ce^{3+} ions. Therefore, we speculated that the phytotoxicity of CeO_2 NPs to duckweed was due to the transformation and the high sensitivity of duckweed to the released Ce^{3+} ions.

5. Conclusions

In summary, the effects of CeO_2 NPs and Ce^{3+} ions on the aquatic higher plant duckweed were studied. At concentration higher than 0.1 mg/L, they could reduce the biomass and photosynthetic pigment contents of duckweed, causing significant cell death. Moreover, the toxicity of Ce^{3+} ions to duckweed was greater than that of CeO_2 NPs of the same concentration. SEM observation showed that the frond stomata were deformed, and the parameters of stomatal morphology were significantly changed under the two treatments. The in situ distribution of Ce in duckweed roots suggested that roots played a dominant role in the uptake of Ce. XANES and LCF results confirmed the transformation of Ce from Ce(IV) to Ce(III) in CeO_2-NP-treated duckweed and that the toxicity of CeO_2 NPs to duckweed originated mainly from the sensitively of this species to the released ions. The findings of this work were helpful for assessing the ecotoxicity of CeO_2 NPs and for better understanding the interaction between metal-based NPs and aquatic plants.

Supplementary Materials: The following supporting information can be downloaded at: https://www.mdpi.com/article/10.3390/nano13182523/s1, Figure S1: TEM (A) and SEM (B) images of CeO_2 NPs; Figure S2: Size distribution of CeO_2 NPs (10 mg/L) in deionized water and Steinberg's medium at 0 days by DLS; Figure S3: Size distribution of CeO_2 NPs in Steinberg's medium after exposure; Figure S4: Photosynthetic pigment contents of duckweed under different treatments; Figure S5: The photographs of duckweed in the control (A), CeO_2 NP (B), and Ce^{3+} ion (C) groups; Figure S6: The aperture width (A) and area (B) of stomata on fronds of different treatment groups; Figure S7: Ce contents in duckweed under different treatments; Figure S8: TEM images of root sections of duckweed unexposed (A,B), exposed to 10 mg/L CeO_2 NPs (C,D), and exposed to 10 mg/L to Ce^{3+} ions (E,F); Table S1: BAF of Ce in duckweed after exposure to CeO_2 NPs and Ce^{3+} ions.

Author Contributions: Conceptualization, Y.M. and Z.Z.; methodology, Y.L. and X.Z.; software, W.D.; validation, Y.W. and Z.S.; formal analysis, J.S.; investigation, Y.L.; data curation, X.Z.; writing—original draft preparation, Y.L. and X.Z.; writing—review and editing, Y.M.; visualization, X.H.; supervision, Y.M. and Z.Z.; project administration, F.Y. and Z.Z.; funding acquisition, Y.M. and Z.Z. All authors have read and agreed to the published version of the manuscript.

Funding: This research was funded by the National Natural Science Foundation of China (Grant Nos. 12175263, 11875267, and 12075262) and the Introduced Overseas Students Funding Program of Hebei Province in 2022 (C20220502).

Data Availability Statement: Not applicable.

Conflicts of Interest: The authors declare no conflict of interest.

References

1. Zhang, Y.; Hou, F.; Tan, Y. CeO_2 nanoplates with a hexagonal structure and their catalytic applications in highly selective hydrogenation of substituted nitroaromatics. *Chem. Commun.* **2012**, *48*, 2391–2393. [CrossRef]
2. Johnson, A.C.; Park, B. Predicting contamination by the fuel additive cerium oxide engineered nanoparticles within the United Kingdom and the associated risks. *Environ. Toxicol. Chem.* **2012**, *31*, 2582–2587. [CrossRef] [PubMed]
3. Celardo, I.; Pedersen, J.Z.; Traversa, E.; Ghibelli, L. Pharmacological potential of cerium oxide nanoparticles. *Nanoscale* **2011**, *3*, 1411–1420. [CrossRef] [PubMed]
4. Keller, A.A.; McFerran, S.; Lazareva, A.; Suh, S. Global life cycle releases of engineered nanomaterials. *J. Nanopart. Res.* **2013**, *15*, 1–17.
5. Xu, C.; Qu, X.G. Cerium oxide nanoparticle: A remarkably versatile rare earth nanomaterial for biological applications. *NPG Asia Mater.* **2014**, *6*, e90. [CrossRef]

6. Zhao, X.C.; Yu, M.; Xu, D.; Liu, A.F.; Hou, X.W.; Hao, F.; Long, Y.M.; Zhou, Q.F.; Jiang, G.B. Distribution, Bioaccumulation, Trophic Transfer, and Influences of CeO_2 Nanoparticles in a Constructed Aquatic Food Web. *Environ. Sci. Technol.* **2017**, *51*, 5205–5214. [CrossRef]
7. Servin, A.D.; White, J.C. Nanotechnology in agriculture: Next steps for understanding engineered nanoparticle exposure and risk. *Nanoimpact* **2016**, *1*, 9–12. [CrossRef]
8. Wojcieszek, J.; Jimenez-Lamana, J.; Bierla, K.; Ruzik, L.; Asztemborska, M.; Jarosz, M.; Szpunar, J. Uptake, translocation, size characterization and localization of cerium oxide nanoparticles in radish (*Raphanus sativus* L.). *Sci. Total Environ.* **2019**, *683*, 284–292. [CrossRef]
9. Giese, B.; Klaessig, F.; Park, B.; Kaegi, R.; Steinfeldt, M.; Wigger, H.; von Gleich, A.; Gottschalk, F. Risks, Release and Concentrations of Engineered Nanomaterial in the Environment. *Sci. Rep.* **2018**, *8*, 1565. [CrossRef]
10. Manier, N.; Bado-Nilles, A.; Delalain, P.; Aguerre-Chariol, O.; Pandard, P. Ecotoxicity of non-aged and aged CeO_2 nanomaterials towards freshwater microalgae. *Environ. Pollut.* **2013**, *180*, 63–70. [CrossRef]
11. Xie, C.; Ma, Y.; Zhang, P.; Zhang, J.; Li, X.; Zheng, K.; Li, A.; Wu, W.; Pang, Q.; He, X. Elucidating the origin of the toxicity of nano-CeO_2 to Chlorella pyrenoidosa: The role of specific surface area and chemical composition. *Environ. Sci. Nano* **2021**, *8*, 1701.
12. Radić, S.; Domijan, A.M.; Glavaš Ljubimir, K.; Maldini, K.; Ivešić, M.; Peharec Štefanić, P.; Krivohlavek, A. Toxicity of nanosilver and fumonisin B(1) and their interactions on duckweed (*Lemna minor* L.). *Chemosphere* **2019**, *229*, 86–93. [CrossRef]
13. Rippner, D.A.; Lien, J.; Balla, H.; Guo, T.; Green, P.G.; Young, T.M.; Parikh, S.J. Surface modification induced cuprous oxide nanoparticle toxicity to duckweed at sub-toxic metal concentrations. *Sci. Total Environ.* **2020**, *722*, 137607. [CrossRef]
14. Modlitbová, P.; Hlaváček, A.; Švestková, T.; Pořízka, P.; Šimoníková, L.; Novotný, K.; Kaiser, J. The effects of photon-upconversion nanoparticles on the growth of radish and duckweed: Bioaccumulation, imaging, and spectroscopic studies. *Chemosphere* **2019**, *225*, 723–734. [CrossRef] [PubMed]
15. Tarrahi, R.; Khataee, A.; Movafeghi, A.; Rezanejad, F.; Gohari, G. Toxicological implications of selenium nanoparticles with different coatings along with Se(4+) on Lemna minor. *Chemosphere* **2017**, *181*, 655–665. [CrossRef] [PubMed]
16. Lalau, C.M.; Simioni, C.; Vicentini, D.S.; Ouriques, L.C.; Mohedano, R.A.; Puerari, R.C.; Matias, W.G. Toxicological effects of AgNPs on duckweed (*Landoltia punctata*). *Sci. Total Environ.* **2020**, *710*, 136318. [CrossRef] [PubMed]
17. Stegemeier, J.P.; Colman, B.P.; Schwab, F.; Wiesner, M.R.; Lowry, G.V. Uptake and Distribution of Silver in the Aquatic Plant *Landoltia punctata* (Duckweed) Exposed to Silver and Silver Sulfide Nanoparticles. *Environ. Sci. Technol.* **2017**, *51*, 4936–4943. [CrossRef]
18. Lalau, C.M.; Mohedano, R.d.A.; Schmidt, É.C.; Bouzon, Z.L.; Ouriques, L.C.; dos Santos, R.W.; da Costa, C.H.; Vicentini, D.S.; Matias, W.G. Toxicological effects of copper oxide nanoparticles on the growth rate, photosynthetic pigment content, and cell morphology of the duckweed *Landoltia punctata*. *Protoplasma* **2015**, *252*, 221–229. [CrossRef]
19. Yue, L.; Zhao, J.; Yu, X.; Lv, K.; Wang, Z.; Xing, B. Interaction of CuO nanoparticles with duckweed (*Lemna minor L..*): Uptake, distribution and ROS production sites. *Environ. Pollut.* **2018**, *243*, 543–552. [CrossRef]
20. Zhang, Z.; He, X.; Zhang, H.; Ma, Y.; Zhang, P.; Ding, Y.; Zhao, Y. Uptake and distribution of ceria nanoparticles in cucumber plants. *Metallomics* **2011**, *3*, 816–822. [CrossRef]
21. No, O.T. *221: Lemna sp. Growth Inhibition Test*; OECD Guidelines for the Testing of Chemicals, Section; OECD: Paris, France, 2006; Volume 2, pp. 1–22.
22. Baker, C.J.; Mock, N.M. An Improved Method for Monitoring Cell-Death in Cell-Suspension and Leaf Disc Assays Using Evans Blue. *Plant Cell Tissue Organ Cult.* **1994**, *39*, 7–12. [CrossRef]
23. Dong, G.; Nkoh, J.N.; Hong, Z.N.; Dong, Y.; Lu, H.L.; Yang, J.; Pan, X.Y.; Xu, R.K. Phytotoxicity of Cu^{2+} and Cd^{2+} to the roots of four different wheat cultivars as related to charge properties and chemical forms of the metals on whole plant roots. *Ecotoxicol. Environ. Saf.* **2020**, *196*, 110545. [CrossRef] [PubMed]
24. You, Y.; Liu, L.J.; Wang, Y.; Li, J.X.; Ying, Z.N.; Hou, Z.L.; Liu, H.J.; Du, S.T. Graphene oxide decreases Cd concentration in rice seedlings but intensifies growth restriction. *J. Hazard. Mater.* **2021**, *417*, 125958. [CrossRef]
25. Liu, M.; Feng, S.; Ma, Y.; Xie, C.; He, X.; Ding, Y.; Zhang, J.; Luo, W.; Zheng, L.; Chen, D.; et al. Influence of Surface Charge on the Phytotoxicity, Transformation, and Translocation of CeO2 Nanoparticles in Cucumber Plants. *ACS Appl. Mater. Interfaces* **2019**, *11*, 16905–16913. [CrossRef] [PubMed]
26. Zhang, P.; Ma, Y.; Zhang, Z.; He, X.; Li, Y.; Zhang, J.; Zheng, L.; Zhao, Y. Species-specific toxicity of ceria nanoparticles to *Lactuca* plants. *Nanotoxicology* **2015**, *9*, 1–8. [CrossRef]
27. Cui, D.; Zhang, P.; Ma, Y.; He, X.; Li, Y.; Zhang, J.; Zhao, Y.; Zhang, Z. Effect of cerium oxide nanoparticles on asparagus lettuce cultured in an agar medium. *Environ. Sci. Nano* **2014**, *1*, 459–465. [CrossRef]
28. Zhang, W.; Ebbs, S.D.; Musante, C.; White, J.C.; Gao, C.; Ma, X. Uptake and Accumulation of Bulk and Nanosized Cerium Oxide Particles and Ionic Cerium by Radish (*Raphanus sativus* L.). *J. Agric. Food Chem.* **2015**, *63*, 382–390. [CrossRef]
29. Chaudhary, E.; Sharma, P. Effect of cadmium on nutrients concentration in duckweed: A case of *Lemna minor* and *Lemna gibba*. *Environ. Sustain.* **2021**, *4*, 183–190. [CrossRef]
30. Morkunas, I.; Woźniak, A.; Mai, V.C.; Rucińska-Sobkowiak, R.; Jeandet, P. The role of heavy metals in plant response to biotic stress. *Molecules* **2018**, *23*, 2320. [CrossRef]
31. Shtangeeva, I.; Ayrault, S.J.E.; Botany, E. Effects of Eu and Ca on yield and mineral nutrition of wheat (*Triticum aestivum*) seedlings. *Environ. Exp. Bot.* **2007**, *59*, 49–58. [CrossRef]

32. Parida, A.K.; Das, A.B.; Mittra, B. Effects of salt on growth, ion accumulation, photosynthesis and leaf anatomy of the mangrove, *Bruguiera parviflora*. *Trees-Struct. Funct.* **2004**, *18*, 167–174. [CrossRef]
33. Shen, H.; Ren, Q.G.; Mi, Y.; Shi, X.F.; Yao, H.Y.; Jin, C.Z.; Huang, Y.Y.; He, W.; Zhang, J.; Liu, B. Investigation of metal ion accumulation in *Euglena gracilis* by fluorescence methods. *Nucl. Instrum. Methods Phys. Res. Sect. B-Beam Interact. Mater. At.* **2002**, *189*, 506–510. [CrossRef]
34. Miller, G.W.; Pushnik, J.C.; Welkie, G.W. Iron chlorosis, a world wide problem, the relation of chlorophyll biosynthesis to iron. *J. Plant Nutr.* **1984**, *7*, 1–22. [CrossRef]
35. Terry, N.; Low, G. Leaf chlorophyll content and its relation to the intracellular localization of iron. *J. Plant Nutr.* **1982**, *5*, 301–310. [CrossRef]
36. Hu, Z.; Richter, H.; Sparovek, G.; Schnug, E. Physiological and biochemical effects of rare earth elements on plants and their agricultural significance: A review. *J. Plant Nutr.* **2004**, *27*, 183–220. [CrossRef]
37. Diatloff, E.; Smith, F.; Asher, C. Effects of Lanthanum and Cerium on the Growth and Mineral Nutrition of Corn and Mungbean. *Ann. Bot.* **2008**, *101*, 971–982. [CrossRef] [PubMed]
38. Goodwin, P.; DeVay, J.; Meredith, C. Roles of water stress and phytotoxins in the development of Pierce's disease of the grapevine. *Physiol. Mol. Plant Pathol.* **1988**, *32*, 1–15. [CrossRef]
39. Zhao, J.; Ren, W.; Dai, Y.; Liu, L.; Wang, Z.; Yu, X.; Zhang, J.; Wang, X.; Xing, B. Uptake, Distribution, and Transformation of CuO NPs in a Floating Plant Eichhornia crassipes and Related Stomatal Responses. *Environ. Sci. Technol.* **2017**, *51*, 7686–7695. [CrossRef]
40. Noori, A.; Ngo, A.; Gutierrez, P.; Theberge, S.; White, J.C. Silver nanoparticle detection and accumulation in tomato (*Lycopersicon esculentum*). *J. Nanopart. Res.* **2020**, *22*, 131. [CrossRef]
41. Natarajan, L.; Jenifer, M.; Mukherjee, A. Eco-corona formation on the nanomaterials in the aquatic systems lessens their toxic impact: A comprehensive review. *Environ. Res.* **2021**, *194*, 110669. [CrossRef]
42. Schwabe, F.; Schulin, R.; Rupper, P.; Rotzetter, A.; Stark, W.; Nowack, B. Dissolution and transformation of cerium oxide nanoparticles in plant growth media. *J. Nanopart. Res.* **2014**, *16*, 2668. [CrossRef]
43. Zhang, P.; Ma, Y.; Zhang, Z.; He, X.; Zhang, J.; Guo, Z.; Tai, R.; Zhao, Y.; Chai, Z. Biotransformation of Ceria Nanoparticles in Cucumber Plants. *ACS Nano* **2012**, *6*, 9943–9950. [CrossRef] [PubMed]
44. Schwabe, F.; Tanner, S.; Schulin, R.; Rotzetter, A.C.; Stark, W.J.; von Quadt, A.; Nowack, B. Dissolved Cerium contributes to uptake of Ce in presence of differently sized CeO_2-nanoparticles by three crop plants. *Metallomics* **2015**, *7*, 466–477. [CrossRef] [PubMed]
45. Escolà Casas, M.; Matamoros, V. Linking plant-root exudate changes to micropollutant exposure in aquatic plants (*Lemna minor* and *Salvinia natans*). A prospective metabolomic study. *Chemosphere* **2022**, *287*, 132056. [CrossRef]
46. Wan, D.; Yang, J.; Wang, X.; Xiang, W.; Selvinsimpson, S.; Chen, Y. Wavelength-Dependent Photoreactivity of Root Exudates from Aquatic Plants under UV-LED Irradiation. *ACS ES&T Water* **2022**, *2*, 2613–2622. [CrossRef]

nanomaterials

MDPI

Article

The Pd (II) Reduction Mechanisms in *Bacillus megaterium* Y-4 Revealed by Proteomic Analysis

Yuan Chen [1,2], Jiaxing Wang [3], Daidi Chen [1], Boxi Wang [1], Jinchuan Wu [1], Rongrong Liu [2,*] and Qingxin Li [1,*]

[1] Guangdong Provincial Engineering Laboratory of Biomass High Value Utilization, Institute of Biological and Medical Engineering, Guangdong Academy of Sciences, Guangzhou 510316, China
[2] Key Laboratory for Water Quality and Conservation of the Pearl River Delta, Institute of Environmental Research at Greater Bay Area, Ministry of Education, Guangzhou University, Guangzhou 510006, China
[3] Ministry of Education Key Laboratory of Pollution Control and Ecological Remediation for Industrial Agglomeration Area, School of Environment and Energy, South China University of Technology, Guangzhou 510006, China; 19120332369@163.com
* Correspondence: liurr@gzhu.edu.cn (R.L.); qingxin_li@outlook.com (Q.L.)

Abstract: Many studies have been conducted on the microbial reduction of Pd (II) to palladium nanoparticles (Pd-NPs) due to the environmental friendliness, low cost, and the decreased toxicity of Pd (II) ions. In this study, we investigate the reduction mechanism of Pd (II) by *Bacillus megaterium* Y-4 through proteomics. The data are available via ProteomeXchange with identifier PXD049711. Our results revealed that *B. megaterium* Y-4 may use the endogenous electron donor (NAD(P)H) generated by nirB, tdh, and fabG and reductase to reduce Pd (II) to Pd-NPs. The expression levels of fabG, tdh, gudB, and rocG that generate NAD(P)H were further increased, and the number of reduced Pd-NPs was further increased with the exogenous electron donor sodium formate. Endogenous electron mediators such as quinones and flavins in *B. megaterium* Y-4 can further enhance Pd (II) reduction. The findings provided invaluable information regarding the reduction mechanism of Pd (II) by *B. megaterium* Y-4 at the proteome level.

Keywords: palladium; biological reduction; protein expression; proteomics; *Bacillus megaterium* Y-4

1. Introduction

Palladium (Pd) is a valuable metal and has been widely applied in pharmaceutics, electroplating, and precision manufacturing due to its excellent catalytic performance, excellent effectiveness, and good stability [1]. However, palladium is extremely rare and expensive [2]. At the same time, the discharge of palladium-containing waste impacts the environmental ecosystem and threatens human health [3]. Therefore, researchers have been motivated to develop diverse methods to effectively recover and recycle Pd from different sources [4]. Some microorganisms (e.g., *Geobacter sulfurreducens* [4,5], *Shewanella oneidensis* [6,7], *Enterococcus faecalis* [8], and *Enterobacter cloacae* [9]) can convert Pd (II) through a reduction reaction under room temperature and pressure [10]. Compared to conventional physical and chemical methods, the microbial reduction of Pd (II) is renewable and low-cost without introducing contamination [11]. Pd (0) nanoparticles (Pd-NPs) can be bio-synthesized in periplasmic, intracellular, or extracellular space [12]. Biosynthesized palladium nanoparticles can play a role in the reduction and removal of Cr (VI) [13], the dehalogenation of polychlorinated dioxins [14], the Mizoroki–Heck reaction, etc. [15].

Pd (II) reduction can be achieved through several pathways. In the reduction reaction, Pd (II) serves as a respiratory terminal electron acceptor when dissimilatory metal-reducing bacteria (DMRB) oxidize organic matter and then transfer the released electrons to Pd (II) for respiration, resulting in the bio-reduction to Pd-NPs [9,16]. The role of the NADH dehydrogenases and hydrogenases (especially HyaB) of *Shewanella oneidensis* MR-1 was demonstrated to be responsible for the formation of bio-Pd on the outer membrane and in the periplasmic space, respectively [17]. Transcriptomic analyses revealed that the genes encoding NADH–quinone oxidoreductase, dehydrogenases, cytochrome c reductase, cytochrome c oxidase, the quinone cycle, and ribE in *Bacillus thuringiensis* Y9 have a strong positive relationship with palladium reduction [18]. The genes encoding oxidoreductase, metabolic enzymes, and cell-surface proteins were mainly identified to be responsible for the reduction of Pd (II) [19]. Further investigation of the microbial reduction mechanism of Pd (II) and the identification of related proteins will provide insights into the microbial reduction of palladium.

Our previous research demonstrated that the aerobic bacteria *B. megaterium* Y-4 can reduce Pd (II) [12], but it is unclear which enzymes and proteins are involved in this process. Proteomics is a powerful tool to monitor protein levels and has been applied to elucidate the cellular molecular mechanism of organisms responding to environmental stress induced by different conditions. Analyzing differentially expressed proteins (DEPs) is a strategy used to identify proteins involved in the response [20]. Here, to elucidate the Pd (II) reduction mechanisms in *B. megaterium* Y-4, we analyzed the proteomes of *B. megaterium* Y-4 after Pd (II) exposure under anaerobic conditions in the absence (N_2) and presence of 5 mM sodium formate (N_2-SF). The DEPs were identified by high-resolution liquid chromatography–tandem mass spectrometry (LC-MS/MS). Our results showed that the nirB, tdh, and fabG that are responsible for NAD(P)H production were upregulated under both conditions. *B. megaterium* Y-4 can use NAD(P)H to provide electrons for Pd (II) reduction. The higher expression level of gudB and zwf proteins that produce NAD(P)H in N_2-SF than in N_2 may partially explain the higher Pd (II) reduction in N_2-SF than N_2. Pd (II) may also be reduced by the ssuE, trx-1, and flavin reductase family proteins. The reduction of Pd (II) can be enhanced through the presence of endogenous electron mediators such as the menaquinone produced by the qoxB protein and flavin mononucleotide (FMN) produced by ssuE and ssuD. In addition, the ribosome-relevant proteins can maintain the translation, protein metabolism, and RNA process. There are upregulated flagellar motility, ATP-related proteins, and pentose-phosphate-pathway-related proteins, which can generate more energy to maintain normal metabolism, benefit-tending, and harm avoidance. These processes are conducive to enhancing the reduction of Pd (II). In this study, the mechanisms by which *B. megaterium* Y-4 produces biogenic Pd-NPs were elucidated through analyzing proteomics under two conditions.

2. Materials and Methods

2.1. Materials

TMT® mass tagging kits and reagents were purchased from Thermo (Shanghai, China). A Bradford protein quantification kit was purchased from Beyotime (Shanghai, China). Dithiothreitol (DTT), iodoacetamide (IAM), ammonium bicarbonate, ammonium hydrogen triethylcarbate buffer (TEAB), ammonia solution, and trifluoroacetic acid (TFA) were purchased from Sigma. Sodium dodecyl sulfate (SDS) and carbamide were purchased from National Pharmaceutical Holdings Limited. Mass spectrometry grade pancreatic enzyme was purchased from Promega. LC-MS-grade ultrapure water and LC-MS-grade acetonitrile were purchased from Thermo Fisher Chemical (Shanghai, China). LC-MS-grade formic acid was purchased from Thermo Fisher Scientific (Shanghai, China), and acetone was purchased from Beijing Chemical Factory. A ProteoMiner low-abundance protein enrichment kit was purchased from Bio-Rad (Shanghai, China). To prepare the Pd (II) solution, Na_2PdCl_4 (Aladdin Industrial Corporation, Shanghai, China) was dissolved in distilled water.

2.2. Synthesis and Analysis of Microbial Pd (0) Nanoparticle

The *B. megaterium* Y-4 (MH472619) used in this study was isolated from the electronics waste factory. *B. megaterium* Y-4 was activated by inclined cultivation for 1–2 days before it was aerobically cultured in 100 mL Luria Broth medium in a 250 mL conical flask at 30 °C and 150 rpm. The log-growth *B. megaterium* Y-4 cells were obtained through centrifugation (4000 g, 5 min) and washed with 10 mL phosphate buffer (pH = 7.0) to remove the supernatant medium. Cells with 303 mg·L^{-1} (dry weight) were mixed with 10 mg·L^{-1} Pd (II) under N_2 and N_2-SF conditions, respectively. Then, *B. megaterium* Y-4 cells loaded with Pd-NPs were washed with ultrapure water and subjected to proteomics analysis. The detailed proteomics analysis procedures are described in the Supplementary Information (SI). The mass spectrometry proteomics data were deposited into the ProteomeXchange Consortium via the PRIDE [21] partner repository with the dataset identifier PXD049711. TMT was used for protein labeling and quantification (Table 1).

Table 1. Labeling information of the sample.

Run Name	126	127N	128N	129N	130N	131N	132N	133N	134N
run1	Ctrl-1	Ctrl-2	Ctrl-3	N_2-1	N_2-2	N_2-3	N_2-SF-1	N_2-SF-2	N_2-SF-3

The residual Pd concentration was determined with an inductively coupled plasma mass spectrometer (ICP-MS, Agilent 7900, Santa Clara, CA, USA). All samples were filtered through 0.22 μm glass fiber filters (Tianjin Branch billion Lung Experimental Equipment Co., Ltd., Tianjin, China) before analysis. Ionic palladium (Pd) standard solutions (1000 mg· L^{-1}) were diluted as the calibration solution and internal standard.

The removal efficiency (%) of Pd (II) was calculated by the following formula:

$$\text{The removal efficiency (\%)} = (1 - C_e/C_0) \times 100 \quad (1)$$

where C_e (mg/L) is the Pd (II) residual concentration after reaction and C_0 (mg/L) is the initial Pd (II) concentration.

The microbial Pd-NPs were freeze-dried and characterized by X-ray photoelectron spectroscopy (XPS, PHI X-Tool, DE). The XPS spectra were recorded with the Al Ka line at 15 kV and 51 W. The binding energies were determined at 284.8 eV by referencing the C1s component due to carbon being bound only to carbon or hydrogen. The biological slices used for transmission electron microscopy (TEM) (TECNAI 10, PHILIPS, NED) analysis were prepared following standard protocols [12].

3. Results

3.1. Production of Pd-NPs by B. megaterium Y-4

The bioreduction of palladium by *B. megaterium* Y-4 was investigated under anaerobic conditions with N_2 and in presence of both N_2 and 5 mM sodium formate (N_2-SF). Sodium formate was used as an electron donor for the reduction. The removal efficiencies of Pd (II) reached 92.86% and 93.68% under N_2 and N_2-SF conditions, respectively (Figure S1). The chemical states of palladium after being treated with *B. megaterium* Y-4 under N_2 and N_2-SF conditions were further characterized by TEM and XPS spectroscopy. Firstly, the production of Pd-NPs by bacterial cells was observed: the TEM analysis demonstrated that many Pd-NPs were distributed in the periplasmic space and some Pd-NPs were also released from the live cells into the aqueous solution (Figure 1A,C). Secondly, the changes in the different states of the palladium were elucidated by XPS. As depicted in Figure 1B,D and Table S3, binding energies centered at 340.0–340.7 eV and 334.7–335.4 eV were observed, which was attributed to Pd 3d/2 and Pd 5d/2 of Pd (0). In addition, binding energies centered at 342.3–343.4 eV and 337.1–338.1 eV were observed, which was ascribed to Pd 3d/2 and Pd 5d/2 of Pd (II) [18]. The presence of Pd (0) after *B. megaterium* Y treatment demonstrates the conversion of palladium under the experimental conditions. Lastly, the quantification of the states of the palladium was achieved in the study and the conversion rates of the palladium under different conditions were compared. The XPS analysis (Table S3) indicated that the total peak area of Pd (0) synthesized by live cells was larger under N_2-SF conditions (Figure 1D) than that under N_2 conditions (Figure 1B). These results indicated that *B. megaterium* Y-4 can use the energy storage compounds to reduce Pd (II), and the exogenous electron donor sodium formate can further enhance the reduction of Pd (II) under N_2 conditions.

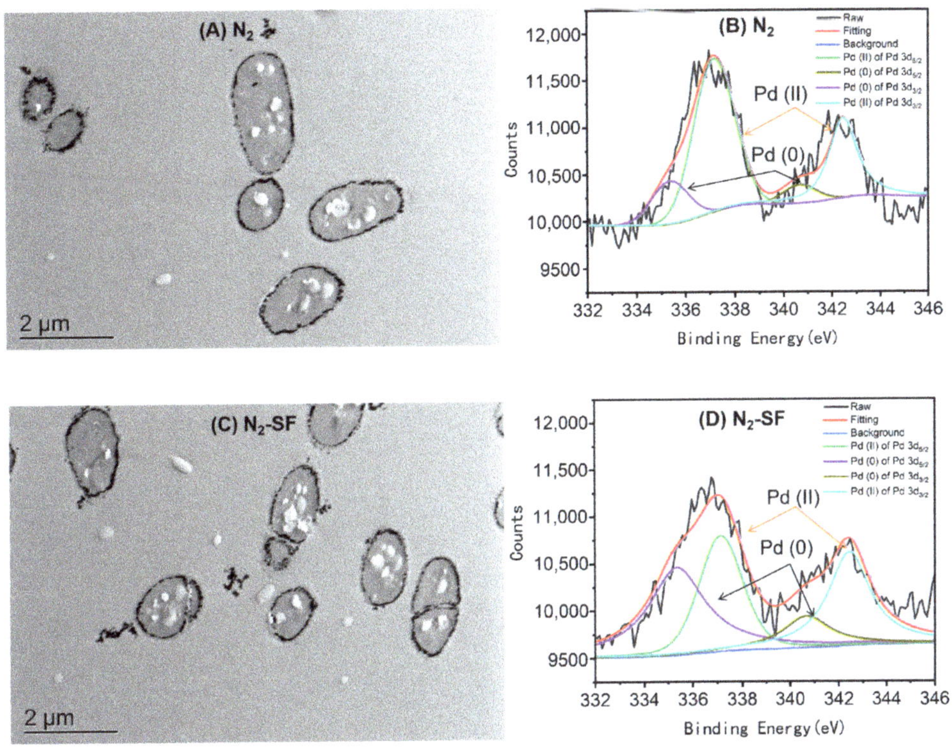

Figure 1. TEM and XPS of Pd-NPs synthesized by live *Bacillus megaterium* Y-4 under N_2 (**A**,**B**) and N_2-SF (**C**,**D**) conditions. Signals corresponding to Pd (II) and Pd (0) are indicated with arrows.

3.2. Protein Identification and Comparison during Palladium Reduction

The protein levels of *B. megaterium* Y-4 were evaluated during palladium reduction to identify proteins that are critical for the reaction. Up to 2560 proteins from *B. megaterium* Y-4 were identified by proteomics analysis, and the parent ion mass tolerance distribution and principal component analysis are shown in Figures S2 and S3. The protein levels before and after treatment were comparted to obtain quantitative ratios and to understand the mechanisms for palladium reduction by this bacterium. A quantitative ratio greater than 1.2 was defined as upregulation, whereas a value below 0.833 was considered downregulation. All quantifiable proteins exhibiting increased (≥1.2-fold) and decreased (≤0.833-fold) expression levels are listed in Table S4 ($p < 0.05$).

Compared with the control sample without treatment, 49 upregulated proteins and 43 downregulated proteins in *B. megaterium* Y-4 were induced after treatment with N_2 (Figure 2A and Figure S4A), 64 upregulated proteins and 64 downregulated proteins were identified after N_2-SF treatment (Figure 2B and Figure S4A), and 45 upregulated proteins and 42 downregulated proteins were observed in samples treated under both conditions (Figure 2C,D and Figure S4B). Among these upregulated proteins, nirB (WP_098786927.1) exhibited the highest level of upregulation. Other upregulated proteins included hxlA (WP_013055599.1), spxA (WP_013055422.1), rpsO (WP_013058831.1), hfq (WP_013058791.1), galU (WP_161515009.1), and dinG (WP_013056100.1). The downregulated proteins included aceA (WP_013055343.1), katA (WP_098534856.1), trpB (WP_013059002.1), trpE (WP_013059006.1), trpA (WP_013059001.1), and trpC (WP_013059004.1) (Table S4). For the N_2-treated sample, four upregulated proteins (WP_013056022.1, WP_013059402.1, WP_025752035.1, and cyoC (WP_053487176.1)) and one downregulated protein (WP_053487046.1) were unique and not observed in the N_2-SF-treated sample (Figure 2C,D and Figure S4B). For the N_2-SF-treated sample, 19 upregulated proteins (such as aroQ (WP_013059157.1), flgL (WP_057275290.1), panD (WP_013056099.1), and zwf (WP_013057959.1)) and 22 downregulated proteins (such as rpmI (WP_013059426.1, rpmB (WP_013058914.1), rpmG (WP_013054911.1), and qoxB (WP_013056572.1)) were unique (Figure 2C,D and Figure S4B). Nearly all of the DEPs exhibited either direct or in direct interactions with other proteins (Figures S5 and S6).

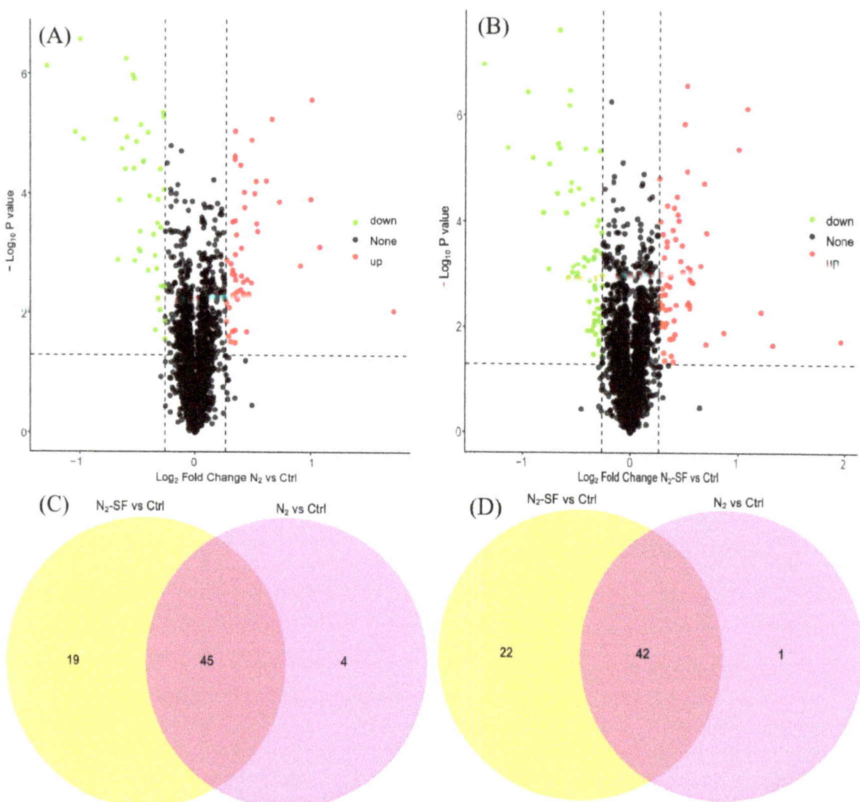

Figure 2. Volcano plots of DEPs in N_2- (**A**) and N_2-SF (**B**)-treated samples; Venn diagram showing the common and unique upregulated (**C**) or downregulated (**D**) expressed proteins. Upregulated and downregulated samples are indicated in yellow and purple, respectively.

3.3. Functional Characterization of Differentially Expressed Proteins

These DEPs under different conditions were classified according to their subcellular locations (Figure S7). For the N_2-treated sample, 47.37%, 26.32%, 15.79%, and 10.53% of the DEPs were located in the cytoplasm, cell membrane, extracell, and cell wall, respectively. For the N_2-SF-treated sample, 41.67%, 33.33%, 16.67%, and 8.33% of the DEPs were located in the cytoplasm, cell membrane, extracell, and cell wall, respectively.

Based on the Gene Ontology (GO) analysis tool, the biological functions affected by DEPs were classified to three functional groups, namely, molecular function (MF), cellular component (CC), and biological process (BP). For the N_2-treated sample, the order of the number of these DEPs enriched in GO functions was as follows: catalytic activity (GO:000382) (11 upregulated and 16 downregulated proteins) > metabolic process (GO:0008152) (11 upregulated and 15 downregulated proteins) > single-organism process (GO:0044699) (3 upregulated and 13 downregulated proteins) > single-organism metabolic process (GO:0044710) (2 upregulated and 12 downregulated proteins) > oxidation-reduction process (GO:0055114) (1 upregulated and 8 downregulated proteins) > oxidoreductase activity (GO:0016491) (2 upregulated and 7 downregulated proteins) (Figure 3). For the N_2-SF-treated sample, the order of the number of these DEPs enriched in GO functions was as follows: metabolic process (GO:0008152) (14 upregulated and 19 downregulated proteins) > single-organism process (GO:0044699) (8 upregulated and 15 downregulated proteins) > single-organism metabolic process (GO:0044710) (5 upregulated and 14 downregulated proteins) > oxidation-reduction process (GO:0055114) (3 upregulated proteins and 9 downregulated proteins) > oxidoreductase activity (GO:0016491) (3 upregulated and 9 downregulated proteins) > organonitrogen compound metabolic process (GO:1901564) (5 upregulated and 5 downregulated proteins) > small molecule metabolic process (GO:0044281) (4 upregulated and 4 downregulated proteins) (Figure 4).

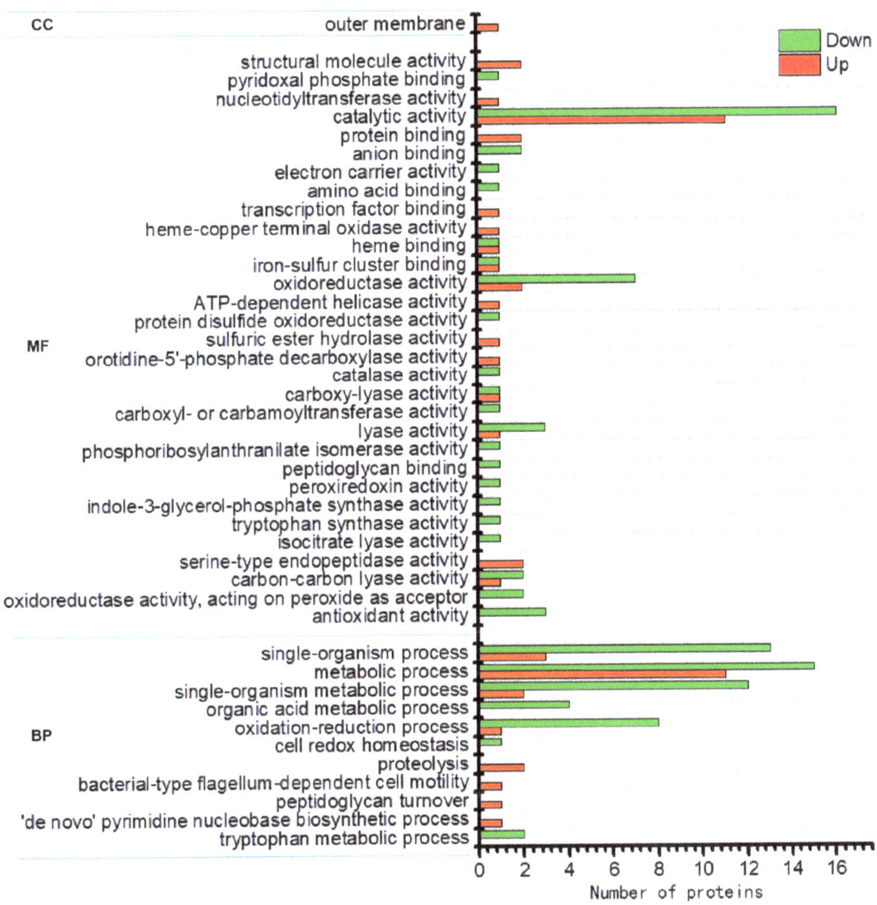

Figure 3. Functional category distribution (CC, cellular component; MF, molecular function; BP, biological process) of differentially expressed proteins in the N_2-treated sample.

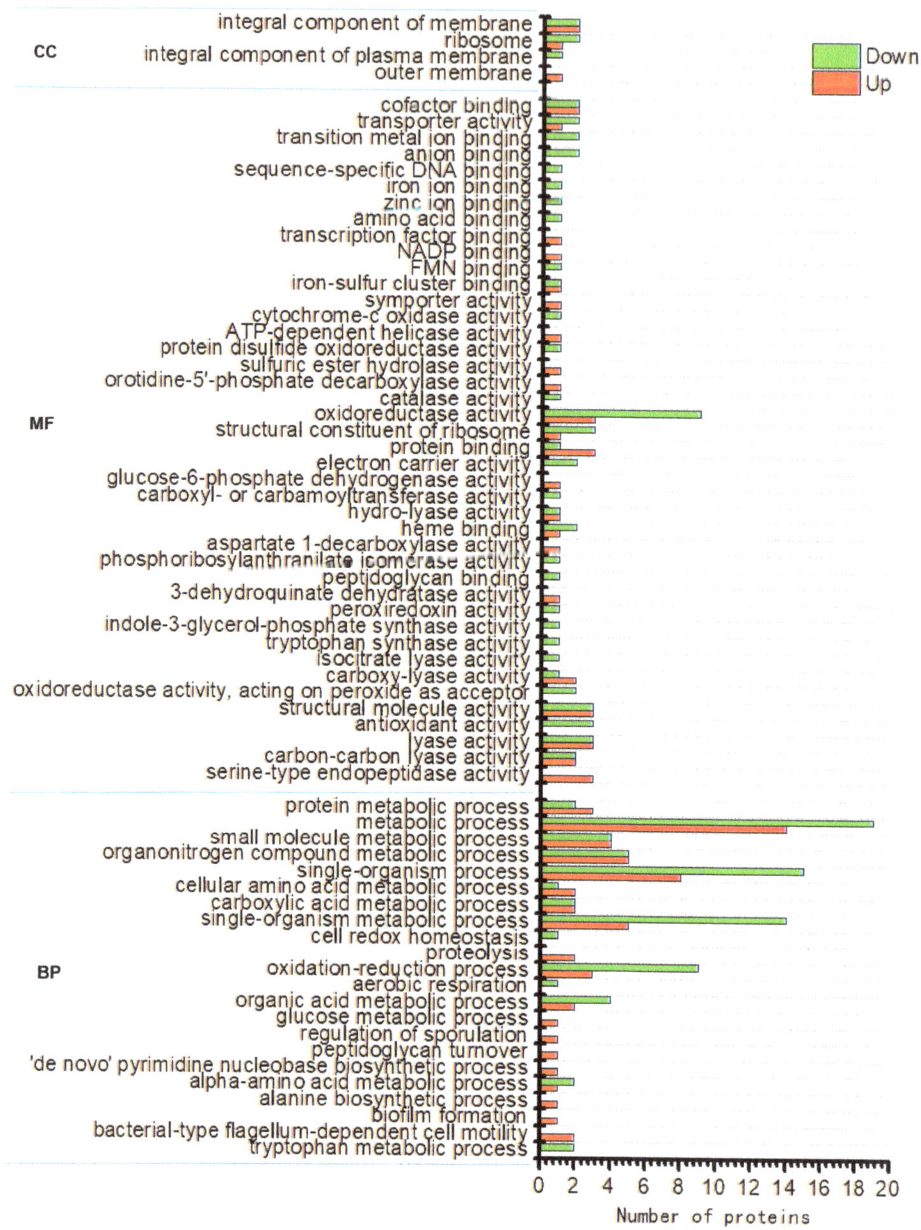

Figure 4. Functional category distribution (CC, cellular component; MF, molecular function; BP, biological process) of differentially expressed proteins in the N_2-SF-treated sample.

3.4. Functional Enrichment of Differentially Regulated Proteins

To investigate the functional differences in the upregulated and downregulated proteins, these proteins were subjected separately for GO, KEGG pathway, and domain enrichment analysis. The GO enrichment analysis results are shown in Figure 5, Figures S8 and S9. For the N_2-treated sample, the upregulated proteins were mainly enriched in serine-type endopeptidase activity (GO:0004252), hydrolase activity (GO:0016787), orotidine-5′-phosphate decarboxylase activity (GO:0004590), sulfuric ester hydrolase activity (GO:0008484), and the 'de novo' pyrimidine nucleobase biosynthetic process (GO:0006207) (Figure S8A). Meanwhile, the downregulated proteins were mainly related to antioxidant activity (GO:0016209), the tryptophan metabolic process (GO:0006568), oxidoreductase activity (GO:0016684), the single-organism metabolic process (GO:0044710), and the oxidation-reduction process (GO:0055114) (Figure S8B). For the N_2-SF-treated sample, the upregulated proteins were mainly involved in serine-type endopeptidase activity (GO:0004252), bacterial-type flagellum-dependent cell motility (GO:0071973), 3-dehydroquinate dehydratase activity (GO:0003855), biofilm formation (GO:0042710), aspartate 1-decarboxylase activity (GO:0004068), the alanine biosynthetic process (GO:0006523), carboxy-lyase activity (GO:0016831), and glucose-6-phosphate dehydrogenase activity (GO:0004345) (Figure S9A). The downregulated proteins mainly significantly participated in antioxidant activity (GO:0016209), the tryptophan metabolic process (GO:0006568), oxidoreductase activity (GO:0016684), the single-organism metabolic process (GO:0004451), and tryptophan synthase activity (GO:0004834) (Figure S9B).

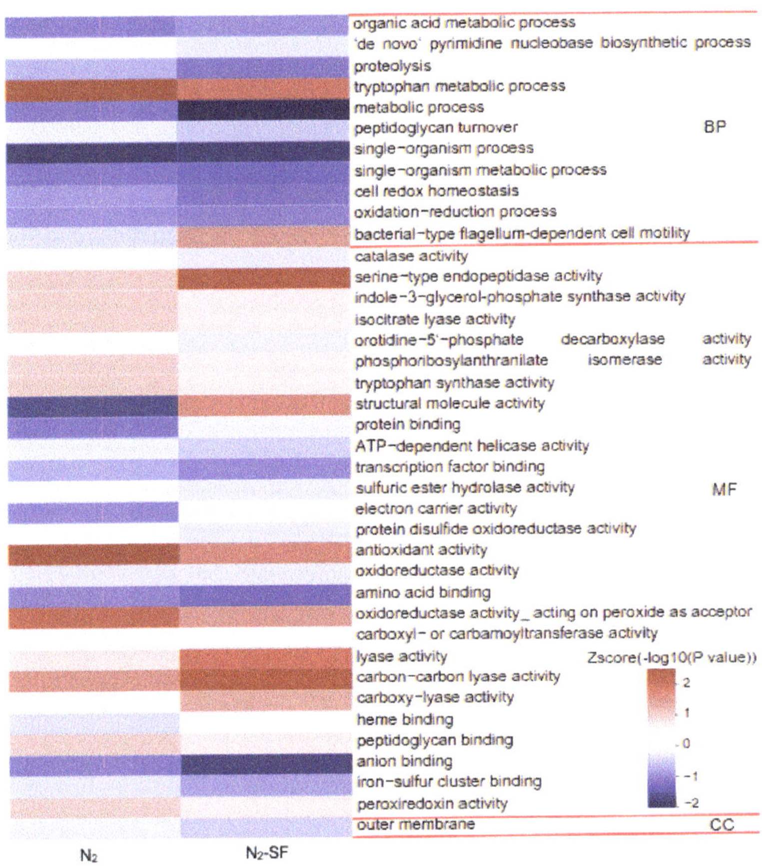

Figure 5. Function of the affected proteins during palladium reduction. A hierarchical clustering analysis was conducted on the differentially expressed proteins based on GO-analysis-based enrichment. The *p*-values were converted into Z-scores, which are displayed in the color legend with a red hue indicating significant enrichment. CC, cellular component; MF, molecular function; BP, biological process.

The KEGG pathway enrichment results are shown in Figure 6, Figures S10 and S11. For the N_2-treated sample, the upregulated proteins were responsible for quorum sensing (map02024) and cationic antimicrobial peptide (CAMP) resistance (map01503) (Figure S10A); the downregulated proteins were critical for phenylalanine, tyrosine, and tryptophan biosynthesis (map00400), carotenoid biosynthesis (map00906), sulfur metabolism (map00920), the biosynthesis of amino acids (map01230), the biosynthesis of secondary metabolites (map01110), ABC transporters (map02010), and phenazine biosynthesis (map00405) (Figure S10B). For the N_2-SF-treated sample, the upregulated DEPs were important for flagellar assembly (map02040), nitrogen metabolism (map00910), and quorum sensing (map02024) (Figure S11B); the downregulated DEPs were important for phenylalanine, tyrosine, and tryptophan biosynthesis (map00400), carotenoid biosynthesis (map00906), and sulfur metabolism (map00920) (Figure S11B).

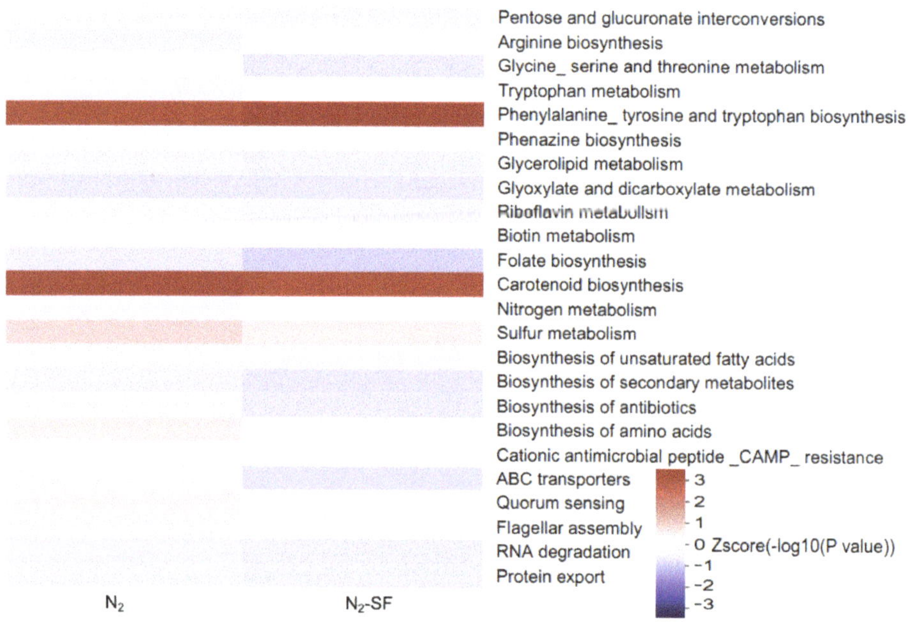

Figure 6. Affected pathways during palladium reduction. KEGG-pathway-based enrichment was employed to conduct a hierarchical clustering analysis on the proteins that were expressed differently. The *p*-values were transformed into Z-scores for hierarchical clustering analysis. The Z-score is shown in the color legend, and the red color represents significant enrichments.

The DEPs in all cases were also examined using protein domain enrichment analysis (Figure 7, Figures S12 and S13). For the N_2-treated sample, the upregulated proteins included protease-associated domain (IPR003137), peptidase S8/S53, subtilisin/kexin/sedolisin (IPR000209), alpha/beta hydrolase, N-terminal (IPR022742), and ribosomal protein S15 (IPR000589) (Figure S12A). The downregulated expressed proteins contained luciferase-like domain (IPR011251), extracellular solute-binding protein, family 3 (IPR001638), amine oxidase (IPR002937), alkyl hydroperoxide reductase subunit C/Thiol-specific antioxidant (IPR000866), and thioredoxin-like fold (IPR012336) (Figure S12B). For the N_2-SF-treated sample, the upregulated proteins included flagellin, D0/D1 domain (IPR001029), protease-associated domain, PA (IPR003137), peptidase S8/S53, subtilisin/kexin/sedolisin (IPR000209), and alpha/beta hydrolase, N-terminal (IPR022742) (Figure S13A); the downregulated proteins included luciferase-like domain (IPR011251), extracellular solute-binding protein, family 3 (IPR001638), NADPH-dependent FMN reductase (IPR005025), amine oxidase (IPR002937), and alkyl hydroperoxide reductase subunit C/Thiol-specific antioxidant (IPR000866) (Figure S13B).

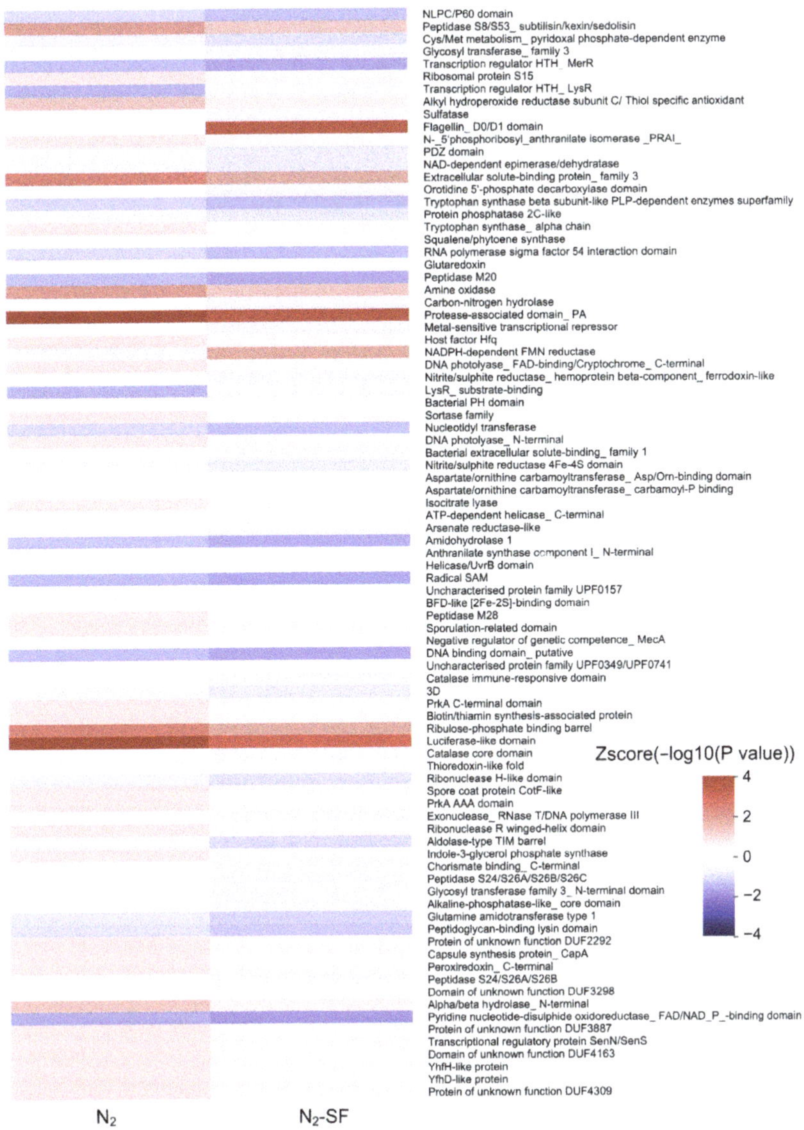

Figure 7. Hierarchical clustering analysis was conducted for the differentially expressed proteins according to protein-domain-based enrichment. The *p*-values were transformed into Z-scores for hierarchical clustering analysis. The Z-score is shown in the color legend, and the red color represents significant enrichments.

3.5. Metabolic Pathways Identified for Palladium Reduction

Based on the above proteome results, several metabolic pathways of palladium bioreduction by *B. megaterium* Y-4 were proposed (Figure 8, Table S4). Firstly, NAD(P)H with strong reducibility can directly reduce Pd (II) [22,23]. In *B. megaterium* Y-4, the NAD(P)H-generating proteins included nirB, tdh, fabG, gudB, zwf, and ssuE. For the N_2-SF-treated sample, only ssuE was downregulated, while nirB, tdh, fabG, gudB, and zwf were upregulated. For the N_2-treated sample, the following expression pattern was observed: only ssuE downregulated, gudB and zwf nonsignificant, and nirB, tdh, and fabG upregulated. Furthermore, the expression of tdh and fabG in the N_2-SF-treated sample was higher than that in the N_2-treated sample, which was consistent with the trend of the Pd (II) reduction amount. Figure S14 and Table S4 summarize the metabolic pathways involved by some proteins. The nirB and gudB proteins were involved in nitrogen metabolism (map00910) (Figure S14A). The fabG (WP_098239123.1) protein played a role in biotin metabolism (map00780) (Figure S14B). The ssuE protein was critical for riboflavin metabolism (map00740) (Figure S14C). The reduction of Pd (II) may be achieved through these metabolic processes. A study showed that NADH dehydrogenases of *Shewanella oneidensis* MR-1 that can produce NADH contribute to the reduction of Pd (II) [17]. The NADH also contributes to Se (VI) bioreduction through regulating the electron transfer [24]. Secondly, reductase plays significant roles in a series of intracellular redox reactions and is involved in various electron transfer cycles. In the N_2-SF-treated sample, four reductase proteins (trx-1, flavin reductase family protein, and two crtIs) were downregulated (Table S4). In the N_2-treated sample, three proteins (trx-1 and two crtIs) were downregulated (Table S4). Pd (II) may be reduced as electron acceptors under the action of these reductases. The crtI proteins may reduce Pd (II) through carotenoid biosynthesis (map00906) and the biosynthesis of secondary metabolites (map01110) (Table S4). Thirdly, previous studies have reported that Q-circle [25] and riboflavin synthase [26] can act as endogenous electron mediators and accelerate the electron transfer process in the bacterial cells, enhancing the efficiency of Pd (II) bioreduction. For the N_2-SF-treated samples, the qoxB Q-cycle protein and three riboflavin-related proteins (LLM-class flavin-dependent oxidoreductase, ssuD, ssuE) were downregulated. For the N_2-treated samples, the cyoC Q-cycle protein was upregulated, and three riboflavin-related proteins (the LLM-class flavin-dependent oxidoreductase, ssuD and ssuE) were downregulated. The ssuD and ssuE proteins are significantly enriched in sulfur metabolism (map00920) (Table S4). These results suggested that *B. megaterium* Y-4 may defend Pd (II) stress by reducing sulfate uptake. Chromate also as a competitive inhibitor of sulfate uptake [27]. Fourthly, The DinG, lysC and ATP-binding cassette domain-containing protein can generate energy to maintain normal metabolism for reduction of Pd (II). For the N_2-treated sample, the DinG and lysC were both up-regulated. The lysC proteins was significantly enriched in glycine, serine and threonine metabolism (map00260) (Figure S15). In the N_2-SF-treated sample, the DinG- and ATP-binding cassette-domain-containing proteins were both upregulated. ABC transporters are transmembrane proteins that utilize energy to carry substrates into the cells [28]. The N_2-treated sample and N_2-SF-treated sample exhibited a decrease in the transporter substrate-binding-domain-containing protein, amino acid ABC transporter substrate-binding protein, and extracellular solute-binding protein. The iron–hydroxamate ABC transporter substrate-binding protein was also downregulated in the N_2-treated sample. These results indicate that the energy generated can be increased and the transport proteins are inhibited under N_2 and N_2-SF conditions. Fifthly, the ribosome-relevant proteins (e.g., rpsO, rpmG, rpmB, and GNAT family N-acetyltransferase) participate in translation, protein metabolism, and RNA processes. As a consequence, more proteins are synthesized to maintain the cellular structure and function, to repair damaged proteins, and to produce new proteins against Pd (II) stress. In the N_2-treated sample, the rpsO was upregulated and the GNAT family N-acetyltransferase was downregulated. For the N_2-SF-treated sample, the rpsO was upregulated while the GNAT family N-acetyltransferase, rpmG, and rpmB were downregulated. Lastly, flagellar motility is very important to allow bacteria to move toward favorable conditions, form biofilms, and acquire nutrients [29,30]. *B. megaterium* Y-4 could exhibit benefit-tending and harm-avoiding behaviors through the flagellum and motility under Pd (II) stress. The fliC (FC = 2.1) was upregulated in the N_2-treated sample. The FlgL (FC = 1.22) and fliC (FC = 2.3) were upregulated in the N_2-SF-treated sample. The fliC protein was significantly enriched in flagellar assembly (map02040).

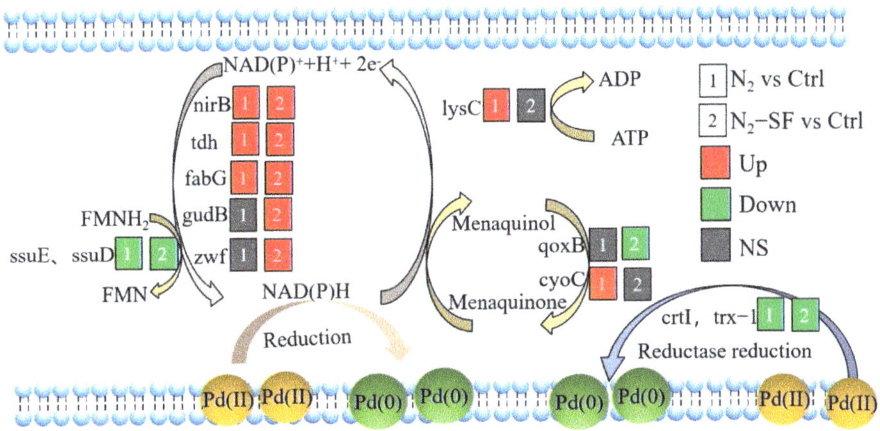

Figure 8. Proposed reduction mechanism of Pd (II) by *Bacillus megaterium* Y-4. Colored boxes indicate the differential expression profiles (upregulated, red; downregulated, green; no significance (NS), gray) for the reduction of Pd (II) by *B. megaterium* Y-4 cells under anaerobic conditions in the presence of N_2 (1), and both N_2 and 5 mM sodium formate (N_2-SF) (2) compared with pure cells without palladium. The information of genes involved in these cellular metabolisms is shown in Supplementary Table S4.

4. Discussion

The recycling of palladium is an important task due to its importance in industry and potential harm to the environment. The microbial reduction of palladium has been used in treating palladium-containing water and it is an economic and green strategy that does not pollute the environment. As metal reduction is a complicated process, many proteins are involved in this process. Understanding the pathways of proteins during palladium reduction will provide valuable insights into more efficient processing and provide useful information for understanding the response of a bacterium under stress. In this study, we used *B. megaterium* Y-4 to reduce palladium under anaerobic conditions in the absence and presence of sodium formate. The palladium removal efficiency can be improved in the presence of an external electron donor. The Pd nanoparticles were observed both inside and outside of the bacterial cells (Figure 1). Our XPS experiment also confirmed the reduction of palladium by *B. megaterium* Y-4.

Proteomics analysis was employed to explore the changes in the protein levels in the bacterium during palladium reduction. A number of proteins were identified (Figure 2), which are critical for understanding the pathways that are critical for palladium reduction. The locations of the DEPs include the cell wall, extracell, cell membrane, and cytoplasm, which suggests that the microbial reduction of palladium in this study may include the transportation of palladium across the cell wall into the cytoplasm, which requires proteins at these locations. The upregulation of proteins suggests that palladium may stimulate certain pathways to reduce the toxic effect caused by metal taking. The downregulation of proteins might arise from the toxic effect of palladium. Nonetheless, based on GO and KEGG analysis, the following pathways were affected during the microbial reduction of palladium. NAD(P)H may play important roles in palladium reduction as several related proteins were upregulated [17]. Interestingly, the downregulation of reductase and oxidoreductase was observed under both conditions, which suggests that these enzymes play important roles in the reduction reaction [19], while the lower protein level might be due to their involvement in reduction and palladium did not induce their production. In addition, proteins related to energy synthesis were stimulated and the levels of proteins for transportation were reduced. The results suggest that the uptake of palladium may result in the denature of some transportation proteins and a reduction in the required additional energy, which may contribute to the restoration of the function of proteins [28], the response to the toxic effect of the metal, and the synthesis of more proteins. Indeed, some ribosome-related proteins were also affected during palladium reduction. The Dep data also showed that palladium also affected the motion of the bacterium, as proteins related to flagellar motility such as fliC were upregulated.

Based on the data accumulated in this study, *B. megaterium* Y-4 may utilize endogenous electron donors (such as NAD(P)H) or reductase to provide electrons for Pd (II) reduction. Endogenous electron mediators (quinones and flavins) may further enhance Pd (II) reduction [18,25]. The ribosome-relevant proteins can maintain the translation, protein metabolism, and RNA process. More energy is required to maintain metabolisms, benefit-tending, and harm avoidance through the ATP-related protein, flagellum motility, and pentose phosphate pathway. The identification of these proteins demonstrated that palladium reduction is a complicated process involving multiple enzymes and proteins in several pathways. Therefore, our study provides insights into the mechanism of palladium reduction in *B. megaterium* Y-4. Based on the obtained information, further study on individual proteins or pathways will expand our knowledge on the function of these proteins. The information is also helpful for improving metal removal efficiency in microbial treatment.

5. Conclusions

In summary, the microbial reduction of Pd (II) to Pd-NPs by *B. megaterium* Y-4 can be achieved through the synergistic action of multiple proteins involved in several cellular pathways. We found that NirB, tdh, fabG, gudB, and zwf proteins produce NAD(P)H, which can provide electrons for *B. megaterium* Y-4 for the reduction of Pd (II). Additionally, Pd (II) may also be reduced by the reductase produced by the ssuE, trx-1, and flavin reductase family protein. The reduction efficiency can be accelerated through the menaquinone produced by the qoxB protein and FMN produced by ssuE and ssuD. Further investigations are still needed on how these genes or proteins influence the microbial reduction of Pd (II) in future work. Our findings provide a new perspective for understanding the bioreduction mechanisms of Pd (II) to create Pd-NPs.

Supplementary Materials: The following supporting information can be downloaded at: https://www.mdpi.com/article/10.3390/nano14060512/s1, Table S1. peptide fraction separation liquid chromatography elution gradient table. Table S2. Liquid chromatography elution gradient table. Table S3. The Pd (II) and Pd (0) XPS peak area of different samples. Table S4. All the identified quantifiable upregulated and downregulated proteins. Figure S1. The removal efficiency (%) of Pd (II) by *B. megaterium* Y-4 under N_2, and N_2-5 mM sodium formate (N_2-SF) conditions, respectively. Figure S2. Parent ion mass tolerance distribution. The abscissa is the mass deviation, and the ordinate is the parent ion density distribution of the corresponding error. Most peaks close to 0, suggesting small mass deviation. Figure S3. Principal component analysis. The horizontal axis PC1 and vertical axis PC2 represent the scores of the first and second ranked principal components, respectively, and the scatter color represents the experimental grouping of the samples. Figure S4. Number of differentially expressed proteins (A) and venn diagram showing common and unique differentially expressed proteins were compared (B). The DEPs under different conditions are shown in different color. The proteins were observed under both conditions are shown, and 87 proteins were present under both conditions. Figure S5. The protein and protein interaction networks of all differentially expressed proteins in N_2 treated sample. Up-(red) and down-(green) regulated DEPs are indicated. Size of the dot correlates to the absolute value of N_2 treated log2FC. Figure S6. The protein and protein interaction networks of all differentially expressed proteins in the N_2-SF treated sample. Up-(red) and down-(green) regulated DEPs are shown. Size of the dot corelates to the absolute value of N_2-SF treated log2FC. Figure S7. Subcellular location of differentially expressed proteins. N_2 and N_2-SF treated samples were analyzed and the proteins from different locations are shown in different color. Figure S8. GO-based enrichment analysis of up-regulated proteins (A, red) and down-regulated proteins (B, green) in N_2 treated sample. Figure S9. GO-based enrichment analysis of up-regulated (A, red) and down-regulated (B, green) proteins in the N_2-SF treated sample. Figure S10. KEGG-based enrichment analysis of up-regulated (A, red) and down-regulated (B, green) proteins of N_2 treated sample. Figure S11. KEGG-based enrichment analysis of up-regulated (A, red) and down-regulated (B, green) proteins of N_2-SF treated sample. Figure S12. Protein domain-based enrichment analysis of up-regulated (A, red) and down-regulated (B, green) proteins of N_2 treated sample. The protein domains are indicated on the left side and the possibility values are indicated on the right side. Figure S13. Protein domain-based enrichment analysis of up-regulated (A, red) and down-regulated (B, green) proteins of N_2-SF treated sample. The protein domains are indicated on the left side and the possibility values are indicated on the right side. Figure S14. Schematic diagram of nitrogen metabolism, biotin metabolism and riboflavin metabolism. Colored boxes indicate the different expression profiles (upregulated, red; downregulated, green; no significance (NS), gray) under N_2 (1) and N_2-SF (2) conditions compared with cells without exposing to palladium. The information of upregulated or downregulated proteins involved in these cellular metabolisms is shown in the Supplementary Table S3. Figure S15. Schematic diagram of glycine, serine and threonine metabolism. Colored boxes indicate the different expression profiles. Upregulated, downregulated and no significantly (NS) changed pathways are shown in red, green and gray, respectively under N_2 (1) and N_2-SF (2) conditions compared with cells without exposing to palladium. The information of upregulated or downregulated proteins involved in these cellular metabolisms is shown in the Supplementary Table S4. References [31–36] are cited in the Supplementary Materials.

Author Contributions: Conceptualization, Y.C. and Q.L.; methodology, Y.C. and J.W. (Jiaxing Wang); software, B.W.; validation, Y.C., R.L. and Q.L.; formal analysis, Y.C. and D.C.; investigation, Y.C.; resources, Y.C., Q.L., R.L. and J.W. (Jinchuan Wu); data curation, Y.C. and J.W. (Jiaxing Wang); writing—original draft preparation, Y.C. and D.C.; writing—review and editing, Q.L. and R.L.; visualization, Y.C.; supervision, Q.L. and R.L.; project administration, Y.C., Q.L., R.L. and J.W. (Jinchuan Wu); funding acquisition, Y.C., Q.L., R.L. and J.W. (Jinchuan Wu). All authors have read and agreed to the published version of the manuscript.

Funding: This work was supported by Guangzhou Science and Technology Plan Project (2024A04J0407), Guangdong Basic and Applied Basic Research Foundation (2020A1515111168, 2022A1515010366), the "Hundred Talent Program" (2020GDASYL-20200102009, 2020GDASYL-20200102010), GDAS' Project of Science and Technology Development (2022GDASZH-2022010110, 2023GDASZH-2023010102), the leading talents of Guangdong province "Pearl River Talent Program" (2021CX020301), and the Guangdong modern agricultural industrial technology system innovation team construction project with agricultural products as the unit (2023KJ104).

Data Availability Statement: Data available on reasonable request.

Conflicts of Interest: The authors declare no conflicts of interest.

References

1. Zhou, L.; Yang, J.; Ma, F.; Pi, S.; Tang, A.; Li, A. Recycling of Pd (0) catalysts by magnetic nanocomposites—Microbial extracellular polymeric substances@Fe_3O_4. *J. Environ. Manag.* **2021**, *280*, 111834. [CrossRef] [PubMed]
2. Sharma, S.; Rajesh, N. Augmenting the adsorption of palladium from spent catalyst using a thiazole ligand tethered on an amine functionalized polymeric resin. *Chem. Eng. J.* **2016**, *283*, 999–1008. [CrossRef]
3. Chen, Y.; Chen, Y. Difference in toxicity of Pd (II) and mechanism of action before and after reduction by *Bacillus wiedmannii* MSM. *Environ. Sci. Pollut. R* **2021**, *29*, 1824–1835. [CrossRef]
4. Yates, M.D.; Cusick, R.D.; Logan, B.E. Extracellular palladium nanoparticle production using *Geobacter sulfurreducens*. *ACS Sustain. Chem. Eng.* **2013**, *1*, 1165–1171. [CrossRef]
5. Tuo, Y.; Liu, G.; Zhou, J.; Wang, A.; Wang, J.; Jin, R.; Lv, H. Microbial formation of palladium nanoparticles by *Geobacter sulfurreducens* for chromate reduction. *Bioresour. Technol.* **2013**, *133*, 606–611. [CrossRef]
6. De Windt, W.; Aelterman, P.; Verstraete, W. Bioreductive deposition of palladium (0) nanoparticles on *Shewanella oneidensis* with catalytic activity towards reductive dechlorination of polychlorinated biphenyls. *Environ. Microbiol.* **2005**, *7*, 314–325. [CrossRef]
7. De Windt, W.; Boon, N.; Van den Bulcke, J.; Rubberecht, L.; Prata, F.; Mast, J.; Hennebel, T.; Verstraete, W. Biological control of the size and reactivity of catalytic Pd (0) produced by *Shewanella oneidensis*. *Anton. Leeuw. Int. J. G* **2006**, *90*, 377–389. [CrossRef] [PubMed]
8. Cui, J.; Zhu, N.; Kang, N.; Ha, C.; Shi, C.; Wu, P. Biorecovery mechanism of palladium as nanoparticles by *Enterococcus faecalis*: From biosorption to bioreduction. *Chem. Eng. J.* **2017**, *328*, 1051–1057. [CrossRef]
9. You, L.; Pan, D.; Chen, N.; Lin, W.; Chen, Q.; Rensing, C.; Zhou, S. Extracellular electron transfer of *Enterobacter cloacae* SgZ-5T via bi-mediators for the biorecovery of palladium as nanorods. *Environ. Int.* **2019**, *123*, 1–9. [CrossRef]
10. Asada, T.; Edamitsu, Y.; Wakashiba, K.; Nomura, T. Bioreduction of aqueous palladium ions by *Shewanella algae* under atmospheric conditions. *Hydrometallurgy* **2024**, *224*, 106244. [CrossRef]

11. Bunge, M.; Sobjerg, L.S.; Rotaru, A.E.; Gauthier, D.; Lindhardt, A.T.; Hause, G.; Finster, K.; Kingshott, P.; Skrydstrup, T.; Meyer, R.L. Formation of palladium(0) nanoparticles at microbial surfaces. *Biotechnol. Bioeng.* **2010**, *107*, 206–215. [CrossRef]
12. Chen, Y.; Hu, K.; Chen, Y. The effect of biotic and abiotic environmental factors on Pd (II) adsorption and reduction by *Bacillus megaterium* Y-4. *Chemosphere* **2019**, *220*, 1058–1066. [CrossRef]
13. Zhang, Y.; Zhao, Q.; Chen, B. Reduction and removal of Cr(VI) in water using biosynthesized palladium nanoparticles loaded *Shewanella oneidensis* MR-1. *Sci. Total Environ.* **2022**, *805*, 150336. [CrossRef] [PubMed]
14. Schlüter, M.; Hentzel, T.; Suarez, C.; Koch, M.; Lorenz, W.G.; Böhm, L.; Düring, R.; Koinig, K.A.; Bunge, M. Synthesis of novel palladium(0) nanocatalysts by microorganisms from heavy-metal-influenced high-alpine sites for dehalogenation of polychlorinated dioxins. *Chemosphere* **2014**, *117*, 462–470. [CrossRef] [PubMed]
15. Rybochkin, P.V.; Perchikov, R.N.; Ya Karlinskii, B.; Kamanina, O.A.; Arlyapov, V.A.; Kashin, A.S.; Ananikov, V.P. Aerobic bacteria-supported biohybrid palladium catalysts for efficient cross-coupling reactions. *J. Catal.* **2024**, *429*, 115238. [CrossRef]
16. Wu, R.; Tian, X.; Xiao, Y.; Ulstrup, J.; Christensen, H.; Zhao, F.; Zhang, J. Selective electrocatalysis of biofuel molecular oxidation using palladium nanoparticles generated on *Shewanella oneidensis* MR-1. *J. Mater. Chem. A* **2018**, *6*, 10655–10662. [CrossRef]
17. Yang, Z.; Hou, Y.; Zhang, B.; Cheng, H.; Yong, Y.; Liu, W.; Han, J.; Liu, S.; Wang, A. Insights into palladium nanoparticles produced by *Shewanella oneidensis* MR-1, Roles of NADH dehydrogenases and hydrogenases. *Environ. Res.* **2020**, *191*, 110196. [CrossRef]
18. Chen, Y.; Chen, Y.; Wang, B.; Mao, C.; Liu, C.; Liu, R.; Yan, B. From trash to treasure: Microbial conversion of palladium contaminants into valuable Pd nanoparticles by *Bacillus thuringiensis* Y9. *J. Clean. Prod.* **2022**, *366*, 132880. [CrossRef]
19. Matsumoto, T.; Kamino, M.; Yamada, R.; Konishi, Y.; Ogino, H. Identification of genes responsible for reducing palladium ion in *Escherichia coli*. *J. Biotechnol.* **2020**, *324*, 7–10. [CrossRef]
20. Gang, H.; Xiao, C.; Xiao, Y.; Yan, W.; Bai, R.; Ding, R.; Yang, Z.; Zhao, F. Proteomic analysis of the reduction and resistance mechanisms of *Shewanella oneidensis* MR-1 under long-term hexavalent chromium stress. *Environ. Int.* **2019**, *127*, 94–102. [CrossRef]
21. Perez-Riverol, Y.; Bai, J.; Bandla, C.; García-Seisdedos, D.; Hewapathirana, S.; Kamatchinathan, S.; Kundu, D.J., Prakash, A.; Frericks-Zipper, A.; Eisenacher, M.; et al. The PRIDE database resources in 2022, a hub for mass spectrometry-based proteomics evidences. *Nucleic Acids Res.* **2022**, *50*, D543–D552. [CrossRef] [PubMed]
22. Choi, Y.; Park, T.J.; Lee, D.C.; Lee, S.Y. Recombinant *Escherichia coli* as a biofactory for various single- and multi-element nanomaterials. *Proc. Natl. Acad. Sci. USA* **2018**, *115*, 5944–5949. [CrossRef]
23. Dundas, C.M.; Graham, A.J.; Romanovicz, D.K.; Keitz, B.K. Extracellular electron transfer by *Shewanella oneidensis* controls palladium nanoparticle phenotype. *ACS Synth. Biol.* **2018**, *7*, 2726–2736. [CrossRef]
24. Li, L.; Yan, W.; Zhang, B.; Zhang, H.; Geng, R.; Sun, S.; Guan, X. Coupling of selenate reduction and pyrrhotite oxidation by indigenous microbial consortium in natural aquifer. *Water Res.* **2023**, *238*, 119987. [CrossRef] [PubMed]
25. Pat-Espadas, A.M.; Razo-Flores, E.; Rangel-Mendez, J.R.; Cervantes, F.J. Direct and quinone-mediated palladium reduction by *Geobacter sulfurreducens*: Mechanisms and modeling. *Environ. Sci. Technol.* **2014**, *48*, 2910–2919. [CrossRef] [PubMed]
26. Brutinel, E.D.; Gralnick, J.A. Shuttling happens: Soluble flavin mediators of extracellular electron transfer in Shewanella. *Appl. Microbiol. Biot.* **2012**, *93*, 41–48. [CrossRef]
27. Ohtake, H.; Cervantes, C.; Silver, S. Decreased chromate uptake in *Pseudomonas fluorescens* carrying a chromate resistance plasmid. *J. Bacteriol.* **1987**, *169*, 3853–3856. [CrossRef]
28. Bulut, H.; Moniot, S.; Licht, A.; Scheffel, F.; Gathmann, S.; Saenger, W.; Schneider, E. Crystal Structures of Two Solute Receptors for l-Cystine and l-Cysteine, Respectively, of the Human Pathogen Neisseria gonorrhoeae. *J. Mol. Biol.* **2012**, *415*, 560–572. [CrossRef]
29. Qin, Q.L.; Li, Y.; Zhang, Y.J.; Zhou, Z.M.; Zhang, W.X.; Chen, X.L.; Zhang, X.Y.; Zhou, B.C.; Wang, L.; Zhang, Y.Z. Comparative genomics reveals a deep-sea sediment-adapted life style of *Pseudoalteromonas* sp. SM9913. *ISME J.* **2011**, *5*, 274–284. [CrossRef]
30. Wang, L.; Huang, L.; Su, Y.; Qin, Y.; Kong, W.; Ma, Y.; Xu, X.; Lin, M.; Zheng, J.; Yan, Q. Involvement of the flagellar assembly pathway in *Vibrio alginolyticus* adhesion under environmental stresses. *Front Cell. Infect Microbiol.* **2015**, *5*, 59. [CrossRef]
31. Marx, H.; Minogue, C.E.; Jayaraman, D.; Richards, A.L.; Kwiecien, N.W.; Siahpirani, A.F.; Rajasekar, S.; Maeda, J.; Garcia, K.; Del Valle-Echevarria, A.R.; et al. A proteomic atlas of the legume *Medicago truncatula* and its nitrogen-fixing endosymbiont *Sinorhizobium meliloti*. *Nat. Biotechnol.* **2016**, *34*, 1198. [CrossRef] [PubMed]
32. Nie, L.; Zhang, H.; Wu, Z.; Wang, Y.; Liu, H.; Wu, X.; Wang, W. Modified TCA/acetone precipitation of plant proteins for proteomic analysis. *PLoS ONE* **2018**, *13*, e0202238. [CrossRef] [PubMed]
33. Bradford, M.M. A rapid and sensitive method for the quantitation of microgram quantities of protein utilizing the principle of protein-dye binding. *Anal. Biochem.* **1976**, *72*, 248–254. [CrossRef]
34. Kachuk, C.; Stephen, K.; Doucette, A. Comparison of sodium dodecyl sulfate depletion techniques for proteome analysis by mass spectrometry. *J Chromatogr. A* **2015**, *1418*, 158–166. [CrossRef]
35. Wisniewski, J.R.; Zougman, A.; Nagaraj, N.; Mann, M. Universal sample preparation method for proteome analysis. *Nat. Methods* **2009**, *6*, 359–360. [CrossRef]
36. Gillette, M.A.; Satpathy, S.; Cao, S.; Dhanasekaran, S.M.; Vasaikar, S.V.; Krug, K.; Petralia, F.; Li, Y.Z.; Liang, W.W.; Reva, B.; et al. Proteogenomic Characterization Reveals Therapeutic Vulnerabilities in Lung Adenocarcinoma. *Cell* **2020**, *182*, 200. [CrossRef]

MDPI AG
Grosspeteranlage 5
4052 Basel
Switzerland
Tel.: +41 61 683 77 34

Nanomaterials Editorial Office
E-mail: nanomaterials@mdpi.com
www.mdpi.com/journal/nanomaterials

www.ingramcontent.com/pod-product-compliance
Lightning Source LLC
Chambersburg PA
CBHW061605210926
51850CB00009B/2220
9783725876952